Topics in Geobiology

Volume 47

Series Editors
Neil H. Landman
American Museum of Natural History, New York, NY, USA

Peter J. Harries
North Carolina State University, Raleigh, NC, USA

The **Topics in Geobiology** series covers the broad discipline of geobiology that is devoted to documenting life history of the Earth. A critical theme inherent in addressing this issue and one that is at the heart of the series is the interplay between the history of life and the changing environment. The series aims for high quality, scholarly volumes of original research as well as broad reviews.

Geobiology remains a vibrant as well as a rapidly advancing and dynamic field. Given this field's multidiscipline nature, it treats a broad spectrum of geologic, biologic, and geochemical themes all focused on documenting and understanding the fossil record and what it reveals about the evolutionary history of life. The Topics in Geobiology series was initiated to delve into how these numerous facets have influenced and controlled life on Earth.

Recent volumes have showcased specific taxonomic groups, major themes in the discipline, as well as approaches to improving our understanding of how life has evolved.

Taxonomic volumes focus on the biology and paleobiology of organisms – their ecology and mode of life – and, in addition, the fossil record – their phylogeny and evolutionary patterns – as well as their distribution in time and space.

Theme-based volumes, such as predator-prey relationships, biomineralization, paleobiogeography, and approaches to high-resolution stratigraphy, cover specific topics and how important elements are manifested in a wide range of organisms and how those dynamics have changed through the evolutionary history of life.

Comments or suggestions for future volumes are welcomed.

More information about this series at http://www.springer.com/series/6623

Carrie L. Tyler • Chris L. Schneider
Editors

Marine Conservation Paleobiology

Springer

Editors
Carrie L. Tyler
Department of Geology and Environmental Earth Science
Miami University
Oxford, OH, USA

Chris L. Schneider
Department of Earth and Atmospheric Sciences
University of Alberta, Edmonton
Alberta, Canada

ISSN 0275-0120
Topics in Geobiology
ISBN 978-3-030-08859-0 ISBN 978-3-319-73795-9 (eBook)
https://doi.org/10.1007/978-3-319-73795-9

Printed on acid-free paper

This Springer imprint is published by the registered company Springer International Publishing AG part of Springer Nature.
The registered company address is: Gewerbestrasse 11, 6330 Cham, Switzerland

Preface

We believe that a book of this nature is timely and important in highlighting the contributions of conservation paleobiology in solving modern ecological crises. The consequences of millennia of human impacts on ecosystems including extinction, biodiversity loss, and changes in community structure and species composition remain largely unknown. These trends may yet be reversible, but developing effective strategies for conservation, remediation, restoration, and policy calls for a thorough understanding of long-term ecological processes and the ability to examine ecosystems before, during, and after perturbation. There is thus an urgent need for a cross-disciplinary synthesis of modern and historical perspectives. Although paleontological and modern data have traditionally been viewed as disparate data types, fossils can yield high resolution data suitable for community analysis, comparable to modern assemblage data.

We are pleased to present research from some of the foremost and upcoming conservation paleobiologists. Contributions range from case studies with direct application to reviews and meta-analyses, providing broader implications to conservation efforts, and to training of future generations. In this volume, we aim to:

1. Define the goals of conservation paleobiology and our role in conservation science.
2. Highlight how conservation paleoecology can be used to identify and understand ecosystem crises.
3. Provide case studies demonstrating applications to modern communities.
4. Stimulate novel applications of paleontological approach to neontological data.
5. Encourage cross-disciplinary dialogue and *application* of research to management and conservation.
6. Explore the future of conservation paleoecology.

We have the capability, technology, and innovation to enact positive change in the biosphere. There are a multitude of efforts undertaken by societies and governments to mitigate the effects of global climate change. Marine protected areas and reserves, for example, provide safe havens for biodiversity and contribute to sustainable populations for harvested resources. However, conservation biology is a crisis-driven discipline, with little time and often insufficient financial resources to respond. Furthermore, conservation science studies systems *already undergoing change*. In almost all cases, the baseline condition of the ideal ecosystem is gone, and in many of those cases, given the trajectory of climate change and related ocean deterioration, the baseline may no longer be achievable. Conservation paleobiology can provide timely and critical insights to conservation, management, and restoration efforts, providing data on timescales otherwise unavailable and of true pre-impact conditions.

We must collaborate if we are going to maintain a productive marine biosphere for future humanity, and we challenge conservation paleobiology as a field to truly be interdisciplinary not just on paper, but in action. We continue to work towards a better future, to which conservation paleobiology can make significant contributions.

Oxford, OH, USA Carrie L. Tyler
Edmonton, AB, Canada Chris L. Schneider

Acknowledgements

We are deeply grateful to all of the reviewers who devoted their time, expertise, and experience to improving each chapter, despite their busy schedules. We wish to thank a few of the advocates of this new frontier of research, particularly Gregory Dietl, Karl Flessa, Jeremy Jackson, Patricia Kelley, and Sue Kidwell. To all of the authors, thank you for your contributions; we are thrilled to share such high-quality work with the scientific community. We are particularly grateful to Greg Dietl for thoughtful discussions, which influenced the shaping and content of this volume. We also thank Lindsey Leighton for his advice on editing a special volume such as this one. Thank you also to the editorial staff at Springer, without whom this volume could not have been published.

Contents

Contributors

Gregory P. Dietl Paleontological Research Institution, Ithaca, NY, USA

Department of Earth and Atmospheric Sciences, Cornell University, Ithaca, NY, USA

Ashley A. Dineen Invertebrate Zoology and Geology, Institute for Biodiversity Science and Sustainability, California Academy of Sciences, San Francisco, CA, USA

Stephen R. Durham Department of Earth and Atmospheric Sciences, Cornell University, Ithaca, NY, USA

Paleontological Research Institution, Ithaca, NY, USA

Karl W. Flessa Department of Geosciences, University of Arizona, Tucson, AZ, USA

Amanda N. Grant School of Earth Sciences and Environmental Sustainability, Northern Arizona University, Flagstaff, AZ, USA

Patricia H. Kelley Department of Earth and Ocean Sciences, University of North Carolina Wilmington, Wilmington, NC, USA

Kristopher M. Kusnerik Division of Invertebrate Paleontology, Florida Museum of Natural History, Gainesville, FL, USA

Lindsey R. Leighton Department of Earth and Atmospheric Sciences, University of Alberta, Edmonton, Alberta, Canada

Rowan Lockwood Department of Geology, The College of William and Mary, Williamsburg, VA, USA

Matthew J. Pruden Department of Earth and Atmospheric Sciences, University of Alberta, Edmonton, Alberta, Canada

Peter D. Roopnarine Invertebrate Zoology and Geology, Institute for Biodiversity Science and Sustainability, California Academy of Sciences, San Francisco, CA, USA

Michael Savarese Marine and Ecological Sciences, Florida Gulf Coast University, Ft. Myers, FL, USA

Chris L. Schneider Department of Earth and Atmospheric Sciences, University of Alberta, Edmonton, Alberta, Canada

Jansen A. Smith Department of Earth and Atmospheric Sciences, Cornell University, Ithaca, NY, USA

Carrie L. Tyler Department of Geology and Environmental Earth Science, Miami University, Oxford, OH, USA

Christy C. Visaggi Department of Geosciences, Georgia State University, Atlanta, GA, USA

An Overview of Conservation Paleobiology

Carrie L. Tyler and Chris L. Schneider

Abstract The field of conservation paleobiology was formally established in the early 2000s, as a growing body of literature substantiating the fidelity of paleontological data on a variety of spatial and temporal scales emerged, and paleontologists became increasingly aware of the potential insights that the fossil record could provide into the current biodiversity crisis. Conservation paleobiology contributes a temporal scope and historical perspective lacking from the relatively short time spans covered by modern ecological studies, progressively in demand in the face of changing climate and environmental degradation. The increasing number of conservation paleobiology studies in the past decade validates the potential contributions of this field to conservation efforts, and fall within a range of temporal categories (e.g., "near-time" and "deep-time"). Data are not restricted to fossils, but can also include historical reports and archeological evidence (conservation archeobiology). Although temporal resolution often declines with increased assemblage age, ancient ecosystems document responses to, and recoveries from, global crises. Thus, the field of conservation paleobiology, when considered in concert with historical ecology and conservation biology, has the potential to positively affect future ecosystems and biodiversity.

Keywords Conservation science · Conservation biology · Geohistorical data · Environmental history · Conservation archeobiology

C. L. Tyler (✉)
Department of Geology and Environmental Earth Science, Miami University, Oxford, OH, USA
e-mail: tylercl@miamiOH.edu

C. L. Schneider
Department of Earth and Atmospheric Sciences, University of Alberta, Edmonton, Alberta, Canada
e-mail: clschnei@ualberta.ca

C. L. Tyler, C. L. Schneider (eds.), *Marine Conservation Paleobiology*, Topics in Geobiology 47, https://doi.org/10.1007/978-3-319-73795-9_1

1 Defining and Establishing Conservation Paleobiology as a Discipline

The importance of records documenting previous ecosystem states is increasingly recognized in conservation and ecosystem management (e.g., Willis and Birks 2006; Knowlton and Jackson 2008; Louys et al. 2012), particularly in identifying restoration targets for degraded or ailing ecosystems (Jackson and Hobbs 2009). Ideally, baseline conditions should be established prior to human disturbance; however, human activities have been altering the Earth since the Late Pleistocene (Jackson et al. 2001; Boivin et al. 2016), increasing in scope and intensity with the advent of agriculture 10,000 years ago, and later by the rise of large-scale industrialization. Marine ecosystems have arguably been altered far more than those on land (Jackson et al. 2001; Knowlton and Jackson 2008; Ban and Alder 2008), and as a result, "pristine" ocean habitats have effectively vanished and may not be recoverable. Furthermore, global and local environmental changes continue to occur, with shifting baselines creating a moving target for conservation scientists and resource management (Pauly 1995; Jackson et al. 2001). Yet few modern ecological studies consider the historically unaltered state of ecosystems (Jackson et al. 2001; Willis and Birks 2006; Froyd and Willis 2008), thereby limiting our understanding of marine ecosystem functioning predating anthropogenic effects.

In modern ecosystems, long-term ecological studies are sparse and predominantly focused on terrestrial botanical systems (although this is improving with the rise of programs such as the Long-Term Ecological Research Network). Generally, ecosystem monitoring rarely extends beyond the past few decades (Jackson and Hobbs 2009), and possibly the longest-running example of ecological monitoring, the Park Grass experiment in Rothermstead, southern England, has been running since 1856 (Silvertown et al. 2006). Designed to address agricultural questions, this environmental monitoring experiment has continued for over 160 years. However, many ecological theories require data on far longer timescales to test (Louys 2012), e.g., neutral theory (Hubbell 2001). While broadly applicable and valuable insights into ecological processes can be gained through modern observations, long-term data from other types of habitats and ecosystems are necessary to determine the generality of these findings. Geohistorical and paleontological data can, therefore, provide information essential to conservation science, particularly given the benefits of understanding long-term ecological processes, and that true "pristine" habitats likely no longer exist.

As a nascent discipline, conservation paleobiology is currently ideally poised for deliberate and thoughtful definition, and determination of agreed upon goals through rigorous academic debate (Tyler 2018). Conservation paleobiology provides a deep-time perspective for a biosphere in flux and has been defined as the *"application of the methods and theories of paleontology to the conservation and restoration of biodiversity and ecosystem services"* (Dietl and Flessa 2011; Dietl et al. 2015). As this definition includes some of the goals of conservation science—the conservation and restoration of biodiversity and maintenance of ecosystem services (Soule 1985;

Meine et al. 2006)—Dietl (2016) suggested that it may be beneficial to approach conservation paleobiology as a sub-discipline within conservation science. This approach would likely promote the dissemination of paleontological data, ideas, and information in conservation science (Dietl 2016). As conservation scientists and resource managers remain relatively unaware of potential geohistorical contributions (Smith et al. 2018; Savarese 2018), this paradigm shift would be potentially transformative to the field. It should be noted that not all perceive conservation paleobiology as a value-driven science (see Dietl and Flessa 2018).

The rapid rise of conservation paleobiology is due in part to crucial work validating the ability of paleontology to contribute to ecology and conservation science, and the growing realization that skeletal remains can be used to examine recent changes in ecosystems over years to decades (Dietl et al. 2011; Behrensmeyer and Miller 2012). Taphonomic investigations assessing the quality of the recent fossil record and identifying the limits of its applicability to neontological studies demonstrate high fidelity between live communities and fossil assemblages (see Kidwell and Tomašových 2013, and Kidwell 2015 for review). Likewise, critical autecological and synecological studies, such as functional morphology, community ecology, and predator–prey interactions, are necessary to understand past systems. Discussions and colloquia have produced a plethora of ideas on the relevance of fossil data to crises in the modern biosphere (e.g., Willis and Birks 2006; Dietl and Flessa 2009; Dietl et al. 2011; Dietl and Flessa 2011; Louys 2012; Rick and Lockwood 2013; Gatti et al. 2015; Dietl et al. 2015; Dietl 2016; Fordham et al. 2016). A brief reflection on the growth and development of the discipline illustrates that an awareness of paleontological applications to conservation began just prior to the turn of the century, followed by rapid growth over the last decade, particularly in recent years (Table 1). Research applicable to conservation biology obviously began much earlier, chiefly in the last 20 years (Louys 2012). However, immediately prior to 2001, significant momentum was achieved in the paleontological community introducing the term “conservation paleobiology” (presumably prompting the organization of the 2001 North American Paleontological Convention topical session “New Uses for the Dead: Paleobiological Contributions to Conservation Paleobiology”), as the pursuit of research relevant to biodiversity and conservation became widespread. A search of titles, abstracts, and keywords in the Web of Science citation database (30th October 2016, WoS Core Collections, ts = “conservation” and ts = “paleobiology”) suggests that the phrase “conservation paleobiology” was first used in the title of a peer-reviewed journal article in only 2009 (Simões et al. 2009), reflecting its growing popularity in reference to research applicable to historical ecology and conservation science at research symposia and in discussions among paleoecologists. In 2011, a conservation paleobiology workshop funded by the National Science Foundation appears to have spurred a subsequent burst of research activity as evidenced by the numerous books, symposia, and topical sessions in subsequent years (Table 1).

Table 1 Timeline for the growth and development of conservation paleobiology

Year	Event
2001	North American Paleontological Society Symposium: "New Uses for the Dead: Paleobiological Contributions to Conservation Paleobiology" (Organizers M Droser and KW Flessa)
2005	Committee on the Geological Record of Biosphere Dynamics; National Research Council; "The Geological Record of Ecological Dynamics: Understanding the Effects of Future Environmental Change" (Chair KW Flessa)
2009	Paleontological Society Short Course at the Geological Society of America Annual Conference: "Conservation Paleobiology: Using the Past to Manage for the Future" (Organizers GP Dietl and KW Flessa); First published use of "Conservation Paleobiology" in the title of a peer-reviewed journal article (Simões et al. 2009) published in the journal *Historical Biology*
2011	NSF-Funded Workshop (Paleontological Research Institution, Ithica, New York): "Conservation Paleobiology: Opportunities for the Earth Sciences"
2012	Publication of the book "Paleontology in Ecology and Conservation" (Author and Editor J Louys)
2013	Geological Society of America Symposium: "Conservation Paleobiology—The Microfossil Record" (Organizers S Goldstein and JH Lipps)
2013	Ecological Society of America Symposium: "Resilience, Disturbance, and Long-Term Environmental Change-Integrating Paleoecology into Conservation Management in the Anthropocene" (Organizer AW Ireland)
2014	North American Paleontological Society Symposia: "The Microfossil Record: The Past is the Key to the Future (or Present) in Conservation Paleobiology" (Organizers P Hallock and LS Collins); "Conservation Paleobiology: Ecosystem, Community, and Species Response to Environmental Change" (Organizers CL Tyler, SN Casebolt, R Terry)
2015	Global Change Biology Annual Symposium (UC Berkeley): "Integrating biology and Paleobiology to Enhance Conservation of Terrestrial Ecosystems on a Rapidly Changing Planet"; Society for Vertebrate Paleontology Conservation Paleobiology Symposium – Annual Meeting
2016	4th International Sclerochronology Conference: "Environmental Monitoring and Conservation Paleobiology" (Organizer GP Dietl)
2017	AAAS Annual Meeting Symposium "Conservation Paleobiology: Finding Solutions in the Fossil Record" (Organizer S Kidwell)

Significant professional and academic activities expressly dedicated to the topic of conservation paleobiology

2 Data in Conservation Paleobiology

Given the very recent establishment of the discipline, the accomplishments of the last decade are particularly impressive. A large body of literature already exists, and research with applications to conservation continues to grow. Geohistorical data applied to conservation science include a wide range of data types and continuity with sub-fossil and modern assemblages, employed to understand ancient and modern ecosystems (Rick and Lockwood 2013). Particular emphasis has been

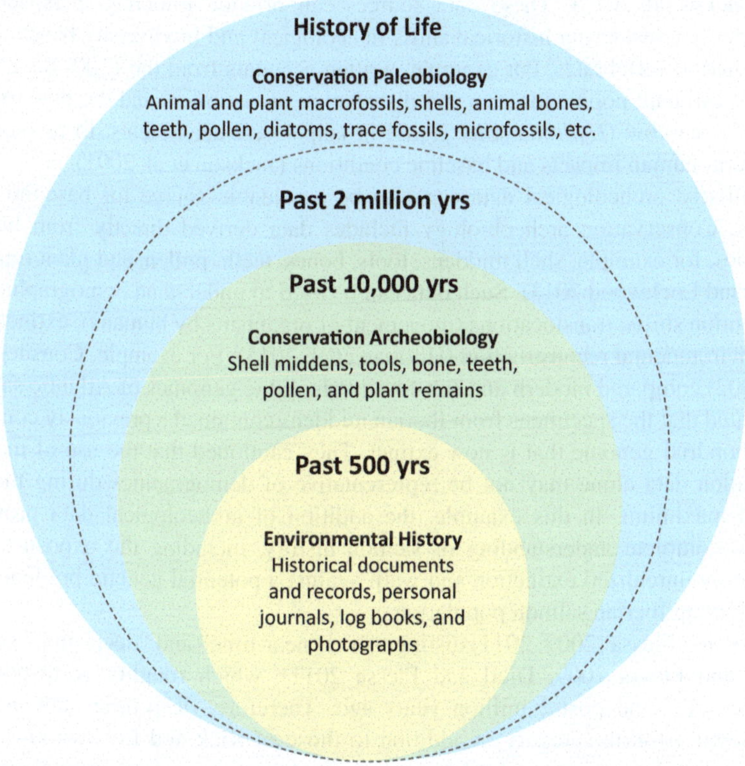

Fig. 1 Range of data types within historical ecology that are commonly applied to conservation science. If conservation paleobiology data are considered a sub-discipline within conservation science, historical ecology must be extended beyond human origins. Conservation paleobiology can then further be divided into "near-time" considering the past 2 million years (within the dashed line), and "deep-time," potentially including all of the history of life (Dietl and Flessa 2011; Rick and Lockwood 2013). Overlapping spheres illustrate overlapping temporal scales, for example, a conservation paleobiology study may compare fossil species abundance and biodiversity from the early Pleistocene to the modern

placed on human–environment interactions (Balée 2006; Szabó and Hédl 2011) and natural variation pre- and post-human activities (Dietl and Flessa 2011). Data can generally be grouped into three categories (Rick and Lockwood 2013): (1) environmental history covering the past 500 years, (2) conservation archeobiology, encompassing the past 10,000 years, and (3) conservation paleobiology, generally from the Pleistocene and Holocene, but potentially including the entire history of life (Fig. 1).

Historical records can provide valuable insights into ecosystem conditions prior to European colonization, and baseline data for restoration targets. Written historical records include a wide range of species and ecosystem data preserved mainly in personal journals, log books, and photographs (Szabó and Hédl 2011; Rick and Lockwood 2013). These data sources can provide important perspectives, particularly considering historical shifts in ecological and biodiversity baselines of large marine vertebrates. For example, written accounts from the 1700s have been used to estimate population size and distribution of sea otters and the now extinct Stellar's sea cow (*Hydrodamalis gigas*) in kelp forest ecosystems, to reconstruct long-term human impacts and baseline conditions (Jackson et al. 2001).

Similarly, archeological data can also be a valuable source for baseline conditions. Conservation archeobiology includes data derived directly from human activities, for example, shell middens, tools, bones, teeth, pollen, and plant remains (Rick and Lockwood 2013). Such data can be used to understand demographic and distribution shifts, translocations (movement of organisms by humans), extinctions, and environmental reconstruction (Hofman et al. 2015). For example, Consuegra et al. (2002) compared modern and sub-fossil Paleolithic genomes of Atlantic salmon and found that the specimens from Iberian middens contained a previously common mitochondrial genome that is now extinct. They cautioned that the use of modern population data alone may not be representative of demographics during the last glacial maximum. In this example, the addition of archeological data provided a more complete understanding of salmon history, including the exposure of a previously unrealized extinction and warn against a potential genetic bottleneck in the declining Iberian salmon population.

Dietl and Flessa (2009, 2011) distinguished "near-time" and "deep-time" studies (Dietl and Flessa 2009; Dietl and Flessa 2011), which roughly correspond to intervals pre- and post-2 million years ago. Therefore, deep-time data may be considered a fourth category in addition to those of Rick and Lockwood (2013), as these fossil remains are likely to represent less similar communities, and to have undergone more extensive alteration than their younger counterparts (Fig. 1). Paleontological information from the past 2 million years includes similar species and ecosystems to that of today's oceans and may, therefore, be directly relevant and applicable to current conservation (Smith et al. 2018; Dietl and Flessa 2011; Dietl et al. 2015). These studies include establishing baseline conditions and capture the response to human arrival (e.g., decline in oyster sizes between Late Pleistocene, colonial, and modern times along the US Atlantic coast; Kusnerik et al. 2018), and impacts of human-induced stressors (the history of overfishing and habitat loss since pre-Roman times in the Mediterranean; Sala et al. 2012; Boada et al. 2015). Studies of deep-time, prior to 2 million years ago, are differentiated by the greater dissimilarity between ancient and modern ecosystems and species (for a thorough review, see Dietl and Flessa 2009; Dietl and Flessa 2011; Dietl et al. 2015). A few examples of deep-time studies include: (1) The varied responses of ecosystems and species to perturbations, especially in terms of extinction, survivorship, and recovery; (2) The range of conditions leading to stable ecosystem states; (3) The conditions under which stable states persist; (4) Evolutionary and ecological processes in the absence of human interference; and (5) The evolutionary and ecological consequences of biotic invasions.

Although deep-time data come with some caveats that are comparatively negligible in near-time studies (for instance, the temporal resolution of paleobiological data generally declines with age before present), the resulting "big-picture" perspective can provide valuable information on how the marine biosphere has responded to, and recovered from, global crises in the past, such as major climate shifts and mass extinction events. In this regard, deep-time studies alone can answer questions about the marine biosphere before, during, and after a global catastrophe (e.g., applying the characteristics of successful refugia in past mass extinctions to marine reserves; Schneider 2018). Ancient ecosystems and species may also differ significantly from those of the modern, and an exact correspondence with present taxa is often not possible. This is perhaps best emphasized through the recognition of the Cambrian, Paleozoic, and Modern evolutionary faunas put forth by Sepkoski (1984), in which the more ancient the assemblage, the generally more dissimilar from modern systems in taxonomic dominants. However, assuming uniformitarianism, ecological processes and evolutionary forces acting within the biosphere generally followed the same principles. Therefore, even in ancient ecosystems that differ substantially in composition from those of the present, valuable insights can still be gained. For example, there have been two prior periods with major barrier reef systems: the past few million years, and the Devonian Period. Conservation efforts in modern reef systems may, therefore, benefit not only from the research of Pleistocene interglacial reefs (e.g., Greenstein and Pandolfi 2008), but also the future investigation of the causes and process of global reef demise in the Late Devonian with a focus on applicability to modern crises in tropical oceans.

3 Looking Forward

Conservation is defined in the Merriam-Webster's Dictionary as *"a careful preservation and protection of something; especially: planned management of a natural resource to prevent exploitation, destruction or neglect."* If we wish to continue to declare ourselves "conservation paleobiologists," our research goals must contribute to preservation and planned management. Paleobiological studies frequently include only a brief phrase or statement about the potential relevance of their findings to modern systems and conservation science (Burnham 2001; Kelley et al. 2018; Savarese 2018). Conservation paleobiology must move beyond *potential* and towards *applications* to narrow the implementation gap. Relevant paleobiological information needs to be presented, and made available, to the appropriate audiences, i.e., resource managers, government agencies, and stakeholders (Smith et al. 2018; Kelley et al. 2018; Savarese 2018). Furthermore, as conservation paleobiology and conservation science have developed in parallel as two disparate disciplines, managers are typically unaware of what conservation paleobiology has to offer. Savarese (2018) provides a thorough discussion on establishing connections and communication with decision-making groups. Some of these issues could also be mitigated by enacting conservation paleobiology as a sub-discipline within conservation science, as suggested by Dietl (2016).

Numerous principal objectives of conservation paleobiology have been proposed by Dietl and others (Dietl and Flessa 2009; Dietl and Flessa 2011; Dietl et al. 2015) and can be summarized as follows:

1. Establish baselines to compare pre- and post-anthropogenic disturbance conditions
2. Examine response to perturbations, both natural and anthropogenic in origin
3. Document historical range of variability within ecosystems, and predict potential geographic range shifts in response to climate change
4. Provide information for the determination of realistic restoration targets
5. Distinguish between anthropogenic and non-anthropogenic change
6. Recognition of ecological legacies only explainable by prior events and conditions
7. Identify vulnerable species in urgent need of protection and evaluating their extinction risk
8. Identify invasive species
9. Detect recent shifts in species abundance
10. Assess changes in genetic diversity
11. Identify future refugia in anticipation of climate change

Ultimately, we must strive to contribute to conservation in a changing world. Global and local environmental change will continue to occur, and conservation scientists and ecologists are therefore trying to quantify and predict biotic changes using systems either already altered or currently in flux. As paleontologists, we have the unique and critical perspective of *time* and *change* that is lacking in the study of modern systems. As conservation paleobiologists, we can provide crucial baselines, insights into pre-impact conditions and ecosystems states, and perspectives on the "hows" and "whys" of biosphere resilience through global catastrophes. Change is inevitable. Whether or not the oceans will be able to provide services in the future is uncertain. Thus, conservation paleobiology may prove essential for maintaining the marine biosphere along with its resources and ecosystem services, and ultimately, the survivorship of our own species.

References

Balée W (2006) The research program of historical ecology. Annu Rev Anthropol 35:75–98

Ban N, Alder J (2008) How wild is the ocean? Assessing the intensity of anthropogenic marine activities in British Columbia, Canada. Aquat Conserv Mar Freshwat Ecosyst 18:55–85

Behrensmeyer AK, Miller JH (2012) Building links between ecology and paleontology using taphonomic studies of recent vertebrate communities. In: Louys J (ed) Paleontology in ecology and conservation. Springer, New York, pp 69–91

Boada J, Arthur R, Farina S et al (2015) Hotspots of predation persist outside marine reserves in the historically fished Mediterranean Sea. Biol Conserv 191:67–74

Boivin NL, Zeder MA, Fuller DQ et al (2016) Ecological consequences of human niche construction: examining long-term anthropogenic shaping of global species distributions. Proc Natl Acad Sci 113:6388–6396

Burnham RJ (2001) Is conservation biology a paleontological pursuit? PALAIOS 16:423

Consuegra S, García de Leániz C, Serdio A et al (2002) Mitochondrial DNA variation in Pleistocene and modern Atlantic salmon from the Iberian glacial refugium. Mol Ecol 11:2037–2048

Dietl GP (2016) Brave new world of conservation paleobiology. Front Ecol Evol 4:21

Dietl GP, Flessa KW (eds) (2009) Conservation paleobiology: using the past to manage for the future. The Paleontological Society Papers vol. 15, Lubbock, TX

Dietl GP, Flessa KW (2011) Conservation paleobiology: putting the dead to work. Trends Ecol Evol 26:30–37

Dietl GP, Flessa KW (2018) Should conservation paleobiologists save the world on their own time? In: Tyler CL, Schneider CL (eds) Marine conservation paleobiology. Springer, Cham, pp 11–22

Dietl GP, Kidwell SM, Brenner M et al (2011) Convservation paleobiology: opportunities for Earth science. Ithica, New York

Dietl GP, Kidwell SM, Brenner M et al (2015) Conservation paleobiology: leveraging knowledge of the past to inform conservation and restoration. Annu Rev Earth Planet Sci 43:79–103

Fordham DA, Akçakaya HR, Alroy J et al (2016) Predicting and mitigating future biodiversity loss using long-term ecological proxies. Nat Clim Chang 6:909–916

Froyd CA, Willis KJ (2008) Emerging issues in biodiversity and conservation management: the need for a palaeoecological perspective. Quat Sci Rev 27:1723–1732

Gatti G, Bianchi CN, Parravicini V et al (2015) Ecological change, sliding baselines and the importance of historical data: lessons from combing observational and quantitative data on a temperate reef over 70 years. PLoS One 10:e0118581

Greenstein BJ, Pandolfi JM (2008) Escaping the heat: range shifts of reef coral taxa in coastal Western Australia. Glob Chang Biol 14:513–528

Hofman CA, Rick TC, Fleischer RC, Maldonado JE (2015) Conservation archaeogenomics: ancient DNA and biodiversity in the Anthropocene. Trends Ecol Evol 30:540–549

Hubbell SP (2001) The unified neutral theory of biodiversity and biogeography. Princeton University Press, Princeton

Jackson ST, Hobbs RJ (2009) Ecological restoration in the light of ecological history. Science 325:567–569

Jackson JBC, Kirby MX, Berger WH et al (2001) Historical overfishing and the recent collapse of coastal ecosystems. Science 293:629–638

Kelley PH, Dietl GP, Visaggi CC (2018) Training tomorrow's conservation paleobiologist. In: Tyler CL, Schneider CL (eds) Marine conservation paleobiology. Springer, Cham, pp 207–223

Kidwell SM (2015) Biology in the Anthropocene: challenges and insights from young fossil records. Proc Natl Acad Sci U S A 112:4922–4929

Kidwell SM, Tomašových A (2013) Implications of time-averaged death assemblages for ecology and conservation biology. Annu Rev Ecol Evol Syst 44:22.1–22.25

Knowlton N, Jackson JBC (2008) Shifting baselines, local impacts, and global change on coral reefs. PLoS Biol 6:1–6

Kusnerik KM, Lockwood R, Grant AN (2018) Using the fossil record to establish a baseline and recommendations for oyster mitigation in the mid-Atlantic U.S. In: Tyler CL, Schneider CL (eds) Marine conservation paleobiology. Springer, Cham, pp 75–103

Louys J (2012) Paleontology in ecology and conservation: an introduction. In: Louys J (ed) Paleontology in ecology and conservation. Springer, Berlin Heidelberg, pp 1–8

Louys J, Wilkinson DM, Bishop LC (2012) Ecology needs a paleontological perspective. In: Louys J (ed) Paleontology in ecology and conservation. Springer, New York, pp 23–38

Meine C, Soule M, Noss RF (2006) "A mission-driven discipline": the growth of conservation biology. Conserv Biol 20:631–651

Pauly D (1995) Anecdotes and the shifting baseline syndrome of fisheries. Trends Ecol Evol 10:430

Rick TC, Lockwood R (2013) Integrating paleobiology, archeology, and history to inform biological conservation. Conserv Biol 27:45–54

Sala E, Ballesteros E, Dendrinos P et al (2012) The structure of Mediterranean rocky reef ecosystems across environmental and human gradients, and conservation implications. PLoS One 7:e32742

Savarese M (2018) Effectively connecting conservation paleobiological research to environmental management: examples from Greater Everglades' restoration of southwest Florida. In: Tyler CL, Schneider CL (eds) Marine conservation paleobiology. Springer, Cham, pp 55–73

Schneider CL (2018) Marine refugia, past, present, and future: lessons from ancient geologic crises for modern marine ecosystem conservation. In: Tyler CL, Schneider CL (eds) Marine conservation paleobiology. Springer, Cham, pp 161–206

Sepkoski JJ (1984) A kinetic model of Phanerozoic taxonomic diversity. III. Post-Paleozoic families and mass extinctions. Paleobiology 10:246–267

Silvertown J, Poulton P, Johnston E et al (2006) The park grass experiment 1856-2006: its contribution to ecology. J Ecol 94:801–814

Simões MG, Rodrigues SC, Kowalewski M (2009) *Bouchardia rosea*, a vanishing brachiopod species of the Brazilian platform: taphonomy, historical ecology and conservation paleobiology. Hist Biol 21:123–137

Smith JA, Durham SR, Dietl GP (2018) Conceptions of long-term data among marine conservation biologists and what conservation paleobiologists need to know. In: Tyler CL, Schneider CL (eds) Marine conservation paleobiology. Springer, Cham, pp 23–54

Soule ME (1985) What is conservation biology? Bioscience 35:727–734

Szabó P, Hédl R (2011) Advancing the integration of history and ecology for conservation. Conserv Biol 25:680–687

Tyler CL (2018) A conceptual map of conservation paleobiology: visualizing a discipline. In: Tyler CL, Schneider CL (eds) Marine conservation paleobiology. Springer, Cham, pp 225–252

Willis KJ, Birks HJB (2006) What is natural? The need for a long-term perspective in biodiversity conservation. Science 314:1261–1265

Should Conservation Paleobiologists Save the World on Their Own Time?

Gregory P. Dietl and Karl W. Flessa

Abstract Conservation paleobiology is an applied science. Academically based conservation paleobiologists can put their science to work in field applications, informing environmental policy and management, or both. This essay provides some thoughts on how conservation paleobiologists can engage with stakeholders and policymakers. Advocacy is one option, but it is not required. The decision to advocate for a particular conservation policy or goal is a personal one and drawing on one's scientific expertise and experience is part of such advocacy. Marshalling facts in support of one's values can strengthen advocacy if one is explicit about one's values and committed to the ethical norms of science. Styles of engagement with the policy world will vary: from arbiters of scientific information to advocates for particular solutions to conservation problems. Engaging with policymakers and implementing policy also means accepting that not all facts are equally important and that compromise is often necessary. There are ample resources available to those who decide that they want to get involved: take advantage of workshops, formal programs, and publications about science communication and the science–policy interface. Making decisions based on evidence is not a partisan issue—don't be afraid to engage.

Keywords Conservation biology · Conservation science · Objectivity · Science advocacy · Science communication · Values

G. P. Dietl (✉)
Paleontological Research Institution, Ithaca, NY, USA

Department of Earth and Atmospheric Sciences, Cornell University, Ithaca, NY, USA
e-mail: gpd3@cornell.edu

K. W. Flessa
Department of Geosciences, University of Arizona, Tucson, AZ, USA
e-mail: kflessa@email.arizona.edu

C. L. Tyler, C. L. Schneider (eds.), *Marine Conservation Paleobiology*,
Topics in Geobiology 47, https://doi.org/10.1007/978-3-319-73795-9_2

1 Always Academicize?

Stanley Fish's 2008 book, *Save the World on Your Own Time*, made an unabashed public defense of what he thought academics could "legitimately (as opposed to presumptuously)" claim to be able to do as part of their job. Fish argued that it was imperative always to "academicize" issues. By this he meant that we should always separate an issue from contexts that pose a choice of what to do—describe a specific policy option, not advocate for or against it. According to Fish (p. 169), once you start engaging in "discussions designed to produce action in the world, you are surely doing something, but it is not academic, even if you give it that name."

Fish made a valid point, but one that seems at odds with the job description of conservation scientists (including conservation paleobiologists) whose research sits at the nexus of science and politics. Today it is increasingly accepted by public and policymakers alike that scientists should become more closely involved in the search for solutions to major environmental problems (Baron 2016). These requests for increased participation, however, raise important questions about how a conservation paleobiologist could engage with the public in the process of solving conservation problems.

Here, our goal is to raise awareness in the conservation paleobiology community about the options available in making the leap from quietly describing the facts (Fish's academicizing) to entering the policy world to inform decision-makers and help implement policy. We also offer practical advice for advocating responsibly and how to begin. We felt compelled to share our thoughts on the obviously deeply personal decision to venture into the policy world because we believe fear of "crossing an illusory line into advocacy" (Garrard et al. 2016) might stop many conservation paleobiologists from participating in the process.

2 To Advocate, or Not to Advocate

Are scientists uniquely qualified to offer trustworthy advice about which policies might best help conserve biodiversity or ecosystem services? Only a decade ago, many scientists would have answered that such active advocacy should be avoided (Lach et al. 2003), fearing that doing so might damage one's scientific credibility, objectivity, and impartiality, while at the same time eroding the public's confidence in their work (e.g., Lackey 2007; Nelson and Vucetich 2009). But, the times are changing. The last decade has witnessed a radical shift in the mindset of many scientists (Baron 2016). Discussing the value of outreach by scientists to the public and policymakers, Nancy Baron quipped that the question is no longer "Should I do this?, but rather, How do I do this?" (Baron 2016, p. 518). For instance, Singh et al.'s (2014) survey of more than 500 scientists from more than 30 countries worldwide who attended ecological conferences, such as the annual Ecological Society of America meeting, showed that many of them wanted to engage more with policymakers as advocates for political action. Nelson and Vucetich (2009) have also shown that many traditional arguments against science advocacy (e.g., it conflicts with science because the purpose of science is to remain neutral and value-free; Nielsen 2001), really say more about how you should advocate responsibly than whether you should advocate in the first place.

Our own view is that there is nothing inappropriate about wanting to speak out about the very real conservation implications of your research. Because conservation paleobiologists (and all scientists) are citizens first and scientists second, they should have the right to advocate[1] (or not at all) as long as it is done responsibly.

3 Speaking Honestly to Power

What does it mean to be a responsible advocate? For a conservation paleobiologist—or any scientist—to sustain a credible and effective presence in the public sphere, we believe that they must, above all else, be honest—to their science and to their values. To paraphrase the late climate scientist Stephen Schneider, first you need to make your own values and biases "conscious," then, you have to make them explicit, and not assume that others hold them. Some introspection will help. For starters, try asking yourself: What are my personal values and prior beliefs and how do they bias my choice of methods and interpretations (Garrard et al. 2016)? Let's say that you value biodiversity for its own sake and believe that it is currently threatened. Given these personal positions you should be mindful that: (1) your methodological decisions (e.g., sampling design, analytical approach) may enhance your ability to interpret your results as "positive"—that is, to find what you expected (or wanted) to find; and, (2) because biodiversity issues are often emotionally charged, your feelings may come into play when interpreting your data (e.g., when choosing to focus on (or ignore) data that are consistent/inconsistent with a preferred policy option) (Wiens 1997). "No one is exempt from prejudices and values, but the people who know when they are bringing in values and make their biases explicit are more likely to provide balanced assessments—and to be able to single out those who do not" (Schneider n.d.).

When speaking honestly to power, it's also good to be conscious of who you are speaking for. A lot of environmental decision-making involves "stakeholders"—those individuals, groups, agencies, landowners, industries, etc., whose interests will be affected by the decisions that must be made. Academic scientists have a long tradition of advising stakeholders, but are they stakeholders themselves?—not in the traditional sense. Our livelihoods are rarely affected by the pending decisions over

[1]We also prefer to use the term "advocate" and not Sara Goldrick-Rab's (2014) related term "scholar activist." In our opinion, the distinction matters because being a conservation paleobiologist—unlike a scholar activist—means that we start with a goal in mind (e.g., the conservation of biodiversity and ecosystem services), and work for change to achieve this goal (Dietl 2016).

land use, emissions, recovery plans, protected areas, etc. We are usually careful to avoid speaking for our employers, funding sources, or professional societies. We speak for ourselves (after all each of us is a citizen entitled to preferences), but must protect the credibility of scientists everywhere by advocating for a policy option only when there is a rigorous and objective scientific basis for taking that position.

4 From Pure Scientist to Honest Broker

The way that you advocate is your choice. Suppose a politician asks you whether she should support a proposed regulation to stop the spread of an invasive species. What should you do?—try to convince her to support the proposal; only tell her about the current scientific findings on the issue; answer her questions about the science of the issue; or, explain the merits of the proposed regulation as well as several alternatives to it? This is not an easy choice to make, made even more difficult because most scientists are politically naïve and not aware that they even have a choice.

In his 2007 book, *The Honest Broker: Making Sense of Science in Policy and Politics*, Roger Pielke stressed that scientists have choices in what role they play in the policy process. His taxonomy is not prescriptive; its utility lies in illustrating many, though likely not all, of the roles available to scientists in the policy realm.

Pielke argued that when there is widespread public agreement and low scientific uncertainty concerning a potential decision the "linear model" of science and politics, in which science advice comes before and compels political action (Jasanoff 2008), applies and scientists can provide input as "Pure Scientists" who simply provide knowledge or as "Science Arbiters" who serve as advisors in developing policy. A pure scientist makes no effort to engage in advocacy or to promote certain policies outside of how they might communicate by way of peer-reviewed scientific papers. As Pielke explained, "...research results in findings that are placed into a reservoir of knowledge where they will be available to all decision-makers" (p. 15). Like the pure scientist, the science arbiter expresses no opinion on policy preferences, nor does the science arbiter proactively discuss policy at all. Objectivity is the scientist's primary goal. Yet the science arbiter departs from the pure scientist in that they make themselves available to policymakers by providing answers to questions that can be addressed using the tools of science. We suspect that many conservation paleobiologists would feel most comfortable as a pure scientist or science arbiter, as these roles reflect entrenched dichotomies between science and society. Pielke, however, argues that far from keeping politics out of science, acting in accord with the linear model encourages the mapping of political interests onto science, i.e., it leads to a politicization of science—whether the issue is climate change or biodiversity loss—when partisan lobbyists "cherry-pick" scientific results as a means to negotiate for a desired political outcome (see Pielke 2006).

When conflicts in values and a high degree of scientific uncertainty are contributing to disagreements, scientific findings alone cannot lead to a decision or action. For example, with many conservation issues, high scientific uncertainty makes it difficult to exclude a wide range of future outcomes. In such contexts, the expectations that science will resolve political conflict almost always fall short because science provides an "excess of objectivity" that can be used to support a range of "subjective, political positions on complex issues such as climate change . . . or endangered species" (Sarewitz 2000, p. 90). According to Pielke, scientists thus face the question of whether to try to broaden or reduce the range of policy options on the table. If scientists choose to act as "Issue Advocates," they aim to reduce the scope of choices toward one preference or a small set of related preferences. Issue advocates thus attempt to steer policymakers and public opinion to promote a favored issue position (e.g., support for the Aichi biodiversity targets[2]), one they believe is best according to their objective scientific judgment. On the other hand, if scientists seek to broaden the range of options, they become "Honest Brokers of Policy Alternatives" that clarify existing options and identify new options. By expanding thinking, the honest broker encourages decision-makers to think beyond personal values and ideologically driven biases to consider alternatives that may make sense from diverse viewpoints.

Inevitably, to have any real-world impact, most conservation paleobiologists will find themselves filling one (or a combination) of Pielke's advocacy roles. And you may fit one category on one issue (or day) and another category on another issue (or day). We highlight two case studies from among our own ranks. Both are lessons on how to embrace serendipity.

In the media blitz that followed the publication of a *Nature* paper (Barnosky et al. 2012) reporting that our planet faced a "planetary-scale tipping point," conservation paleobiologists—Tony Barnosky and Liz Hadly—found themselves unexpectedly being catapulted into the political fray when the governor of the state of California—Jerry Brown—called. He wanted to ask them questions about the science in the paper, but, importantly, also to translate the science in the paper into something that policymakers could use (Gewin 2014; Barnosky et al. 2014). The resulting consensus document (Barnosky et al. 2014) informed climate policy discussions (Fig. 1) and helped facilitate two international agreements to reduce greenhouse-gas emissions and develop green technology—a memorandum of understanding between two of the world's largest economies, China and California,[3] and the Pacific Climate Pact between the states of California, Oregon, Washington, and British Columbia[4] (see Barnosky et al. 2014 for further details). Throughout the process, Barnosky and Hadly had to make personal decisions on how they would get involved. They have been open about their preferred form of advocacy, what they call "informative advocacy" (Hadly and Barnosky 2014), which is similar in spirit to Peilke's honest broker role. According to Hadly and Barnosky, informative advocacy:

[2]https://www.cbd.int/sp/targets/.

[3]http://gov.ca.gov/news.php?id=18205.

[4]http://gov.ca.gov/news.php?id=18284.

Fig. 1 California's governor, Jerry Brown, presenting consensus statement at NASA's Ames Research Center in 2013. Reprinted by permission from Macmillan Publishers Ltd.: *Nature*, vol. 511: 402–404, copyright 2014

> ... inject[s] ... scientific realities into the many different categories of information that decision makers must take into account when formulating policy. Informative advocacy also has a second goal that is critical: learning from decision makers about the kind of information they need. This back-and-forth dialog ultimately opens new doors for decision makers to formulate solutions to complex problems, and new doors for scientists to understand how their science is socially relevant. (Hadly and Barnosky 2014)

Informative advocacy attempts to expand the range of options, recognizing that there are multiple "scientifically sound paths to desired destinations" (Hadly and Barnosky 2014).

Note that Barnosky and Hadly depended on the media to make the connection between the science and politics of the issue and that initial connection was made at a high level. Their experience contrasts with our second example of how conservation paleobiologists are more likely to become engaged.

Karl Flessa's story starts at a particular place: the mudflats of the Colorado River delta. His lab's work documented, in a scientifically rigorous fashion, what everyone should have expected: that turning off the supply of Colorado River water had a profoundly deleterious effect on its estuary in the upper Gulf of California (Calderon-Aguilara and Flessa 2009). Others had already documented the damage to the river's riparian zone. Along the way, Flessa facilitated collaborations among Mexican and US scientists from universities, conservation groups, and agencies. He would have likely been happy to continue in his roles as conservation paleobiologist and science-facilitator.

But drought happened. Southwestern water agencies wanted to see if a dormant desalination plant at the US-Mexico border could help supplement dwindling supplies. But even a test run of the plant would increase the salinity and decrease the amount of water that supplied an ecologically important Mexican wetland. Conservation groups and agencies from the USA and Mexico came up with a possible solution and asked Flessa to organize a binational science team to monitor the wetland during the desalination plant's trial run. No harm resulted. The conservation groups were pleased, the scientists got to publish papers (Glenn 2013), and the water agencies now include some of the costs of environmental protection in their estimates of the economic feasibility of this desalination plant.

But then an earthquake happened in 2010. It damaged Mexico's system of irrigation canals and the USA agreed to store Mexico's annual allocation of water in a US reservoir, Lake Mead. This was a good deal for the USA as well. Keeping the water level high in Lake Mead provided some insurance against a shortage that would decrease the delivery of water to Arizona, Nevada, and eventually, California. A five-year agreement, called Minute 319 (IBWC 2012) (a "Minute" is an explicit agreement within the confines of the 1944 water treaty between the USA and Mexico) allowed Mexico to continue to store part of its allocation in Lake Mead, stipulated the degree to which Mexico and the USA would share in any future water shortages or surpluses and, importantly for this story, arranged for a transboundary environmental flow (Fig. 2) to see what its effects might be on the hydrology and environment of the Colorado River delta's riparian zone (Flessa et al. 2013). Flessa and others organized another binational science team to monitor the effects of this flow. The flow benefited the trees, birds, groundwater, and people of the riparian zone (Flessa et al. 2016; Bark et al. 2016). Scientists deployed to respond to the policymakers' needs to account for the effects of the environmental flows. And the delta science team is now playing a major role in designing scenarios for future flows. Policy is shaping the science that gets done; science is shaping the policy that develops.

Did analyses of dead clams (i.e., conservation paleobiology) result in the first transboundary environmental flow to the Colorado River delta? No. Did it help scientists take a more active role in decision-making over transboundary water? Yes. Did it move Flessa closer to the policy sphere? Yes, but that was his choice. In the context of Pielke's taxonomy, did Flessa act as a science arbiter, an issue advocate or an honest broker?—a little of each, depending on the circumstances.

These two examples illustrate three essential points. First, exactly which type of advocacy you feel comfortable with, and competent at, will vary among individuals. That is fine—it's a personal decision. You need to find your comfort zone (Flessa 2017). Pielke's taxonomic scheme makes it clear that you have a range of options to consider—all of which have their roles in informing policy decisions—about what best suits your personal situation. The important point to remember is that your choice matters. Second, pathways to engagement with policymakers vary. A few will be high-profile, media-intense events. More are likely to involve on-the-ground collaborations with agencies or conservation groups. And, third, because natural resource issues are often politically charged, all productive engagements will require reciprocity: that scientists understand the needs and limitations of the policymakers, and that policymakers understand the capabilities and limits of the scientists.

Fig. 2 Gates of Morelos Dam opening on March 23, 2014, sending the first transboundary environmental flow into the Colorado River delta. Photo Credit: Karl Flessa

5 Keeping It Real

Decisions about environmental policies and their implementation are rarely—if ever—made on the basis of scientific facts alone. This reality is not a matter of "facts" vs. "feelings." Facts occur in a social context and they never speak for themselves. "If you come from an angle of 'science is the truth you just don't understand the facts', people will not be willing to listen to you" (Xavier Basurto, in Gewin 2017, p. 426). Accept that policy decision-making is not always rational and some facts are more important than others. The facts of climate change will not win over the denialists. The facts of evolution will not win over the creationists. And, in some cases, it's not even a deficit of scientific facts that stands in the way of action. Ideology and competing stakeholder interests can stand in the way. But, this does not mean that scientists should not engage with the public or with policymakers. It simply means that you can't just assume that facts will always win the day.

In addition, there may be many legitimate "facts" at play. For example, the Colorado River's simulated spring flood of 130 million cubic meters was much less than the 320 million cubic meters called for by the conservation-group scientists. Constraints on delivery mechanisms were also facts: the amount delivered was the maximum that the channel and canals could provide without flooding fields and urban areas. The farmer's irrigation contracts were also facts. The water was delivered in late March/early April—months before the spring floods of the pre-dam past. Delivering environmental water later was not possible because channel and canal capacity was already dedicated toward meeting irrigation needs at that time. As a "new" user of the river's water, the environmental groups needed to accommodate the facts on the ground. They decided to avoid letting the perfect be the enemy of the good. They compromised.

6 Overcoming the Fear Factor

By this point, many of you who want to bridge the gap between your conservation paleobiology research and the "real world" may be thinking to yourself that you don't know where to begin. Baron (2016, p. 518) evocatively called this gap—between desire and action—a "valley of death" that can stop scientists from engaging in policy issues. As academic scientists, you probably have little formal training in public engagement (Kelley et al. 2018) and may find the prospect of trying to inject your scientific findings into a public debate a little terrifying—or even a lot. Fortunately, there are resources to turn to for help to get you started.

Nancy Baron's 2010 book, *Escape from the Ivory Tower: A Guide to Making Your Science Matter*, and Lee Badgett's 2016 book, *The Public Professor: How to Use Your Research to Change the World*, both offer accessible and practical advice to scientists looking to engage with the policy world outside of academia. For instance, Badgett boils down the things that effective "public professors" do well, and that everyone else—from senior scientists who have reached a point in their careers when they decide to focus on channeling their scientific expertise toward the public good to newly minted PhDs who share their generation's commitment to community service—should learn how to do better: (1) seeing the bigger picture of a policy debate to identify a role for your research; (2) developing personal relationships with users of scientific information—such as journalists, policymakers, and business leaders—who will be more willing to trust what you have to say if they already know you; and (3) communicating[5] your research outside academic circles (e.g., volunteering your expertise to an advocacy group, writing an op-ed piece, getting your institution to write a media release about your work, reaching out to the agency personnel responsible for implementing environmental policies, participating in legislative visit days, testifying at legislative hearings). Honing and enhancing these public engagement skills should be a high priority for all conservation paleobiologists.

Two widely respected organizations that provide training opportunities to bridge the gap are COMPASS and the Leopold Leadership Program. The mission of COMPASS is "to help scientists effectively share their knowledge in the public discourse and decision-making."[6] According to co-founder Jane Lubchenco the Leopold Leadership program[7] was envisioned as "a leadership and communication training program to help environmental scientists become more effective communi-

[5]A review of the extensive literature on the effective communication of science is beyond the scope of this essay. Nevertheless, how you deliver the facts can be as important as the facts themselves. Randy Olson's 2009 book, *Don't Be Such a Scientist: Talking Substance in an Age of Style*, and Cornelia Dean's 2012 book, *Am I Making Myself Clear?: A Scientist's Guide to Talking to the Public*, provide readable introductions.

[6]http://www.compassonline.org/.

[7]http://leopoldleadership.stanford.edu/.

cators of science to the public and policy makers."[8] Since their founding back in the late 1990s, these programs have helped train thousands of scientists, improving the flow of accurate, credible scientific information to policy makers, and the general public alike on critical conservation issues. Conservation paleobiologists should follow their lead and take advantage of the training opportunities that these programs offer.

Engaging in the policy arena is not easy. If you decide that using your science to shape policy is what you want to do, the opportunities are many and resources are there to help. Don't let fear paralyze your desire to get involved.

7 Later Is Too Late

In the coming years, conservation paleobiologists will (and should) continue to academicize conservation issues, big and small. But, if conservation paleobiologists ever had the luxury of remaining above the political fray, they no longer have it. We know, as scientists, that the status quo is producing the sixth mass extinction event. We believe that conservation paleobiologists who desire to inform public policy to help solve this problem must do more than academicize the issues involved and increasingly engage in the process. This need is greater now than ever before. The examples we have highlighted illustrate that conservation paleobiologists can emerge as powerful agents of change. More of us, however, still need to learn how to bridge the gap between our scholarly work and the public that so urgently needs it. Never underestimate the worth of your contribution.

Acknowledgments We thank Carrie Tyler and Chris Schneider for their invitation to contribute to this volume and their thoughtful suggestions to improve the paper.

References

Badgett MVL (2016) The public professor: how to use your research to change the world. New York University Press, New York, p 256

Bark RH, Robinson J, Flessa KW (2016) Tracking cultural ecosystem services: water chasing the Colorado River restoration pulse flow. Ecol Econ 127:165–172

Barnosky AD, Hadly EA, Bascompte J et al (2012) Approaching a state-shift in Earth's biosphere. Nature 486:52–56

Barnosky AD, Hadly EA, Dirzo R, Fortelius M, Stenseth NC (2014) Translating science for decision makers to help navigate the Anthropocene. Anthr Rev 1:160–170

Baron N (2010) Escape from the ivory tower: a guide to making your science matter. Island Press, Washington, p 272

[8]http://oregonstate.edu/ua/ncs/archives/1998/aug/new-aldo-leopold-program-help-scientists-reach-out.

Baron N (2016) So you want to change the world? Nature 540:517–519
Calderon-Aguilera LE, Flessa KW (2009) Just add water? Transboundary Colorado River flow and ecosystem services in the upper Gulf of California. In: López-Hoffman L, McGovern ED, Varady RG, Flessa KW (eds) Conservation of shared environments: learning from the United States and Mexico. University of Arizona Press, Tucson, pp 154–169
Dean C (2012) Am I making myself clear?: A scientist's guide to talking to the public. Harvard University Press, Cambridge, p 288
Dietl GP (2016) Brave new world of conservation paleobiology. Front Ecol Evol 4:21
Fish S (2008) Save the world on your own time. Oxford University Press, New York, p 189
Flessa KW (2017) Putting the dead to work: translational paleoecology. In: Dietl GP, Flessa KW (eds) Conservation paleobiology: science and practice. University of Chicago Press, Chicago, pp 283–289
Flessa KW, Glenn EP, Hinojosa-Huerta O, de la Parra-Renteria CA, Ramirez-Hernandez J, Schmidt JC, Zamora-Arroyo F (2013) Flooding the Colorado River Delta: a landscape-scale experiment. Eos 94:485–486
Flessa KW, Kendy E, Schlatter K (2016) Minute 319 Colorado River delta environmental flows monitoring interim report, International Boundary and Water Commission. http://www.ibwc.state.gov/Files/Minutes%20319/2016_EFM_InterimReport_Min319.pdf.
Garrard GE, Fidler F, Wintle BC, Chee YE, Bekessy SA (2016) Beyond advocacy: making space for conservation scientists in public debate. Conserv Lett 9:209–212
Gewin V (2014) Hello, Governor. Nature 511:402–404
Gewin V (2017) Communication: post-truth predicaments. Nature 541:425–427
Glenn EP (ed) (2013) Special issue: Colorado delta wetlands. Ecol Eng 59:1–184
Goldrick-Rab S (2014) On scholarly activism. Contexts (blog). https://contexts.org/blog/on-scholarly-activism/. Accessed 4 Dec 2016
Hadly EA, Barnosky AD (2014) Beyond science communication: informative versus prescriptive advocacy. ConsenusForAction Blog. http://consensusforaction.stanford.edu/blog/beyond-science-communicatio.html. Accessed 5 Jan 2016
International Boundary and Water Commission [IBWC] (2012) Interim international cooperative measures in the Colorado River Basin through 2017 and extension of Minute 318 cooperative measures to address the continued effects of the April 2010 earthquake in the Mexicali Valley, Baja California. http://www.ibwc.state.gov/Files/Minutes/Minute_319.pdf.
Jasanoff S (2008) Speaking honestly to power. Sci Am 96:240
Kelley PH, Dietl GP, Visaggi CC (2018) Training tomorrow's conservation paleobiologists. In: Tyler CL, Schneider CL (eds) Marine conservation paleobiology. Springer, Cham, pp 207–223
Lach D, List P, Steel B, Shindler B (2003) Advocacy and credibility of ecological scientists in resource decisionmaking: a regional study. Bioscience 53:170–178
Lackey RT (2007) Science, scientists, and policy advocacy. Conserv Biol 21:12–17
Nelson MP, Vucetich JA (2009) On advocacy by environmental scientists: what, whether, why, and how. Conserv Biol 23:1090–1101
Nielsen LA (2001) Science and advocacy are different—and we need to keep them that way. Hum Dimens Wildl 6:39–47
Olson R (2009) Don't be such a scientist. Talking substance in an age of style. Island Press, Washington, p 206
Pielke RA (2006) When scientists politicize science. Regulation 29:28–34
Pielke RA (2007) The honest broker: making sense of science in policy and politics. Cambridge University Press, Cambridge, p 188
Sarewitz D (2000) Science and environmental policy: an excess of objectivity. In: Frodeman R (ed) Earth matters: the earth sciences, philosophy, and the claims of community. Prentice Hall, Upper Saddle River, pp 79–98
Schneider, SH (n.d.) The Scientist-Advocate. In: "Mediarology". https://stephenschneider.stanford.edu/Mediarology/mediarology.html. Accessed 31 Jan 2017

Singh GC, Tam J, Sisk TD, Klain SC, Mach ME, Martone RG, Chan KMA (2014) A more social science: barriers and incentives for scientists engaging in policy. Front Ecol Environ 12: 161–166

Wiens JA (1997) Scientific responsibility and responsible ecology. Conserv Ecol 1:art16

Conceptions of Long-Term Data Among Marine Conservation Biologists and What Conservation Paleobiologists Need to Know

Jansen A. Smith, Stephen R. Durham, and Gregory P. Dietl

Abstract Marine conservation biologists increasingly recognize the value of long-term data and the temporal context they can provide for modern ecosystems. Such data are also available from conservation paleobiology, but the enormous potential for integration of geohistorical data in marine conservation biology remains unrealized. The lack of a common language for data integration and a tendency in each field to measure different variables, at scales that may differ by orders of magnitude, make integration difficult. To better understand how conservation paleobiology can maximize its potential, we conducted a survey of marine conservation biologists working in the United States.

The respondent population included 90 marine conservation biologists from a variety of workplaces (e.g., governmental, academic) and experience levels (<5 years to >25 years). Survey responses indicated that our fields share common conservation goals (e.g., conservation of biodiversity and ecosystem services) and use long-term data in similar ways (e.g., to establish baselines and elucidate trends and patterns). Respondents, however, mostly considered "long term" to refer to decadal timescales and rarely mentioned geohistorical data.

Overall, the survey results suggest conservation paleobiologists have much work to do before geohistorical data are regularly accepted and applied in marine conservation biology. We highlight four takeaways from the results of

J. A. Smith (✉)
Department of Earth and Atmospheric Sciences, Cornell University, Ithaca, NY, USA
e-mail: jas933@cornell.edu

S. R. Durham
Department of Earth and Atmospheric Sciences, Cornell University, Ithaca, NY, USA

Present address: Paleontological Research Institution, Ithaca, NY, USA
e-mail: durham@priweb.org

G. P. Dietl
Paleontological Research Institution, Ithaca, NY, USA

Department of Earth and Atmospheric Sciences, Cornell University, Ithaca, NY, USA
e-mail: gpd3@cornell.edu

C. L. Tyler, C. L. Schneider (eds.), *Marine Conservation Paleobiology*, Topics in Geobiology 47, https://doi.org/10.1007/978-3-319-73795-9_3

our survey that can help conservation paleobiologists integrate their data into marine conservation practice. (1) Conservation paleobiologists must improve their communication with marine conservation biologists inside and outside of academia. (2) One of the most promising areas for integration is investigating climate change and its ecological implications. (3) The types of long-term data that marine conservation biologists want and need are deliverables conservation paleobiologists can provide. (4) Conservation paleobiologists must be proactive in addressing the barriers that hinder the application of long-term data in conservation practice.

Keywords Conservation barriers · Conservation paleobiology · Cross-disciplinary communication · Environmental stressors · Geohistorical data · Survey

1 What is "Long Term"?

In recent years, long-term data (LTD) have been increasingly recognized for their valuable contributions to improving the understanding of ecology, environmental change, ecosystem service management, and conservation of biodiversity (Strayer et al. 1986; Elliott 1990; Starzomski et al. 2004; Lindenmayer et al. 2012; Krebs 2015; Reed et al. 2016). In particular, LTD can help to elucidate the many important ecological processes that occur slowly or infrequently, are temporally variable, are dependent on historical events, and exhibit time lags (Strayer et al. 1986; Davis 1989; Likens 1989; Carpenter and Turner 2001; Starzomski et al. 2004; Lindenmayer et al. 2012). Even so, there is no consensus on how to define "long term." No fewer than seven definitions exist (Starzomski et al. 2004), with the most simplistic among them defining the term as a distinct period of time (e.g., >10 years; Lindenmayer et al. 2012). Others, however, have applied more restrictive, context-specific definitions (Strayer et al. 1986; Starzomski et al. 2004; Lindenmayer et al. 2012). For instance, Franklin et al. (1990) considered long-term to be the recurrence interval of a disturbance (e.g., floods, wildfires, volcanic eruptions) whereas others have used the organism-specific scale of generation time, which can vary from days for a bacterial community to centuries for trees in a forest (Strayer et al. 1986; Carpenter and Turner 2001; Lindenmayer et al. 2012). Although a single definition is likely overly restrictive, it is critical to communicate clearly what is meant by "long term" between studies and, more importantly, between disciplines, as the future success of restoration and conservation efforts likely lies in our ability to assess and integrate information across multiple timescales (Starzomski et al. 2004).

No larger discrepancy exists in the definition of "long term" than that between neontological and geohistorical studies. In a geohistorical sense, "long term" may mean millions of years (Dietl et al. 2015) whereas neontologically the term is rarely used to imply temporal scales beyond years or decades (Lindenmayer et al. 2012). Although rarely given equal consideration to neontological data, when geohistorical data are used they can greatly improve our understanding of long-acting processes and dynamics (Dietl et al. 2015). For example, in understanding the effects of climate change, geohistorical data have played a central role in understanding the changes that have occurred during the last century (Dietl et al. 2015), including biotic responses to climate change (e.g., changes to population size, de Bruyn et al. 2009; range shifts, Dawson et al. 2011).

Despite this potential, the integration of geohistorical data into conservation practice has been slow (Fritz et al. 2013; Wolkovich et al. 2014; Holmes 2015), even with calls for integration from paleontologists and neontologists alike (Dietl et al. 2015; Price and Schmitz 2016; and references therein). For instance, even as geohistorical case studies are becoming more common, Reed et al. (2016) reported a general paucity of ecological data on the implications of climate change. That is, ecological data predating recent climate change are generally unavailable from traditional sources (e.g., ecosystem monitoring, experimental studies). This disjunction is driven by various factors including a proliferation of discipline-specific jargon, a general lack of interaction between researchers in different disciplines, and, traditionally, a tendency in each discipline to measure different variables at timescales that may differ by orders of magnitude (Polly et al. 2011; Fritz et al. 2013; Dietl et al. 2015; Price and Schmitz 2016). These barriers are beginning to break down, but much work remains to achieve full cross-disciplinary integration.

Here we report the results of a survey of marine conservation biologists that we conducted in an attempt to quantify and qualify their conception of "long term" and to identify opportunities for collaboration between marine conservation biologists and conservation paleobiologists. We wanted to know: How is "long term" defined within the marine conservation community? What types of long-term records do marine conservation biologists use? And, what barriers limit the application of LTD in conservation practice? Addressing questions such as these is important for enabling conservation paleobiologists to identify types of long-term data that marine conservation biologists need and to develop corresponding outreach and research directions in conservation paleobiology that are likely to improve the integration of geohistorical and long-term ecological data into ongoing conservation practices.

2 Survey Implementation

The web survey (Appendix 1) was open for responses during three periods. The initial solicitation (September–November 2015) targeted marine laboratories and researchers in the United States with academic, governmental, and non-governmental affiliations. In response to the distribution of respondents from the first solicitation, which included few academics, two additional solicitations were sent out during January 2016 and April 2016, to the Ecological Society of America listserv (ECOLOG-L) and to members of the National Association of Marine Laboratories, respectively, to reach more academics (see Appendix 2 for a detailed description of survey population selection and distribution). We report the combined results of all three solicitations.

The survey consisted of 23 questions on six broad topics: demographics, conservation goals, long-term data, environmental stressors, baselines, and the challenges of applying LTD in marine conservation. Several question types were used, ranging from multiple-choice to short written responses. Survey responses were categorized for each question and all three authors reviewed each of the categorizations (see Appendix 3 for details on response categorization). Analyses using subsets of the data (e.g., gender, workplace) were not conducted due to small sample sizes. A table of the survey responses can be downloaded at http://doi.org/10.7298/X4VM4965.

3 Survey Responses and What They Mean for Conservation Paleobiologists

A total of 90 marine conservation researchers and professionals completed the survey and they represent a broad cross-section of the marine conservation biology community in the United States.[1] The respondent population consisted of a roughly equal number of men ($n = 46$, 51%) and women ($n = 39$, 43%; Fig. 1a) and the majority of respondents identified as white ($n = 78$, 87%; Fig. 1b). Most respondents worked in *Governmental* ($n = 39$, 43%) or *Academic* ($n = 31$; 34%; Fig. 1c) positions and have graduate degrees ($n = 64$; 71%; Fig. 1d). All respondents who indicated their level of education completed at least 1 year of college ($n = 85$, 94%; Fig. 1d). Respondents identified with fields such as *Fisheries*

[1]The results presented here are focused on the entire population of survey respondents and do not consider subsets that may lead to subtle differences in conservation ideologies and predispositions, such as workplace (Braunisch et al. 2012; Laurance et al. 2012; Cook et al. 2013; Pietri et al. 2013) and gender (Kellert and Berry 1987; Czech et al. 2001; Dougherty et al. 2003; Bremner and Park 2007; Mobley and Kilbourne 2013). Preliminary analyses show minor differences between subsets, but these differences were not statistically comparable due to small sample sizes. When differences did occur, they were relatively minor and did not change the interpretation of the survey results as a whole. For instance, respondents identifying their workplace as *Governmental* tended to select shorter timescales (e.g., months) for LTD compared to those who selected *Academic* (e.g., millennia), but both groups chose the decadal scale most often. Similarly, when asked to rank the importance of environmental stressors, women were more likely to give individual stressors higher importance ranks than men, but both genders agreed on the overall order of importance.

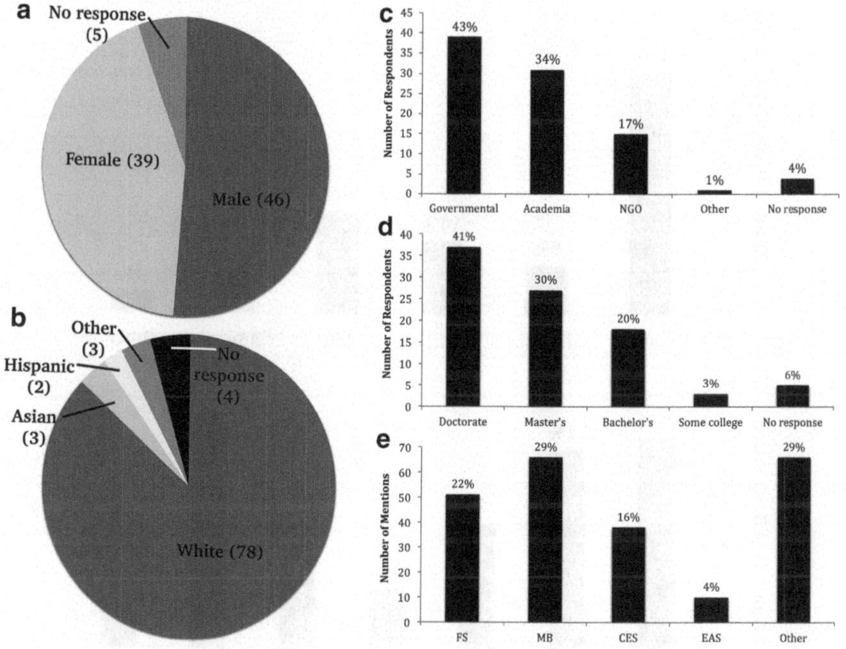

Fig. 1 (**a**) Respondent gender. (**b**) Respondent ethnicity. (**c**) Respondent workplace. (**d**) Highest level of education completed by respondents. (**e**) Respondents' fields of study; *FS* Fisheries science, *MB* Marine biology, *CES* Conservation and environmental sciences, *EAS* Earth and atmospheric sciences. Values in parentheses in (**a**) and (**b**) equal number of responses

science ($n = 51$; 22%) and *Marine biology* ($n = 66$; 29%; Fig. 1e) and filled a variety of conservation-related roles, ranging from *Research* ($n = 49$; 46%) to *Resource management* ($n = 21$, 20%) and project *Administration* ($n = 8$, 7%; Fig. 2a). The survey population also included professionals, across all levels of experience (e.g., <5 years, $n = 20$, 22%; >25 years, $n = 22$, 24%; Fig. 2b), whose work focused on a range of levels of biological organization (e.g., ecosystem, $n = 36$, 40%; Fig. 2c).

Specific responses were variable but, in general, respondents agreed on what overarching goals drive marine conservation biology, what constitutes "long term", the importance of LTD, how to use LTD, and the challenges faced when applying LTD. Perhaps unsurprisingly, conservation paleobiology parallels this consensus in many ways, and the survey results highlight several opportunities for conservation paleobiologists to enhance and expand on the application of LTD in marine conservation.

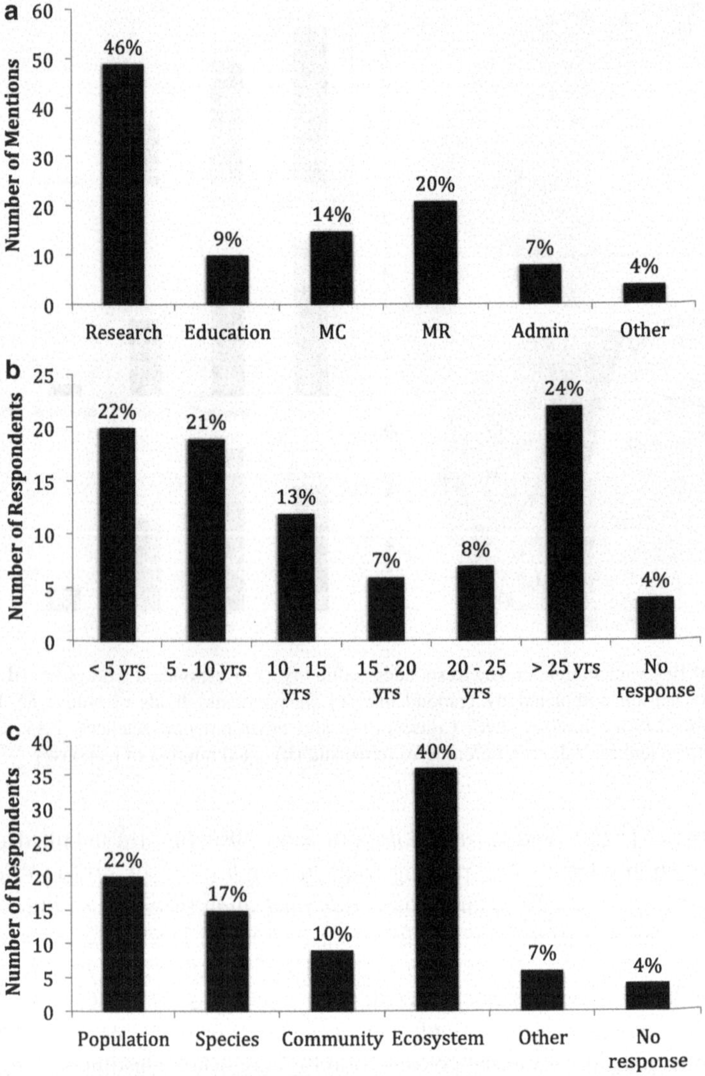

Fig. 2 (**a**) Respondents' conservation related roles within their employing organization; *MC* Management (conservation), *MR* Management (resources). Percentages do not sum to 100% due to rounding. (**b**) Years of experience working in marine conservation biology. (**c**) Level of biological organization at which respondents work or study

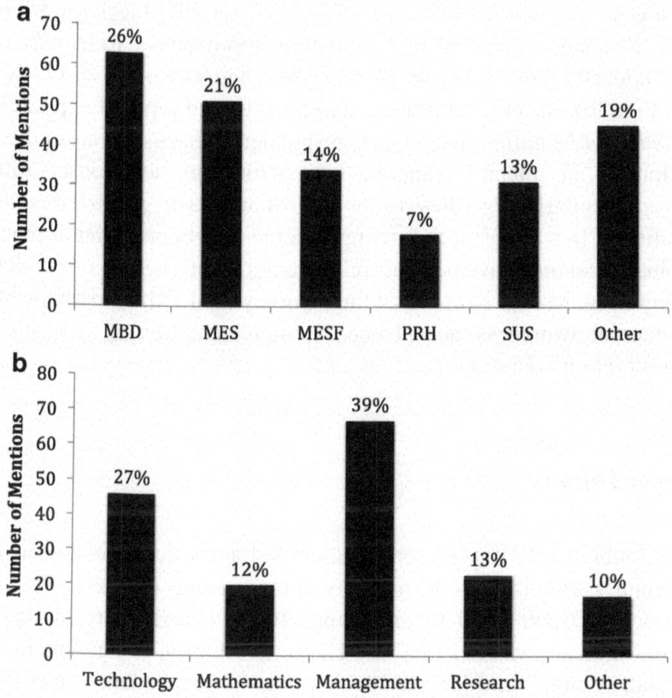

Fig. 3 (**a**) Types of marine conservation biology goals identified by respondents; *MBD* Maintenance of biodiversity, *MES* Maintenance of ecosystem services, *MESF* Maintenance of ecosystem structure and function, *HP* Habitat protection, *SUS* Sustainability. (**b**) Categories of approaches described by respondents to achieve their conservation goals. Percentages do not sum to 100% due to rounding

Conservation Goals

The two conservation goals most commonly mentioned by survey respondents were the *Maintenance of biodiversity* (n = 62, 26%) and *Maintenance of ecosystem services* (n = 51, 21%; Fig. 3a). These goals are well matched by the overarching goal of conservation paleobiology, which is to conserve and restore biodiversity and ecosystem services (Dietl et al. 2015, Dietl 2016). Additionally, geohistorical data have been applied in attempts to achieve the remaining three types of conservation goals mentioned by survey respondents: *Maintenance of ecosystem structure and function* (n = 34, 14%; e.g., Smith et al. 2016), *Protection or restoration of habitats* (n = 18, 7%; e.g., Volety et al. 2009), and *Sustainability* (n = 31, 13%; e.g., Pandolfi et al. 2003). The prevalence of these types of conservation goals in responses highlights the shared principles and philosophical underpinnings of the two disciplines (Dietl 2016) and should form the basis for future collaboration.

To achieve their conservation goals, respondents indicated several types of approaches that they apply, including *Management* (n = 67, 39%; e.g., marine protected areas), *Technology* ($n = 46$, 27%; e.g., satellite tagging), *Mathematics* ($n = 20$, 12%; e.g., statistical and modeling approaches), *Research* ($n = 23$, 13%; e.g., understanding ecosystem processes), and *Other* ($n = 17$, 10%; e.g., education; Fig. 3b). In several instances and for different types of approaches (e.g., *Management, Mathematics, Research*), respondents discussed cross-disciplinary collaboration as an integral component in achieving their conservation goals ($n = 12$, 7%). Similarly, regardless of the type of approach, geohistorical data were rarely mentioned ($n = 5$, 3%). Surprisingly, some novel conservation practices for which geohistorical data have been useful were not mentioned by respondents (e.g., assisted migration, Marris 2011; rewilding, Burney et al. 2012), perhaps indicating a lack of general awareness, acceptance, or implementation of these ideas in the marine conservation community.

Long-Term Data

There is no doubt that the marine conservation biologists we surveyed consider LTD to be important, particularly as the majority of respondents ($n = 60$, 67%) indicated that they use LTD compared to only nine (10%) who do not (21 respondents, 23%, did not indicate whether they use LTD). In addition to agreement that LTD are "essential," "vital," and "incredibly important," respondents most commonly indicated that "long term" refers to the decadal scale ($n = 69$, 49%; Fig. 4a). Relatively few respondents selected timescales of centuries ($n = 21$, 15%) or longer ($n = 18$, 13%) when asked to define "long term" (Fig. 4a), confirming the documented difference in conceptions of the term between neontological and geohistorical disciplines (e.g., Lindenmayer et al. 2012; Dietl et al. 2015). Some respondents did indicate that longer-term temporal data (i.e., beyond decadal timescales) could be useful, particularly when considering humans and their effects on ecosystems. For example, one respondent stated, "... to set goals about the future state of the ecosystem, we need to be aware of conditions of the past (i.e., before industrialization, overfishing activities, etc.)." Another respondent added: "Marine ecosystems work on different time scales than human systems..." Given these responses, and others like them, marine conservation professionals may be receptive to the integration of geohistorical data (Durham and Dietl 2015), particularly as they are often the only source of perspectives beyond the limited time frame of direct human observation (NRC 2005; Dietl et al. 2015).

The distribution of timescales on which respondents considered their sources of LTD to be useful in conservation practice—primarily decades ($n = 204$; 34%), years ($n = 157$, 26%), and months ($n = 68$, 11%; Fig. 4b)—suggests an opening for the integration of geohistorical data. Sources of data on the scale of centuries ($n = 63$, 10%) and longer ($n = 47$, 8%; Fig. 4b) were rare. Furthermore, the

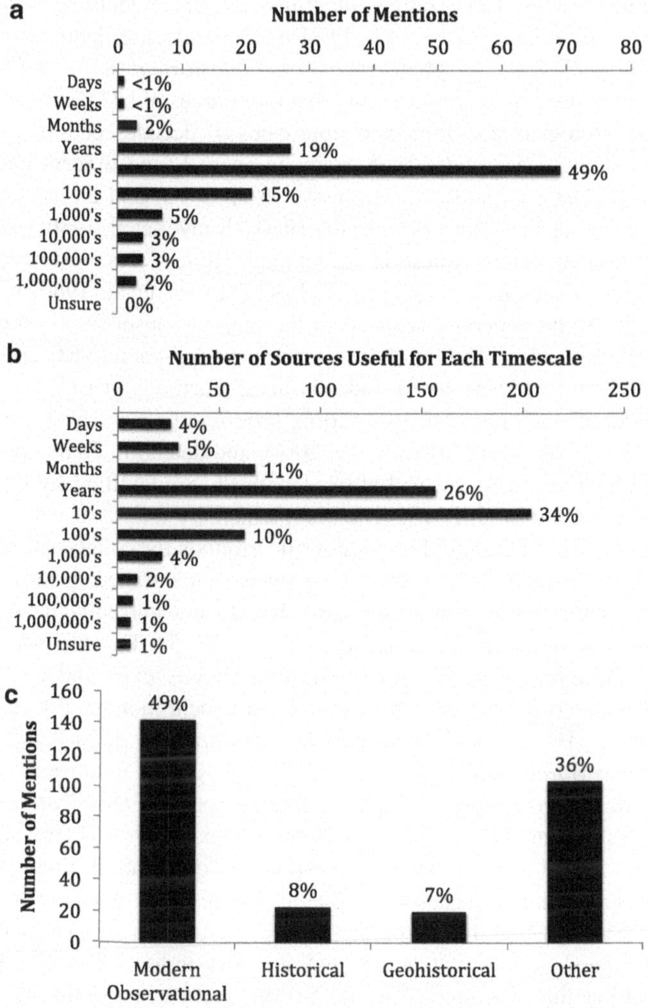

Fig. 4 (**a**) Histogram of respondents' selections of timescales to which the term "long term" applies. (**b**) Histogram of respondents' selections of timescales on which their sources of long-term data can provide information. Percentages do not sum to 100% due to rounding. (**c**) Types of long-term data sources mentioned by respondents

distribution of timescales on which LTD sources were considered by respondents to be informative (Fig. 4b) was shifted towards shorter timescales relative to how they defined "long term" (Fig. 4a). This discrepancy is likely due to the types of data being used. Most LTD sources identified by respondents were categorized as *Modern observational* ($n = 142$; 49%; e.g., monitoring data) as opposed to *Historical* ($n = 23$; 8%; e.g., photographs) or *Geohistorical* ($n = 20$; 7%; Fig. 4c; e.g., sediment cores).[2] It is not surprising that the sources of LTD familiar to marine conservation biologists are limited to timescales of decades or less, particularly given the challenges associated with maintaining real-time observational studies (e.g., funding, data continuity; Lindenmayer et al. 2012) and the general paucity of long-term studies in ecology prior to the 1980s (Strayer et al. 1986; Likens 1989; Franklin et al. 1990). Thus, "questions about longer time scales must be approached by means other than direct long-term studies" (Strayer et al. 1986, p. 4). What does not seem to be generally realized in the marine conservation community is that the application of geohistorical data is one such approach (NRC 2005; Dietl et al. 2015; Durham and Dietl 2015). Indeed, the applications of LTD (regardless of definition) identified by respondents—setting *Baselines* ($n = 50$, 52%), establishing the *Natural range of variability* ($n = 19$, 20%), and detecting *Trends and patterns* ($n = 25$, 26%; Fig. 5a)—are well-aligned with those identified by conservation paleobiologists (Dietl et al. 2015). The complementary nature of our methods for obtaining and using LTD should provide fertile grounds for collaboration.

Conservation paleobiologists should be further encouraged because the types of LTD that marine conservationists want, but do not currently have access to, including *Abiotic* ($n = 15$, 21%) and *Biotic* ($n = 38$, 54; Fig. 5b) variables, map neatly onto the types of geohistorical data that already exist and continue to be collected. Considering mollusks, for example, sclerochronological techniques can generate data on *Abiotic* variables such as *Temperature* ($n = 3$, 20%; e.g., Goodwin et al. 2001) and *Water chemistry* ($n = 10$, 67%; e.g., Cintra-Buenrostro et al. 2012; Fig. 5c), as well as on *Biotic* variables including species' *Distributions* ($n = 11$, 29%; e.g., Smith and Dietl 2016) and *Interactions* ($n = 4$, 11%; e.g., Cintra-Buenrostro et al. 2005; Fig. 5d). Separately, or in combination, these data can be applied to assess human impact on ecosystems. For instance, geohistorical evidence of millennial-scale variability in predator-prey interactions suggested that fishing pressure pre-dated pollution stress in benthic ecosystems of Long Island Sound, USA, indicating that ongoing efforts to restore these ecosystems by addressing nitrogen pollution alone may not succeed if harvest pressure is not also further controlled (Casey et al. 2014).

[2] Three of the four major types of data sources identified in marine historical ecology (sensu Lotze and McClenachan 2014; Jackson and McClenachan 2017), a sister field of conservation paleobiology, were also conspicuously absent. Data types largely absent were geological (e.g., sediment cores), archaeological (e.g., middens), and historical narrative (e.g., accounts of explorers), whereas the fourth, modern scientific and fisheries data (i.e., *Modern observational data* of this study), was mentioned commonly.

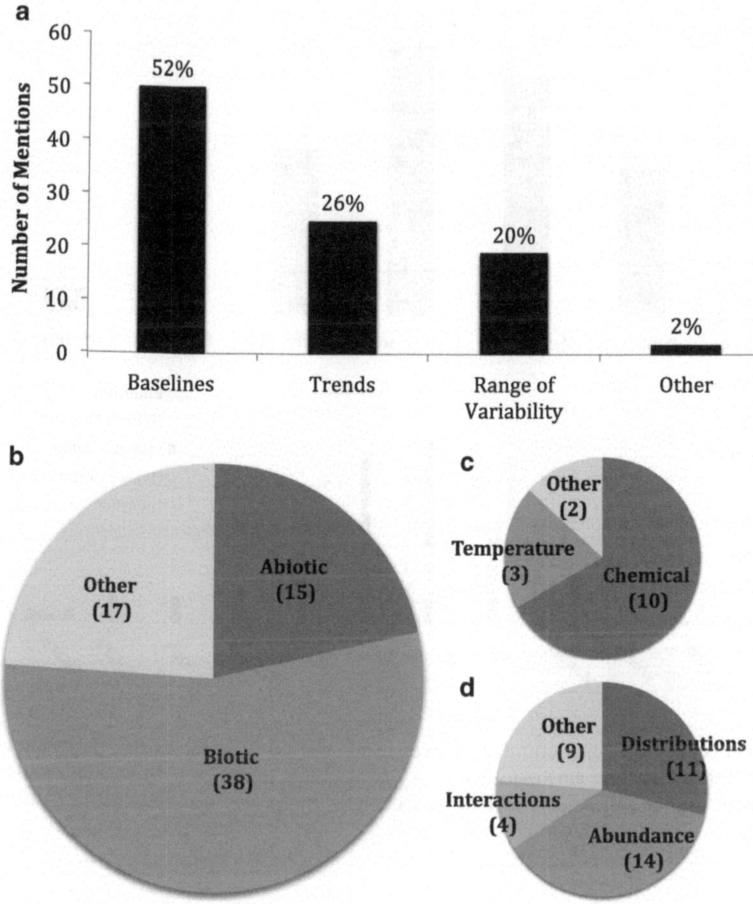

Fig. 5 (**a**) Types of reasons given by respondents in describing how long-term data are used. (**b**) Types of data that respondents would like more of or that are not currently available. (**c**) Breakdown of *Abiotic* data types from 5b. (**d**) Breakdown of *Biotic* data types from 5b. Values in parentheses in (**b**), (**c**), and (**d**) equal number of responses

Environmental Stressors

When respondents were asked to rank the five major stressors identified in the 2005 Millennium Ecosystem Assessment report (pollution, habitat change, climate change, overexploitation, and invasive species; MEA 2005) in order of importance, they most often gave climate change a rank of one (highest importance; $n = 36$ #1 ranks, mean $= 2.18$) and also ranked habitat change as highly important ($n = 32$ #1 ranks, mean $= 2.27$; Table 1). The reported distribution of stressor importance matches well with the distribution of respondents' inclinations for how the LTD sources they provided can be used to address the stressors (Fig. 6a). Whereas

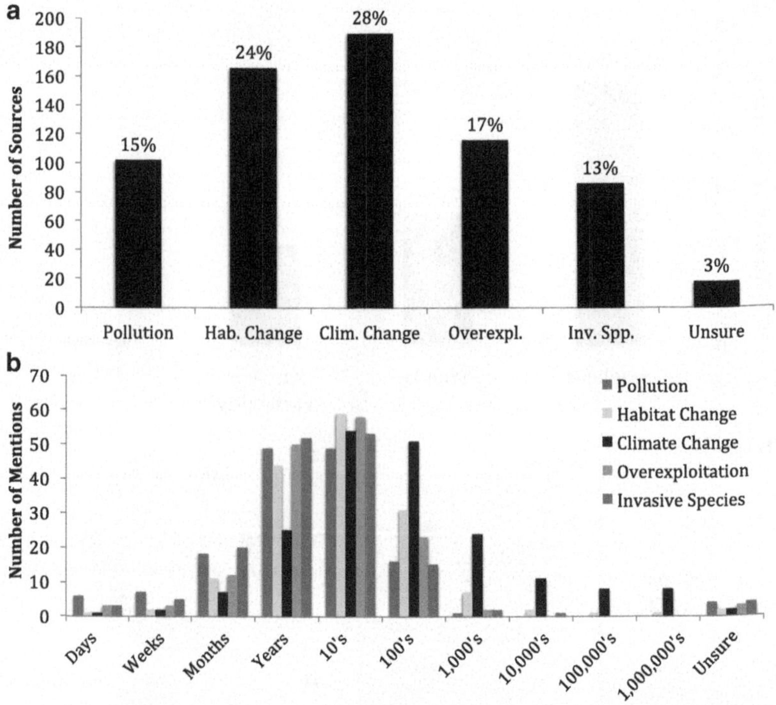

Fig. 6 (**a**) Number of long-term data sources mentioned by respondents that they believe can be applied to each of the environmental stressors identified in the 2005 Millennium Ecosystem Assessment Report (MEA 2005). (**b**) Distribution of timescales identified by respondents as necessary to address each environmental stressor

respondents indicated that data on timescales of decades or shorter would suffice to understand the effects of pollution, habitat change, overexploitation, and invasive species, they recognized that data on timescales of centuries, millennia, or longer are needed to address climate change (Fig. 6b). For pollution, habitat change, overexploitation, and invasive species, the responses approximated a bell curve with a median between years and decades. With climate change responses, however, the median was between decades and centuries and the distribution was skewed towards longer timescales (e.g., thousands to millions of years; Fig. 6b).

The shape of the climate change distribution suggests that marine conservation biologists may welcome data on longer timescales (i.e., centuries, millennia) to address climate change, particularly because the majority of LTD sources provided by respondents were considered to be informative on shorter timescales (i.e., decades or less; Fig. 4b). In light of the importance placed on climate change by respondents (Table 1), the perceived need for data on timescales longer than decades to address biotic responses to climate change (Fig. 6b), and the limited number of LTD sources provided by respondents for timescales greater than decades (Fig. 4b),

Table 1 Respondent ranks of environmental stressor importance (1 = most important, 5 = least important)

	1	2	3	4	5	Unsure	Mean
Pollution	19	25	19	11	8	1	2.60
Habitat change	32	20	14	8	8	0	2.27
Climate change	36	21	9	9	8	0	2.18
Overexploitation	27	19	13	10	12	2	2.60
Invasive species	6	28	23	16	9	1	2.96

conservation paleobiologists have an immediate and unmistakable opportunity to provide highly relevant geohistorical data (MacDonald et al. 2008; Willis et al. 2010). Unfortunately, based on their descriptions of the timescales of data needed to address pollution, habitat change, overexploitation, and invasive species (i.e., predominantly decades or shorter; Fig. 6b), respondents were seemingly unaware of the valuable contributions that geohistorical data can make to understanding their effects (Dietl and Flessa 2011; Dietl et al. 2015; and references therein). Thus, the utility of geohistorical data for helping to address these stressors may be more difficult for marine conservation biologists to accept.

Emphasizing the interactions between climate change and the other four stressors may be a viable path to integrate geohistorical data on all of the stressors because respondents did highlight the importance of interactions between each of these stressors and climate change (Table 2). For example, one respondent stated: “All of these problems are difficult to repair because we do not understand the interactions of multiple drivers and stressors at the ecosystem level.” Hence, because climate change is a long-acting process, data must be from equally long timescales to understand these complex interactions (Dietl et al. 2015). In particular, differentiating between causal factors (e.g., human versus natural) is imperative for reaching conservation solutions, especially given the high incidence of non-additive relationships (e.g., synergistic, antagonistic) between stressors (Crain et al. 2008; Darling and Côté 2008). Each interacting stressor can complicate matters further, as a meta-analysis by Crain et al. (2008) revealed that the addition of a third stressor can double the number of synergistic interactions. As such, the rapidly changing climate we now face can amplify the effects of the other stressors, particularly as we move towards an uncertain future with novel ecosystems (Willis et al. 2010). For example, the poleward spread of nonnative pests and pathogens from the tropics (Bebber et al. 2013) may be exacerbated by a warming climate, just as has been documented during climate shifts in the deep past (Labandeira and Currano 2013). Thus, when interacting with marine conservation biologists, conservation paleobiologists should emphasize that it is not only data on climate change and its biotic effects that are required but also data on the remaining four stressors, especially on the effects of those stressors over timescales (i.e., centuries or longer) that are comparable to the temporal footprint of climate change.

Table 2 Total number of mentions for each environmental stressor in interactions and number of mentions with other stressors[a]

	Total mentions (%)	Habitat change	Climate change	Overexploitation	Invasive species
Pollution	29 (16)	25	25	21	17
Habitat change	47 (26)	–	39	21	20
Climate change	51 (28)	–	–	30	18
Overexploitation	33 (18)	–	–	–	17
Invasive species	19 (11)	–	–	–	–

[a]In 16 instances, respondents selected all five environmental stressors

Baselines

The majority of survey respondents gave equal weight to spatial and temporal data for use in setting baselines ($n = 43$, 66%; Fig. 7a), which is encouraging for conservation paleobiology, particularly as respondents who favored one type of data over the other placed higher value on temporal ($n = 9$, 75%) rather than spatial data ($n = 3$, 25%; Fig. 7b). Even so, many of the baselines they described were from *Modern observational data* ($n = 65$, 42%; e.g., monitoring data) and *Reference sites* ($n = 32$, 21%; e.g., "pristine" reference sites) rather than *Historical* ($n = 25$, 16%; e.g., written records) or *Geohistorical* ($n = 7$, 5%; e.g., death assemblages) sources (Fig. 7c). Given the general consensus that pristine habitats no longer exist (Jackson et al. 2012; Wiens and Hobbs 2015), this practice may be problematic as the comparisons being made do not reflect "healthy" systems and are subject to short-term variation (Kopf et al. 2015). Thus, conservation paleobiologists have an opportunity to provide temporal baselines from geohistorical data (Dietl et al. 2015; Dietl et al. 2016), which can be valuable for smoothing out the short-term variation that is intrinsic to "snapshot" modern surveys (Kowalewski et al. 1998). For instance, the time-averaging and taphonomic inertia inherent to death assemblages in live-dead studies (sensu Kidwell 2007) provide a means to collect data on average conditions over decades or centuries, rather than days or weeks (Kidwell and Tomašových 2013). Geohistorical records can undoubtedly provide baseline data for marine conservation (MacDonald et al. 2008; Willis et al. 2010; Dietl et al. 2015), but conservation paleobiologists should not limit themselves to static baselines concerned with past species diversity and abundance.

In application, baselines are used to define targets for restoration actions; they are the "natural" set of conditions on which goals are established. Many baselines tend to be static and habitually fail to consider humans as a part of the ecosystem (Kopf et al. 2015). As aptly described by one respondent: "Conservation often aims at retaining the status quo or restoring some historic state . . . *this mindset is problematic* . . . reaching an idealized historic state may be impractical or impossible" (italics added). Kopf et al. (2015, p. 803) suggested Anthropocene baselines—"a concept to represent an ecological and theoretical shift from a fixed preindustrial-era reference condition to a dynamic point of reference

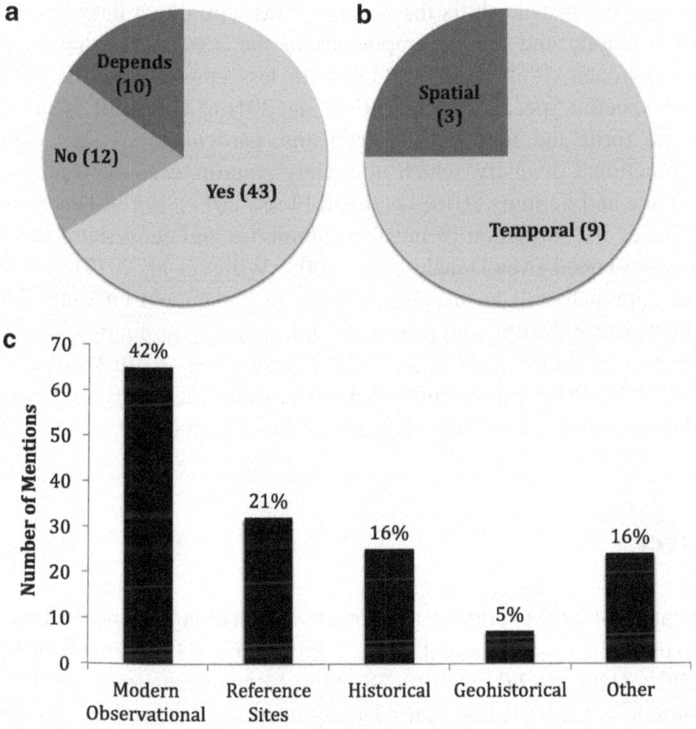

Fig. 7 (**a**) Number of respondents disagreeing (No), agreeing (Yes), or conditionally agreeing (Depends) on the statement that spatial and temporal data have equal value in establishing baseline data or reference conditions. (**b**) Proportion of data types that respondents who disagreed in 7a believe are more valuable for establishing baseline data or reference conditions. (**c**) Types of data used by respondents to establish baselines. Values in parentheses in (**a**) and (**b**) equal number of responses

for human-dominated ecosystems"—as a solution to this problematic mindset. Although this concept may prove to be useful in some management situations, it is reliant on subjective judgment and requires practitioners to recognize the threshold at which ecosystems shift irreversibly to new conditions (i.e., regime shifts). The Anthropocene baselines concept does incorporate human influences and emphasizes a triage approach to conserving biodiversity but the new baselines remain equally static to those from the pre-industrial era. And, for better or worse, many managers and researchers already rely on modern baselines that are human-influenced (Fig. 7c).

Instead, the geohistorical record can provide data on ecosystem function and structure (Eronen et al. 2010; Polly et al. 2011; Fritz et al. 2013; Dietl and Flessa 2017), the maintenance of which is an increasingly important goal for restoration (see Price and Schmitz 2016) and one that was recognized by survey respondents (Fig. 3a). A function-based restoration approach does not discount the importance of biodiversity but instead shifts the focus to traits with direct links to interactions between the abiotic and biotic components of the ecosystem—that is, using an ecometric approach (Polly et al. 2011)—with less emphasis on the presence or absence of specific species (e.g., Dietl et al. 2016). This shift is achieved by studying the form and function of organisms, particularly as they combine to establish functional diversity, which ultimately controls ecosystem processes and services (Price and Schmitz 2016; Dietl and Flessa 2017). For instance, as climate change induces the formation of novel communities and ecosystems that humans have not experienced (MacDonald et al. 2008; Willis et al. 2010), a trait-based, integrative approach will be needed in order to understand ongoing changes in ecosystem structure, health, and processes and to anticipate future changes during the Anthropocene (Starzomski et al. 2004; Eronen et al. 2010; Polly et al. 2011; Fritz et al. 2013; Price and Schmitz 2016; Dietl and Flessa 2017). Conservation paleobiologists can help establish the basis for these evaluations.

Challenges

The application of LTD to conservation practice is challenging and the predominant barriers to the utilization of these data that were identified by respondents (*Funding*, $n = 48$, 36%; *Data availability*, $n = 38$, 28%; *Communication*, $n = 14$, 10%; and *Institutional*, $n = 13$, 10%; Fig. 8) are largely unchanged from those acknowledged decades ago (see Strayer et al. 1986). Even with the establishment of the Long Term Ecological Research program by the National Science Foundation in 1980, these challenges persist. As one respondent explained, "In the project that I work on where we have high quality data sets going back nearly 40 years, we are regularly having to partially refund [sic] the project from new funding sources. When you have a recognized long term data collection program that has these struggles, it is no wonder that there is a paucity of good long term databases directly linked to marine conservation." The utilization of geohistorical data may help overcome these barriers to applying LTD in marine conservation practice.

One of the prominent institutional barriers for marine conservation biologists is the necessity to maintain studies for years or decades to obtain LTD because, as pragmatically described by one respondent, "…it often takes a long time to yield meaningful results." Such studies produce LTD that are valuable, but their longevity requires dedicated leadership, consistent funding, and, not least of all, some good luck (Strayer et al. 1986; Likens 1989; Lindenmayer et al. 2012). The substantial time and effort it takes to collect LTD in real time is a serious obstacle for marine conservation biologists. In particular, several respondents believed that LTD collection is hindered by the perception that "[long-term monitoring] isn't as sexy as some cutting edge work or new venture." Consequently, the relatively rapid and inexpensive assessments (e.g., retrospective studies, Davis 1989;

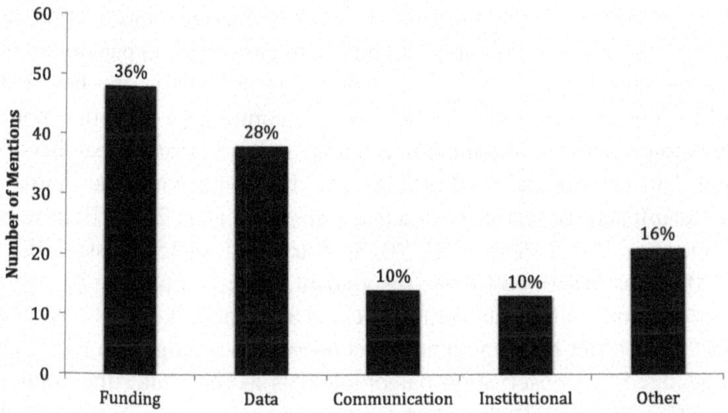

Fig. 8 Types of challenges described by respondents for applying long-term data in conservation practice

live-dead studies, Kidwell 2007) that are possible using geohistorical records may be a useful and resonant selling point for conservation paleobiology. In the right settings (e.g., coastal marine environments), geohistorical records are essentially available on demand. Whereas it is not feasible to monitor an entire coastline, a retrospective geohistorical approach can often provide valuable, location-specific data at relatively low costs (e.g., Dietl and Smith 2016).

Large potential and low costs of geohistorical data notwithstanding, it remains a considerable challenge to secure funding for conservation (Katsanevakis et al. 2015; Fig. 8). Indeed, one respondent noted, "there is a general lack of socio-political will to invest in long-term monitoring, conservation, and management... budgets associated with science and research, particularly climate research, continue to be cut..." Given this troubling trend in funding (see also Kearney et al. 2014; Gauchat 2015; Nadelson and Hardy 2015, Besley 2016; Farrell 2016), conservation science needs innovative solutions, which may be found in cross-disciplinary collaborations (Boulton et al. 2005; Price and Schmitz 2016; Kelley et al. 2018). Marine conservation biologists may not, however, immediately recognize the value of collaborations with conservation paleobiologists, particularly given their lack of awareness of the contributions geohistorical data can make to conservation practice (e.g., on-demand data for all environmental stressors, reduced funding burden). Thus, it will likely be up to conservation paleobiologists to initiate these cross-disciplinary collaborations. In doing so, we should emphasize that, by working together with LTD on complementary timescales, marine conservation biologists may improve their chances of obtaining funding (particularly if they can achieve their conservation goals at lower cost) and, in doing so, simultaneously reduce the previously mentioned institutional barriers (CPW 2012; Boyer et al. 2017; Flessa 2017).

Another barrier to the application of LTD in marine conservation that was discussed by respondents, and applies equally to conservation paleobiology, is the "research-implementation gap" (sensu Arlettaz et al. 2010)—the fact that ideas, data, and techniques generated in the research community are often not applied in conservation practice without explicit outreach efforts from researchers. That is, publishing data in peer-reviewed articles and theorizing within academia are not sufficient to impact conservation practice (Arlettaz et al. 2010; Braunisch et al. 2012; Cook et al. 2013; Pietri et al. 2013; Roux et al. 2015; Flessa 2017; Kelley et al. 2018), or, as described by one respondent, there is a need to be better about "... spreading word of the data being collected and where it can be obtained." This communication barrier is symptomatic of conservation biology as a whole and must be acknowledged by conservation paleobiologists as they enter the applied realm of conservation practice. Rather than wait and hope for practitioners to discover our data, which is uncommon (Braunisch et al. 2012; Griffiths and Dos Santos 2012; Laurance et al. 2012; Boyer et al. 2017), conservation paleobiologists should proactively seek out practitioners as they build cross-disciplinary collaborations and ask what kinds of data are needed (Flessa 2017). Opening this dialogue will allow for geohistorical studies guided by conservation-specific research questions that generate data practitioners will apply.

4 Takeaways for Conservation Paleobiologists

The responses to our survey confirm that "long term" has a different meaning for marine conservation biologists and conservation paleobiologists. Respondents most commonly chose the decadal scale to define "long term" whereas many conservation paleobiologists apply the term to much longer timescales (e.g., centuries and longer; Dietl et al. 2015). Despite this difference, our survey results confirm that our fields are united by the goals that drive them (Dietl 2016) and nearly identical conceptions of the utility of LTD, if not its definition. Adopting a cross-disciplinary approach by integrating data from our complementary sources and timescales is feasible and stands to help us achieve our common goals. We highlight four takeaways from the results of our survey that can help conservation paleobiologists work towards this integration.

1. *Conservation paleobiologists must improve their communication with marine conservation biologists inside and outside of academia.* Respondents rarely mentioned geohistorical data, and were seemingly unaware of their potential to contribute to conservation practice despite the growing number of case studies in conservation paleobiology. Furthermore, respondents indicated that data on decadal timescales are needed to address most ecosystem stressors and the most frequently mentioned barrier to the application of LTD was a lack of funding (often discussed in reference to maintaining costly long-term monitoring studies). This pattern of responses indicates that either marine conservation

biologists are not aware of geohistorical data, or that they do not consider them important for addressing conservation problems.

2. *One of the most promising areas for integration is investigating climate change and its ecological implications.* Although they did not mention geohistorical data, many respondents did recognize that in order to understand climate change, and its effects on ecosystems worldwide, they need data on centennial to millennial scales. Given that their responses also suggested they do not currently have data on these longer timescales, this may be one focus of marine conservation biology where the geohistorical perspectives offered by conservation paleobiology would be immediately welcome.
3. *The types of LTD that marine conservation biologists want and need are deliverables conservation paleobiologists can provide.* Respondents wanted more LTD on temperature, water chemistry, and species abundances, distributions, and interactions. These data are available from marine geohistorical records, so broadcasting our ability to gather them may be beneficial for establishing collaborations with marine conservation biologists.
4. *Conservation paleobiologists can help address the barriers that hinder the application of LTD in conservation practice.* Retrospectively providing LTD on biotic and abiotic variables can help alleviate the challenges with data availability, funding, and institutional barriers that were identified by respondents. Whereas marine conservation biologists often must maintain studies for years or decades to build a long-term dataset, conservation paleobiologists may be able to provide suitable data at considerably lower costs and without the need to gather the data in real time.

5 Moving Forward

Our survey results suggest conservation paleobiologists have much work to do before geohistorical data are regularly accepted and applied in marine conservation biology. We were startled by the lack of awareness of geohistorical data, and their applications in conservation practice, demonstrated by survey respondents. Despite the continued growth of conservation paleobiology as a field over the last decade, our outreach to the marine conservation biology community so far has not been enough. We must redouble our efforts to integrate our data into conservation practice to achieve the full potential of our field. The onus is on conservation paleobiologists to demonstrate the utility of geohistorical data.

Acknowledgments We would like to thank the editors, Carrie Tyler and Chris Schneider, for their invitation to participate in this volume and two reviewers, Michelle Casey and Michael Savarese, whose comments improved the manuscript. We also thank all of those who assisted with survey distribution and those who completed the survey.

Appendix 1: Survey Questions

Demographics and Professional Information

1. With which race/ethnicity do you identify?

   - White
   - Black
   - Asian
   - Hispanic
   - Other: ____________

2. With which gender do you identify?

   - Male
   - Female
   - Other: ____________

3. Please list up to three fields (e.g., fisheries biology, historical ecology, etc.) with which you identify.

4. Which of the following best describes your workplace?

   - Government
   - Nongovernmental organization
   - Academia
   - Other:___________

5. Which of the following best describes your highest completed level of education?

a. Doctorate
b. Master's
c. Bachelor's
d. Other: ____________

6. How many years of experience do you have in marine conservation?

   - <5
   - 5–10
   - 10–15
   - 15–20
   - 20–25
   - >25

7. Please describe your work as it relates to marine conservation in one sentence or less.

8. At what level of biological organization does your work primarily focus?

   - Population
   - Species
   - Community
   - Ecosystem
   - Biome
   - Other: ____________

Goals and Approaches

9. Please list up to three primary goals of the field of marine conservation (e.g., preservation of biodiversity) in your opinion.
10. What are the cutting-edge approaches currently being practiced in marine conservation to achieve the goals you mentioned in the previous question?

Long-term Data: Definitions and Sources

11. In your opinion, to what timescales does the phrase "long-term data" typically refer in the conservation community?

    - Days
    - Weeks
    - Months
    - Years
    - Decades
    - Centuries
    - Thousands of years
    - Tens of thousands of years
    - Hundreds of thousands of years
    - Millions of years
    - Unsure

12. In your opinion, what is the importance of long-term data for achieving the goals of marine conservation?
13. If you use long-term data, how do you use it and why do you use it? Or, if long-term data are not considered in your work, why not?
14. Please list five sources of long-term data and indicate whether you have used each one in your own research.
15. Considering the sources of long-term data you listed previously, at what time scale(s) are data from these sources most useful?

	Days	Weeks	Months	Years	Decades	Centuries	Millennia	10^4 years	10^5 years	10^6 + years	Unsure
Source A	[]	[]	[]	[]	[]	[]	[]	[]	[]	[]	[]
Source B	[]	[]	[]	[]	[]	[]	[]	[]	[]	[]	[]
Source C	[]	[]	[]	[]	[]	[]	[]	[]	[]	[]	[]
Source D	[]	[]	[]	[]	[]	[]	[]	[]	[]	[]	[]
Source E	[]	[]	[]	[]	[]	[]	[]	[]	[]	[]	[]

Long-term Data and Ecological Stressors

16. The Millennium Ecosystem Assessment (2005) identified the following five most-important stressors in ecosystems. Please rate each stressor's importance in marine conservation biology (one being highest importance and five being lowest importance).

	Importance					
	1	2	3	4	5	Unsure
Pollution	[]	[]	[]	[]	[]	[]
Habitat change	[]	[]	[]	[]	[]	[]
Climate change	[]	[]	[]	[]	[]	[]
Overexploitation	[]	[]	[]	[]	[]	[]
Invasive species	[]	[]	[]	[]	[]	[]

17. Given that these stressors interact in complex ways, please identify and briefly describe the interaction that is most pressing to understand in marine conservation, in your opinion (e.g., the additive interaction between invasive species and climate change).
18. Which of the long-term data sources that you identified previously do you believe can be used to address the five stressors or their interactions?

	A	B	C	D	E
Pollution	[]	[]	[]	[]	[]
Habitat change	[]	[]	[]	[]	[]
Climate change	[]	[]	[]	[]	[]
Overexploitation	[]	[]	[]	[]	[]
Invasive species	[]	[]	[]	[]	[]
Unsure	[]	[]	[]	[]	[]

19. Please select one or more timescales of data that would be needed to best address each stressor, in your opinion.

	Days	Weeks	Months	Years	Decades	Centuries	Millennia	10^4 years	10^5 years	10^6+years	Unsure
Pollution	[ ]	[ ]	[ ]	[ ]	[ ]	[ ]	[ ]	[ ]	[ ]	[ ]	[ ]
Habitat change	[ ]	[ ]	[ ]	[ ]	[ ]	[ ]	[ ]	[ ]	[ ]	[ ]	[ ]
Climate change	[ ]	[ ]	[ ]	[ ]	[ ]	[ ]	[ ]	[ ]	[ ]	[ ]	[ ]
Over-exploi-tation	[ ]	[ ]	[ ]	[ ]	[ ]	[ ]	[ ]	[ ]	[ ]	[ ]	[ ]
Invasive species	[ ]	[ ]	[ ]	[ ]	[ ]	[ ]	[ ]	[ ]	[ ]	[ ]	[ ]

Temporal vs. Spatial Data

One important reason for using long-term temporal data is to produce baselines against which current conditions in ecosystems can be compared, but spatial data also are frequently used as references against which to judge current conditions at a specific location. The following two questions are intended to help us understand the balance between use of temporal and spatial data to produce baselines and reference conditions in marine conservation.

20. If you use reference conditions or baselines in your research/conservation work, please list three types of data sources that you use to produce them (e.g., reference sites, monitoring records, etc.).

21. In your opinion, are spatial and temporal data of equal value in establishing reference conditions and baselines? Please explain briefly.

Problems and Challenges with Applying Long-term Data

22. Are there types of long-term data that would be useful, but that aren't currently available or of which you would want more? If so, please give an example.

23. What barriers (e.g., communication, funding, data availability, etc.) have you experienced (or do you perceive to exist) in applying long-term data to marine conservation?

Appendix 2: Survey Population Selection

First Solicitation

In order to establish our survey population, we searched the internet for organizations conducting research or management in marine systems. All institutions, agencies, laboratories, etc. were based in the United States and included National Estuarine Research Reserves (e.g., Chesapeake Bay NERR), Sea Grant programs (e.g., Alaska Sea Grant), governmental departments (e.g., Alaska Department of Fish and Game) and their divisions (Division of Habitat), and academic marine laboratories (e.g., Darling Marine Center, University of Maine). A full list of all organizations contacted for the survey can be downloaded at http://doi.org/10.7298/X4VM4965.

For each organization, we contacted the director, president, or positional equivalent via email prior to the activation of the survey ($n = 202$). If we received a positive response (agreement to distribute the survey) from an organization ($n = 54$), we sent a solicitation to the contact upon activation of the survey. If we did not receive an initial response ($n = 136$), we sent a second email to the contact with a solicitation at the time of survey activation to encourage participation. We did not contact those who responded negatively ($n = 12$) to the initial solicitation. The survey was open September–November 2015.

Second Solicitation

We opened the survey a second time during January 2016. In this period, we sent a solicitation to the Ecological Society of America listserv, ECOLOG-L, in an attempt to reach marine conservation biologists who may not have been reached by our first solicitation. ECOLOG-L is distributed internationally, however the solicitation explicitly requested participation from researchers and managers working in the United States of America.

Third Solicitation

During April 2016, we opened the survey for a third time, with a goal of increasing participation from the academic demographic. We sent a solicitation to the President of the National Association of Marine Laboratories (NAML) who kindly agreed to distribute the survey to the directors of the member laboratories. NAML includes governmental laboratories but its more than 50 members are primarily associated with academia. A full list of all organizations contacted for the survey can be downloaded at http://doi.org/10.7298/X4VM4965. Visit http://www.naml.org/index.php for more information on NAML.

Appendix 3: Categorization of Responses

We categorized the survey responses for all free response and short answer questions prior to analysis. For each question, all three authors reviewed each of the categories to which responses were assigned and when disagreements occurred the categories were discussed until a consensus was reached. Similarly, responses to free response and short answer questions were reviewed collectively and placed within categories after the authors reached agreement. Responses were categorized for the following 11 questions: 3, 7, 9, 10, 12, 14, 17, and 20–23.

Question 3: Please list up to three fields/scientific disciplines (e.g., fisheries biology, historical ecology) with which you identify. Fields and disciplines listed by respondents were grouped into five categories: *Fisheries science*, *Marine biology*, *Conservation and environmental sciences*, *Earth and atmospheric sciences*, and

Other. The use of keywords by respondents facilitated this categorization. For instance, any response including "fishery" (e.g., fisheries biology, fishery management) was categorized as *Fisheries science* and any inclusion of "conservation" or "restoration" (e.g., conservation biology, ecological restoration) was considered *Conservation and environmental sciences*. *Marine biology* was applied generally and was inclusive of responses such as "marine ecology" and "estuarine ecology". The *Earth and atmospheric sciences* category included responses such as "Geography" and "Geology". Responses grouped as *Other* included "biogeochemistry," "genetics," and "molecular biology."

Question 7: Please describe your work as it relates to marine conservation in one sentence or less. Responses to this question were grouped into six categories—*Research*, *Education*, *Management (conservation)*, *Management (resources)*, *Administration*, and *Other*—and were not mutually exclusive. Many respondents indicated that they conducted *Research* (e.g., "I am a marine ecologist studying…") and also filled *Education* (e.g., "Educating bay stewards") or *Management* roles. Keywords were particularly useful when distinguishing between *Management* (*conservation*) and *Management* (*resources*). Responses were grouped under *Management* (*conservation*) when the emphasis was on the preservation or restoration of biodiversity or ecosystems (e.g., "Assess status and trends of ecosystem health in our local estuaries") whereas responses in the *Management* (*resources*) category focused on ecosystem services and fisheries activities (e.g., "...implement resource management actions…"). *Administration* was differentiated from these categories based on the level at which the respondent was working. For example, "Chair of several science or technical advisory committees to coastal policy groups" was considered *Administration* and "…developing strategies to protect and restore salmon habitat" was considered *Management* (*resources*). Responses categorized as *Other* were generally too vague to fit any of the aforementioned categories (e.g., "Working hard today to ensure a better future tomorrow").

Question 9: Please list up to three primary goals in the field of marine conservation (e.g., preservation of biodiversity). Responses were grouped into six categories: *Maintenance of biodiversity*, *Maintenance of ecosystem services*, *Maintenance of ecosystem structure and function*, *Habitat protection*, *Sustainability*, and *Other*. These groupings were not mutually exclusive, as responses such as "Conservation of ecosystem function and services" were considered both *Maintenance of ecosystem services* and *Maintenance of ecosystem structure and function*. *Maintenance of biodiversity*, which was taken to be inclusive of all types of diversity (e.g., genetic, species, ecosystem), *ecosystem services*, and *structure and function* were distinguished based on phrasing and keywords used by respondents—most prominently the category names themselves. *Habitat protection* was applied in a general sense (i.e., not necessarily implying human exclusion from nature). The *Sustainability* category included responses mentioning management practices in a general sense (e.g., "smart management") as well as responses that explicitly mentioned sustainability (e.g., "long-term sustainability"). We acknowledge that the concept of sustainability can be complicated (Callicott and Mumford 1997), but use it here in a broad sense to mean the current and continued coexistence of humans and the ecosystems in which they are embedded. Several respondents also included educational goals (e.g., "education"), focused on research (e.g., "To study the impact we have had . . . "), or gave vague responses (e.g., "Understand marine ecosystems"); these were considered *Other*.

Question 10: What are the cutting-edge approaches currently being practiced in marine conservation to achieve the goals you mentioned in question 9? List no more than three. Responses to this question were variable and ranged from data collection tools (e.g., "drones") to practices (e.g., "adaptive management") and management actions (e.g., "marine protected areas"). Thus, responses were grouped into the broad categories of *Management*, *Technology*, *Mathematics*, *Research*, and *Other*. Many of the responses included multiple types of approaches and others described approaches that spanned more than one of the approach categories (e.g., the response, ". . . statistical and modeling approaches combined with field data from long term studies . . . ", was categorized as *Mathematics* and *Research*). Responses in the *Management* category included decision making (e.g., "utilization of diverse data sets to make management decisions") and management actions (e.g., "Marine Protected Areas") as well as policy changes (e.g., "use laws and politics to control the human activities"). *Technology* approaches referred to improving (e.g., "greater computing power") as well as adapting existing technology to conservation practice (e.g., "use of drones"). *Mathematics* approaches were most commonly related to improved modeling (e.g., "Modeling approaches combined with community-based monitoring . . . ") and analysis (e.g., "spatial analysis"). The *Research* category primarily included descriptions of applying data to conservation practice (e.g., "interdisciplinary collaborative research . . . studying how major river freshwater plumes effect [sic] early life stage survival in marine environments"), some more theoretical considerations (e.g., "ecosystem processes understanding"), and citizen science (e.g., "Developing crowd-sources data [sic] and information products"). The *Other* approaches included responses that were too broad to fit other categories (e.g., "genetics") or did not fit the previous categories (e.g., "education").

Question 12: In your opinion, what is the importance of long-term temporal data for achieving the goals of marine conservation? Responses to this question were classified into one of three commonly described categories (Strayer et al. 1986; Lindenmayer et al. 2012; Dietl et al. 2015)—*Baselines*, *Trends and patterns*, *Range of variability*—and a fourth category, *Other*, for miscellaneous responses. In many cases responses included components of multiple categories and were tallied in each of those categories. Responses in the *Baseline* category typically referred to using LTD to inform decision making in the future (e.g., "To combine with known conditions to be able to model and predict future outcomes"). Responses classified as *Trends and patterns* implied that LTD are important for determining trajectories and removing short-term variation (e.g., "identifies long-term trends in populations or water quality. Eliminates the noise of year-to-year variation . . . "). *Range of variability* most commonly included responses that highlighted the dynamic nature of populations and ecosystems (e.g., "critical for detecting natural dynamics of ecosystems . . . "). The vast majority of responses fell in one of these three categories and two remaining responses were grouped as *Other* (e.g., "Convincing policy makers . . . ").

Question 14: Please list five sources of long-term data and indicate whether you have used each one in your own research. The respondent-provided sources of long-term data were grouped into four categories, *Modern observational*, *Historical*, *Geohistorical*, and *Other*, related to those described for sources of data in marine historical ecology (Lotze and McClenachan 2014; Jackson and McClenachan 2017). In marine historical ecology, "archaeological" is given equivalent status as a data source, however, here it was subsumed under *Geohistorical* due to similarities in timescales and the small number of responses including these data. *Modern observational* included monitoring data and any contemporaneously collected data such as "seabird productivity data," "Weather station data," and "fishery catch data." *Historical* (e.g., "historical documents") was distinguished from *Geohistorical* (e.g., "Paleontological") by its association with records kept by people (e.g., "historical documents"), as opposed to records in nature (e.g., "sediment cores"). The *Other* category included various responses including organizations (e.g., "NOAA") and variables (e.g., "pH") that were too broad to categorize otherwise.

Question 17: Given that these stressors interact in complex ways, please identify and describe the interaction that is most pressing to understand in marine conservation, in your opinion (e.g., the additive interaction between invasive species and climate change)? In 2005, the Millennium Ecosystem Assessment identified five stressors—pollution, habitat change, climate change, overexploitation, and invasive species—as the most important threats to ecosystems and it has subsequently been noted that these stressors often interact in complex ways (Crain et al. 2008; Darling and Côté 2008). Many respondents identified multiple interactions or interactions between three or more stressors they found to be important. Consequently, responses to this question were assessed in two ways. First, the total number of mentions for each stressor was tallied. Second, interactions between stressors were tallied. When three or more stressors were mentioned, each unique pairing was tallied (e.g., a respondent mentioning climate change, habitat change, and pollution resulted in tallies for climate change-habitat change, climate change-pollution, and habitat change-pollution).

Question 20: If you use reference conditions or baselines in your research/conservation work, please list three types of data sources that you use to produce them (e.g., references sites, monitoring records, etc.)? Responses were categorized into five groups: *Modern observational*, *Reference sites*, *Historical*, *Geohistorical*, and *Other*. These categories were chosen to reflect those used in Question 14. Responses classified as *Modern observational* commonly included mentions of monitoring (e.g., "monitoring records"). "Reference sites" was also a frequently given response and formed the basis for the *Reference sites* category; such responses were not considered *Modern observational* because they implied a spatial component rather than continued observation at one or a few sites. Similarly, responses in the *Reference sites* category were distinguished from responses in the *Historical* and *Geohistorical* categories by the mention or implication of spatial rather than temporal data. *Historical* included baselines from human-produced sources including "Literature," "historical data," and "historical accounts." Responses in the *Geohistorical* category often included mentions of paleontological data (e.g., "paleobiology") and geological sources (e.g., "sediment cores"). The *Other* category included responses giving methods (e.g., "Hindcast Circulation and Climate Models") or variables (e.g., "ocean conditions") that could not be linked unequivocally to one of the aforementioned categories.

Question 21: In your opinion, are spatial and temporal data of equal value in establishing reference conditions and baselines? Please explain briefly. Responses to this question were categorized at two levels. First, responses were split into three groups—*Yes*, *No*, and *It depends*—with respect to whether respondents found spatial and temporal data to be of equal value. Second, the *No* category was also subdivided into two groups based on whether respondents found *Temporal* or *Spatial* data to be of greater value for establishing baselines.

Question 22: Are there types of long-term data that would be useful, but that aren't currently available or you would want more of? If so, please give an example. Responses to this question were assessed at three levels. First, responses were divided into those saying *Yes*, *No*, or *Unsure* to the initial question. Second, *Yes* responses were categorized into *Abiotic*, *Biotic*, or *Other* (e.g., "rate or process data") groups. Third, the *Abiotic* and *Biotic* groups were further subdivided into the specific types of data identified by respondents. For the *Abiotic* subgroup, data types included *Temperature* (e.g., "Deep-ocean temperatures."), *Water chemistry* (e.g., "Nutrient concentration of seawater."), and *Other* (e.g., "seismic"). For the *Biotic* subgroup, data types included *Species abundance* (e.g., "abundance of key species"), *Species distribution* (e.g., "species distribution data ..."), *Interactions* (e.g., "predator-prey relationships"), and *Other* (e.g., "species extinction rates").

Question 23: What barriers (e.g., communication, funding, data availability, etc.) have you experienced (or do you perceive to exist) in applying long-term data to marine conservation? Responses to this question were grouped into four categories—*Funding*, *Data availability*, *Communication*, and *Institutional*—similar to those identified by conservation biologists (e.g., Strayer et al. 1986; Lindenmayer et al. 2012) and a fifth category, *Other*, for miscellaneous responses. Many respondents identified multiple barriers and each was tallied under the appropriate category (e.g., "Funding and agency interest" was categorized as *Funding* and *Institutional*). Responses categorized as *Data availability* discussed barriers related to data accessibility or lack of data (e.g., "lack of data availability", "True long-term data is often not available"). The *Communication* category encompassed responses at the level of disciplines (e.g., "...communication may be one barrier, with researchers not recognizing how certain other disciplines might value their contributions") as well as general challenges such as "Communicating long term data can also be difficult if the data is collected on timescales not easily processed by human minds." Responses classified as *Other* included such impediments as education (e.g., "educational barriers") and politics (e.g., "Playing political small ball...").

References

Arlettaz R, Schaub M, Fournier J et al (2010) From publications to public actions: when conservation biologists bridge the gap between research and implementation. Bioscience 60:835–842

Bebber DP, Ramotowski MA, Gurr SJ (2013) Crop pests and pathogens move polewards in a warming world. Nat Clim Chang 3:985–988

Besley JC (2016) The National Science Foundation's science and technology survey and support for science funding, 2006–2014. Public Underst Sci:0963662516649803. https://doi.org/10.1177/0963662516649803

Boulton AJ, Panizzon D, Prior J (2005) Explicit knowledge structures as a tool for overcoming obstacles to interdisciplinary research. Conserv Biol 19:2026–2029

Boyer AG, Brenner M, Burney D et al (2017) Conservation paleobiology roundtable: from promise to application. In: Dietl GP, Flessa KW (eds) Conservation paleobiology: science and practice. University of Chicago Press, Chicago, pp 291–302

Braunisch V, Home R, Pellet J et al (2012) Conservation science relevant to action: a research agenda identified and prioritized by practitioners. Biol Conserv 153:201–210

Bremner A, Park K (2007) Public attitudes to the management of invasive non-native species in Scotland. Biol Conserv 139:306–314

Burney DA, Juvik JO, Burney LP et al (2012) Can unwanted suburban tortoises rescue native Hawaiian plants? The Tortoise 1:104–115

Callicott JB, Mumford K (1997) Ecological sustainability as a conservation concept. Conserv Biol 11:32–40

Carpenter SR, Turner MG (2001) Hares and tortoises: interactions of fast and slow variables in ecosystems. Ecosystems 3:495–497

Casey MM, Dietl GP, Post DM et al (2014) The impact of eutrophication and commercial fishing on molluscan communities in Long Island Sound, USA. Biol Conserv 170:137–144

Cintra-Buenrostro CE, Flessa KW, Dettman DL (2012) Restoration flows for the Colorado River estuary, México: estimates from oxygen isotopes in the bivalve mollusk *Mulinia coloradoensis* (Mactridae: Bivalvia). Wetl Ecol Manag 20:313–327

Cintra-Buenrostro CE, Flessa KW, Guillermo A-S (2005) Who cares about a vanishing clam? Trophic importance of *Mulinia coloradoensis* inferred from predatory damage. PALAIOS 20:296–302

Conservation Paleobiology Workshop [CPW] (2012) Conservation paleobiology: opportunities for the earth sciences. Report to the Division of Earth Sciences, National Science Foundation. Paleontological Research Institution, Ithaca

Cook CN, Mascia MB, Schwartz MW et al (2013) Achieving conservation science that bridges the knowledge-action boundary: achieving effective conservation science. Conserv Biol 27:669–678

Crain CM, Kroeker K, Halpern BS (2008) Interactive and cumulative effects of multiple human stressors in marine systems. Ecol Lett 11:1304–1315

Czech B, Devers PK, Krausman PR (2001) The relationship of gender to species conservation attitudes. Wildl Soc Bull 29:187–194

Darling ES, Côté IM (2008) Quantifying the evidence for ecological synergies. Ecol Lett 11:1278–1286

Davis MB (1989) Retrospective studies. In: Likens GE (ed) Long-term studies in ecology. Springer, New York, pp 71–89

Dawson TP, Jackson ST, House JI et al (2011) Beyond predictions: biodiversity conservation in a changing climate. Science 332:53–58

de Bruyn M, Hall BL, Chauke LF et al (2009) Rapid response of a marine mammal species to Holocene climate and habitat change. PLoS Genet 5:e1000554. https://doi.org/10.1371/journal.pgen.1000554

Dietl GP (2016) Brave new world of conservation paleobiology. Front Ecol Evol 4:21. https://doi.org/10.3389/fevo.2016.00021

Dietl GP, Durham SR, Smith JA et al (2016) Mollusk assemblages as records of past and present ecological status. Front Mar Sci 3:169. https://doi.org/10.3389/fmars.2016.00169

Dietl GP, Flessa KW (2011) Conservation paleobiology: putting the dead to work. Trends Ecol Evol 26:30–37

Dietl GP, Flessa KW (2017) Conservation paleobiology in the Anthropocene In: Dietl GP, Flessa KW (eds) Conservation paleobiology: science and practice. University of Chicago Press, Chicago, pp 303–305

Dietl GP, Kidwell SM, Brenner M et al (2015) Conservation paleobiology: leveraging knowledge of the past to inform conservation and restoration. Annu Rev Earth Planet Sci 43:79–103

Dietl GP, Smith JA (2016) Live-dead analysis reveals long-term response of estuarine bivalve community to water diversions along the Colorado River. Ecol Eng 106:749–756

Dougherty EM, Fulton DC, Anderson DH (2003) The influence of gender on the relationship between wildlife value orientations, beliefs, and the acceptability of lethal deer control in Cuyahoga Valley National Park. Soc Nat Resour 16:603–623

Durham SR, Dietl GP (2015) Perspectives on geohistorical data among oyster restoration professionals in the United States. J Shellfish Res 34:227–239

Elliott JM (1990) The need for long-term investigations in ecology and the contribution of the Freshwater Biological Association. Freshw Biol 23:1–5

Eronen JT, Polly PD, Fred M et al (2010) Ecometrics: the traits that bind the past and present together. Integr Zool 5:88–101

Farrell J (2016) Corporate funding and ideological polarization about climate change. Proc Natl Acad Sci 113:92–97

Flessa KW (2017) Putting the dead to work: translational paleoecology. In: Dietl GP, Flessa KW (eds) Conservation paleobiology: science and practice. University of Chicago Press, Chicago, pp 283–289

Franklin JF, Bledsoe CS, Callahan JT (1990) Contributions of the long-term ecological research program. Bioscience 40:509–523

Fritz SA, Schnitzler J, Eronen JT et al (2013) Diversity in time and space: wanted dead and alive. Trends Ecol Evol 28:509–516

Gauchat G (2015) The political context of science in the United States: public acceptance of evidence-based policy and science funding. Soc Forces 94:723–746

Goodwin DH, Flessa KW, Schone BR et al (2001) Cross-calibration of daily growth increments, stable isotope variation, and temperature in the Gulf of California bivalve mollusk *Chione cortezi*: implications for paleoenvironmental analysis. PALAIOS 16:387–398

Griffiths RA, Dos Santos M (2012) Trends in conservation biology: progress or procrastination in a new millennium? Biol Conserv 153:153–158

Holmes G (2015) What do we talk about when we talk about biodiversity conservation in the Anthropocene? Environ Soc Adv Res 6:87–108

Jackson JBC, Alexander KE, Sala E (2012) Shifting baselines: the past and the future of ocean fisheries. Island Press, Washington

Jackson JBC, McClenachan L (2017) Historical ecology for the paleontologist. In: Dietl GP, Flessa KW (eds) Conservation paleobiology: science and practice. University of Chicago Press, Chicago, pp 87–100

Katsanevakis S, Levin N, Coll M et al (2015) Marine conservation challenges in an era of economic crisis and geopolitical instability: the case of the Mediterranean Sea. Mar Policy 51:31–39

Kearney S, Murray F, Nordan M (2014) A new vision for funding science. Stanford Soc Innov Rev Fall:50–55

Kellert SR, Berry JK (1987) Attitudes, knowledge, and behaviors toward wildlife as affected by gender. Wildl Soc Bull 15:363–371

Kelley PH, Dietl GP, Visaggi CC (2018) Training tomorrow's conservation paleobiologists. In: Tyler CL, Schneider, CL (eds) Marine conservation paleobiology. Springer, Cham, pp 207–223

Kidwell SM (2007) Discordance between living and death assemblages as evidence for anthropogenic ecological change. Proc Natl Acad Sci 104:17701–17706
Kidwell SM, Tomašových A (2013) Implications of time-averaged death assemblages for ecology and conservation biology. Annu Rev Ecol Evol Syst 44:539–563
Kopf RK, Finlayson CM, Humphries P et al (2015) Anthropocene baselines: assessing change and managing biodiversity in human-dominated aquatic ecosystems. Bioscience 65:798–811
Kowalewski M, Goodfriend GA, Flessa KW (1998) High-resolution estimates of temporal mixing within shell beds: the evils and virtues of time-averaging. Paleobiology 24:287–304
Krebs CJ (2015) One hundred years of population ecology: successes, failures and the road ahead. Integr Zool 10:233–240
Labandeira CC, Currano ED (2013) The fossil record of plant-insect dynamics. Annu Rev Earth Planet Sci 41:287–311
Laurance WF, Koster H, Grooten M et al (2012) Making conservation research more relevant for conservation practitioners. Biol Conserv 153:164–168
Likens GE (ed) (1989) Long-term studies in ecology: approaches and alternatives. Springer, New York
Lindenmayer DB, Likens GE, Andersen A et al (2012) Value of long-term ecological studies. Austral Ecol 37:745–757
Lotze HK, McClenachan L (2014) Marine historical ecology: informing the future by learning from the past. In: Bertness MD, Bruno JF, Silliman BR et al (eds) Marine community ecology and conservation. Palgrave Macmillan Sinauer Associates, Sunderland, pp 165–200
MacDonald GM, Bennett KD, Jackson ST et al (2008) Impacts of climate change on species, populations and communities: palaeobiogeographical insights and frontiers. Prog Phys Geogr 32:139–172
Marris E (2011) Rambunctious garden: saving nature in a post-wild world. Bloomsbury USA, New York
Millennium Ecosystem Assessment [MEA] (2005) Ecosystems and human well-being: synthesis. Island Press, Washington
Mobley C, Kilbourne W (2013) Gender differences in pro-environmental intentions: a cross-national perspective on the influence of self-enhancement values and views on technology. Sociol Inq 83:310–332
Nadelson LS, Hardy KK (2015) Trust in science and scientists and the acceptance of evolution. Evol Educ Outreach 8:9
National Research Council [NRC] (2005) The geological record of ecological dynamics: understanding the biotic effects of future environmental change. National Academy Press, Washington
Pandolfi JM, Bradbury RH, Sala E et al (2003) Global trajectories of the long-term decline of coral reef ecosystems. Science 301:955–958
Pietri DM, Gurney GG, Benitez-Vina N et al (2013) Practical recommendations to help students bridge the research-implementation gap and promote conservation: graduate students and the research-implementation gap. Conserv Biol 27:958–967
Polly PD, Eronen JT, Fred M et al (2011) History matters: ecometrics and integrative climate change biology. Proc R Soc B Biol Sci 278:1131–1140
Price SA, Schmitz L (2016) A promising future for integrative biodiversity research: an increased role of scale-dependency and functional biology. Philos Trans R Soc B 371:20150228
Reed DC, Rassweiler AR, Miller RJ et al (2016) The value of a broad temporal and spatial perspective in understanding dynamics of kelp forest ecosystems. Mar Freshw Res 67:14–24
Roux DJ, Kingsford RT, McCool SF et al (2015) The role and value of conservation agency research. Environ Manag 55:1232–1245
Smith JA, Auerbach DA, Flessa KW et al (2016) Fossil clams reveal unintended carbon cycling consequences of Colorado River management. R Soc Open Sci 3:160170
Smith JA, Dietl GP (2016) The value of geohistorical data in identifying a recent human-induced range expansion of a predatory gastropod in the Colorado River delta, Mexico. J Biogeogr 43:791–800

Starzomski BM, Cardinale BJ, Dunne JA et al (2004) Contemporary visions of progress in ecology and thoughts for the future. Ecol Soc 9:14
Strayer D, Glitzenstein JS, Jones CG et al (1986) Long-term ecological studies: an illustrated account of their design, operation, and importance to ecology. The New York Botanical Garden, Institute of Ecosystem Studies, Millbrook
Volety A, Savarese M, Hoye B et al (2009) Landscape pattern: present and past distribution of oysters in South Florida coastal complex (Whitewater Bay/Oyster Bay/Shark to Robert's Rivers). Report to the South Florida Water Management District, Fort Myers, FL, Florida Gulf Coast University
Wiens JA, Hobbs RJ (2015) Integrating conservation and restoration in a changing world. Bioscience 65:302–312
Willis KJ, Bailey RM, Bhagwat SA et al (2010) Biodiversity baselines, thresholds and resilience: testing predictions and assumptions using palaeoecological data. Trends Ecol Evol 25:583–591
Wolkovich EM, Cook BI, McLauchlan KK et al (2014) Temporal ecology in the Anthropocene. Ecol Lett 17:1365–1379

Effectively Connecting Conservation Paleobiological Research to Environmental Management: Examples from Greater Everglades' Restoration of Southwest Florida

Michael Savarese

Abstract Much of the research accomplished by paleontologists can be categorized as conservation paleobiology. Unfortunately, these works often go unrealized or under-appreciated because the environmental professionals best positioned to use those results are detached from the science, as managers or decision-makers, or are completely unaware of the scientific discipline. Commonly, a research program with management implications is undertaken without briefing the relevant agencies in advance and without their input with respect to research objectives and design. Academics typically work within or among their academic institutions with a predestined pathway toward peer-review publication in scholarly journals. Those papers often go unread by agency professionals, and universities at times of promotion rarely value technical reports. Partnership and collaboration with agencies is essential. Agency professionals have very specific environmental management objectives and priorities, often influenced by resource limitations. Academic scientists should adapt their research programs to accommodate agency research needs and priorities, and be willing to adopt research designs that best achieve agency objectives, even if those agencies are unable to financially support the effort. Often agencies can support research through in-kind match for field and laboratory work, or with research staff assistance. When partnerships are strong, "request for proposals" (RFPs) from state and local governmental agencies can be customized for specific talents found among academics.

University scientists and agency professionals in Southwest Florida have developed an appreciation of conservation paleobiology and a culture of cooperation. Effective steps for generating such a productive relationship include: (1) developing and actively participating in "management collaboratives," working groups composed of agency professionals, university scientists, and non-governmental organization (NGO) professionals with stewardship commitments to their region's management and restoration needs; (2) attending and presenting conservation

M. Savarese (✉)
Marine and Ecological Sciences, Florida Gulf Coast University, Ft. Myers, FL, USA
e-mail: msavares@fgcu.edu

C. L. Tyler, C. L. Schneider (eds.), *Marine Conservation Paleobiology*,
Topics in Geobiology 47, https://doi.org/10.1007/978-3-319-73795-9_4

paleobiological work at environmental science and restoration conferences; (3) inviting agency professionals to market and solicit their environmental science priorities and then engaging agency representatives as co-investigators in those studies; and (4) reserving time for professional service for those same agencies and NGOs. Universities can incentivize conservation paleobiological research by valuing technical reports as community-engaged scholarship, particularly if the work results in a management decision or practice that helps the environment. Greater Everglades' restoration efforts have benefitted greatly from conservation paleobiological approaches because of these team-building efforts. A couple of examples are presented.

Keywords Conservation paleobiology · Environmental management · Environmental restoration · Greater Everglades restoration

1 Introduction

Conservation paleobiology is a relatively new discipline that is still inventing itself. Paleontology's value in testing or refining concepts in ecology and evolution is well established, and these applications have obvious relevance to matters in conservation biology. In this mode, paleontological data can provide a deep time, extensive record of life's responses to environmental drivers (Dietl and Flessa 2011; Conservation Paleobiology Workshop 2012). The geohistorical record provides a richer database with which to test ecological processes, particularly when neontological records are not robust. Those same data can be used to validate ecological systems' models; if models withstand testing in post-diction mode (i.e., models that successfully predict situations in the past), their application for prediction is more secure. Finally, that deeper record serves to establish historical precedents: for example, mass extinctions have occurred in the past because of similar drivers; they can happen again.

More importantly, and less appreciated, is paleontology's potential to provide a near-time record of life's response to recent environmental or anthropogenic change (Dietl and Flessa 2011). Society did not have the foresight to characterize natural conditions prior to human development and industrialization. Consequently, restoration managers do not have access to information about those pristine, pre-anthropogenic conditions. Methods and concepts from paleontology, and from the allied disciplines of sedimentology, stratigraphy, and geochemistry, provide a mechanism to infer those pre-anthropogenic conditions. This gives paleontology a direct role in the development of environmental management and restoration strategies.

It's important to recognize that the pre-industrialized world was not necessarily pristine. Significant anthropogenic influences can date back well into aboriginal times (Denevan 1992; Cronon 1996; Knowlton and Jackson 2008). But here too paleontology can directly assist environmental management; the geohistorical record is often of high enough quality to include paleoecological and archaeological data, illuminating shifting baselines throughout human history (e.g., Jackson et al.

2001). With this perspective, environmental managers can better define their restoration target.

As a result, classically trained paleontologists and environmental geoscientists are predisposed to assist in environmental management and restoration. Our professional training and culture, however, inhibit our effectiveness. Our work is often unrecognized, not applied, or not appreciated simply because we are trained in a different academic setting and are employed in different professional networks.

I have been fortunate to be employed at an institution of higher education, Florida Gulf Coast University (FGCU), whose science is genuinely interdisciplinary—I am one of a few geoscientists who do not reside in a Department of Geological Sciences; my colleagues and students come from numerous scientific disciplines; and we are all training students for careers and higher education in environmental science and management (FGCU offers a BA and an MS in Environmental Studies, a BS in Marine Science, an MS in Environmental Science). Additionally, our university was founded with environmental stewardship and literacy as components of its mission, and from its very inception in 1997, the university has been closely partnered with environmental agencies and non-governmental organizations (NGOs) from federal down to local geographic levels. The agencies and organizations with whom we work closely are numerous (see Table 1 for a comprehensive

Table 1 Organizations engaged in Greater Everglades' management and restoration benefitting by conservation paleobiologic research

Agency or Organization	Affiliation
U.S. Army Corps of Engineers (ACOE)	Federal
U.S. Fish and Wildlife Service	Federal
Everglades National Park	Federal
Ten Thousand Islands and Florida Panther National Wildlife Refuges	Federal
Big Cypress National Preserve	Federal
Rookery Bay National Estuarine Research Reserve	Federal and State
Charlotte Harbor National Estuary Program	Federal
U.S. Geological Survey	Federal
National Resources Conservation Services	Federal
Florida Fish & Wildlife Conservation Commission	State
Florida Department of Environmental Protection	State
South Florida Water Management District	State
Collier-Seminole State Park	State
Fakahatchee Strand State Park	State
Lee County	County
Collier County	County
Collier County Coastal Zone Management	County
City of Naples	Municipal
City of Fort Myers Beach	Municipal
Conservancy of Southwest Florida	Not-for-profit
Audubon Florida	Not-for-profit
Florida Wildlife Federation	Not-for-profit

list). Finally, the university serves as the only institution of higher education in Southwest Florida, making us the clearinghouse for the western Everglades. Consequently, we are intimately involved in many aspects of Greater Everglades' restoration.

These, by happenstance, are ideal circumstances under which to be a conservation paleobiologist, but they are rarely united at one university. Consequently, my experiences at FGCU have provided me with insights that are shared in this manuscript and that can serve as a model for other institutions.

2 Defining the Problem

The process of environmental management typically proceeds in this way: an environmental problem requiring resolution is identified; the already existing, relevant science is brought forward; additional science needs are identified; those science efforts are implemented and their results integrated with the previous science; and this collectively informs best management practices. Aspects of this process are captured in crafted and often-revised management plans. Those plans then drive resource (i.e., money, staff time, facilities) allocation and prioritization. Environmental scientists engaged in this process produce the technical reports that fulfill the science needs. They are also sitting at the table with the managers when management plans are developed and when priorities are defined.

As university academics and as scientists outside the environmental profession, we generate peer-reviewed journal articles; those published studies might consider societal and environmental implications, but rarely are they focused on those issues, and the environmental profession is typically not consulted or aware of our activities. As tenure track professors, the currency of scholarly success is the number of peer-reviewed papers and the impact factor held by the journals. Technical reports are not valued similarly, and the societal impact of the environmentally relevant work, assuming the science did affect environmental management and restoration, is not credited. Some universities do define scholarly productivity broadly, sometimes to include "community-engaged scholarship" (Saltmarsh et al. 2009; Whitmer et al. 2010). I'm not aware, however, if promotion and tenure review processes are evaluating this type of scholarship fairly and equitably. Finally, the university scientist is not sitting at the table when science needs are identified and management plans are developed.

Ultimately, we as scientists have to place anticipated and well-received deliverables in the right hands, and our academic institutions must value and reward this type of scholarship.

3 Ensuring Success as a Conservation Paleobiologist

As individual scientists or as working groups within departments, there are a number of practices that can enhance our role in environmental management.

Developing Partnerships and Collaborative Teams

We are accustomed to reaching out to other academics within our discipline for collaboration—teaming up with people that bring needed skills to the effort. We are unaccustomed to partnering with people that have an environmental problem that requires management or correction. As paleontologists, we have unique and helpful skills, but we are not always aware of how our talents can be best applied to environmental science and management. Conservation paleobiologists must engage environmental agencies and the scientists and managers they employ. Regardless of where a faculty member's home institution is located, there are state and local governmental agencies and NGOs that are charged with environmental management and stewardship (for Southwest Florida, see Table 1). Federal environmental agencies may be located at greater distances, but probably still close enough for direct interaction. These agencies should be engaged. Meeting agency professionals at their places of employment, arranging visits to the university for briefings, or presenting a sample of work to those agency professionals are all valuable means to developing collaborative relationships. Most agencies and NGOs employ someone charged with overseeing scientific research (with the term "research" appearing in a position's title). This is the most logical point of first contact.

Assuming you, as a conservation paleobiologist, are flexible in the kinds of scholarship you undertake, soliciting research questions from the agencies is a way to ensure the agency is fully vested in the work you'll pursue. If they have a need that you can fill, then your results will undoubtedly be appreciated and applied in societally useful ways. Often a research question of interest to you is closely related to one of interest to the agency. Combining dual interests may be more cost-effective and allow both to be pursued in concert.

This approach can also result in research support. Agencies often have discretionary funding they can provide outside of competitive bidding and formal competition. These monies could be directed toward projects. Requests for proposals (RFPs) are often crafted by agencies; collaboration ensures you're a recognizable face if and when a call for proposals goes out; and, even without genuine dollars, agencies do have in-kind resources that can facilitate your work (e.g., staff time, field access, supplies).

When applying for funds extramurally and when the work is relevant to an agency's management mission, agency professionals should be invited to become a member of the proposal writing team. Their agency may not permit this; however, even something as simple as a letter of endorsement to append to the proposal can improve the proposal's competitiveness.

Becoming or Engaging a Liaison

At FGCU, many, if not most, of the environmental science faculty members serve as "liaisons" to specific agencies and NGOs. Faculty members serve as points of contact for an agency. I, for example, serve as the liaison to Rookery Bay National Estuarine Research Reserve for FGCU; I sit on Reserve committees, represent the university's interests and capabilities, and pass research needs and opportunities on to our students and faculty. This ensures communication between the agency and university is strong, that opportunities are broadcasted, and that personal and professional relationships develop. Not every university scientist has the right personality for serving as a liaison; some are more predisposed to the role. Establishing a formal plan, one that defines service responsibilities for liaisons within your department or work group, distributes the service workload and guarantees many organizations are engaged. The university and our department value these liaison functions as service, and they are considered at times of review and promotion.

Participate in "Management Collaboratives"

Because environmental problems affect multiple stakeholders, management of those problems requires involvement from multiple agencies and organizations. For this reason, environmental agencies often meet as "management collaboratives," representatives from each of the stakeholder groups participate in regular meetings to coordinate efforts and interpret results. (The U.S. Army Corps calls these "planning charrettes.") This is a mainstay of environmental management: by law, all stakeholder groups must be considered, and, because of resource limitations, agencies must pool their efforts and dollars to remain effective and efficient. For example, both ACOE and the SFWMD, a federal and state partnership, facilitate Everglades' restoration projects. Their oversight process requires that each restoration project be developed by a Project Delivery Team (PDT), which ultimately produces a Project Implementation Report (PIR). Teams are composed of members from all stakeholder groups. Lesser restoration and management efforts, ones not garnering federal and state monies, often employ the same management collaborative model. Serving as a conservation paleobiologist on such collaboratives positions our science well and broadly informs the environmental community of our value and capability.

Compose Technical Reports in Addition to Peer-Reviewed Journal Articles

Perhaps it is unreasonable to assume the university tenure and promotion process will reward technical reports or less-traditional products generated by community-engaged scholarship. That said, it makes good sense to prepare two kinds of products from a conservation paleobiological investigation: the technical report that will be read and utilized by the environmental management community, and a peer-reviewed journal paper that will not be read by the environmental community, but has the respectability of the university community. The two are rarely identical in structure and purpose, but one is readily transformed to the other. Interestingly, some agencies require peer-review validation before scientific results are employed. ACOE, for example, requires its contributing scientists to publish in peer-reviewed journals before results are employed in management or restoration.

Present Your Findings to Stake Holder Groups

As with academic audiences, oral presentation of information to the management community is often more influential than when presented in written form. For this reason, volunteering to give a talk at the agency or to the management collaborative is important. Management decisions are more readily influenced through oral communication than in writing. Oral presentations also generate conversation and personal connection, which increases the likelihood of appreciation and adoption.

Attend and Present at Environmental Science and Restoration Conferences

Conservation paleobiological studies are most commonly given to the wrong professional societies. I too am guilty of this, most often presenting my research at professional meetings catering to geoscientists exclusively. The more appropriate societies (e.g., Society for Ecological Restoration [SER], Association of Environmental Studies and Sciences [AESS], Coastal and Estuarine Research Federation [CERF]) take me out of my comfort zone and introduce me to new people. Most environmental scientists and managers have their academic training in the biological or ecological sciences. Most are unfamiliar with the geological sciences, or assume geology reveals deep time and therefore is irrelevant to modern historical problems. Presenting our work to these audiences informs these disciplines of our science's capabilities.

Train our Students

Most of us teach paleontology in traditional ways, de-emphasizing or excluding entirely the value of conservation paleobiology. Our courses must be restructured to accommodate conservation paleobiology for training of our geoscience students. Though I have no statistics to emphatically make this statement, I suspect there are more employment opportunities, both in the scientific workforce and in academia, for paleontologists doing conservation paleobiology. A geology student with strong training in sedimentology, stratigraphy, geochemistry, and conservation paleobiology makes a strong candidate for the environmental science and management profession.

Additionally, topical units covering conservation paleobiology should appear in courses provided to environmental science and biology students. I would argue that this is the most important audience to affect. A student with a degree in environmental science that enters the workforce and already recognizes the value of conservation paleobiology promotes the incorporation of the discipline. At FGCU, I teach a graduate level course to our MS Environmental Science students entitled "Conservation Paleobiology." At the undergraduate level, taught principally to BS Marine Science students, I offer a paleontology course (titled "Geobiology") that includes conservation paleobiologic examples throughout the semester-long curriculum.

Finally, conservation paleobiological research projects, particularly ones advocated by an agency, make ideal thesis topics for undergraduate and graduate students. Such a project is in essence a scholarly experience and a professional internship packaged together. The student works closely with both a faculty member and one or more environmental professionals—perhaps one of those professionals serves on the student's thesis committee—and the student improves her or his employment opportunities with that agency once the degree is complete.

Reward Faculty for Conducting Community-Engaged Scholarship

In an ideal world, the tenure and promotion process should value community-engaged scholarship, where the research does not necessarily result in a peer-reviewed publication, but where the work's societal effect and usefulness can be documented in other ways. This may be a difficult change to advance in light of institutional tradition, but it is worth exploring. One way to market these accomplishments, even if they cannot be relied upon when being reviewed for promotion, is to develop a well-documented case study within your portfolio that outlines the project's value and how it influenced some environmental management practice. A letter of affidavit should accompany the case study coming from someone within the agency that can speak directly about the work's influence.

Promote and Reward Community Service for Work with Environmental Agencies and NGOs

Developing relationships with environmental agencies is time consuming. Those investments, however, typically result in other, non-scholarly products that serve the greater community. My experience is that these collaborations are valuable service commitments; I'm providing helpful services to the agency by participating in meetings and management collaboratives, and through conversation and presentation. These are deserving of recognition at times of promotion and annual review.

4 Case Studies from Greater Everglades' Restoration

Admittedly, I have benefitted greatly from geographic placement: FGCU sits within the western region of the Greater Everglades, and the counties of Southwest Florida are developing at an alarming rate while available space is limited. This means environmental management must effectively compromise societal expansion with the conservation of natural resources and ecosystem services. Despite the growth and the never-ending desire to live in "paradise," at least during the northern hemisphere winter, the environmental conservation ethic here is strong: citizens recognize the value of our natural systems, if for nothing else than the region's economic prosperity. This means the public is generally agreeable and appreciative though it does not necessarily mean there is adequate financial investment in management and restoration.

Greater Everglades' restoration is principally about "getting the water right" (Sklar et al. 2005; U.S. Department of the Interior Office of Everglades Restoration Initiatives 2016). Our low-lying topography and seasonally intense rainfall translates into an historic proliferation of fresh and brackish water wetlands. As uplands become increasingly limited for development, wetlands are encroached or consumed. Surface hydrology is dominated by sheet flow, meaning that subtle barriers created through development (e.g., roads) can grossly influence water delivery and flow through wetlands. These wetlands historically supported vast and diverse populations of wildlife, with many species currently threatened or endangered. Complicating all of this are the products of water management from the previous century: a vast network of canals, weirs, and impoundments that have effectively drained wetlands, delivering unnatural volumes of freshwater (i.e., freshets) to sensitive estuaries. Greater Everglades' restoration is essentially about replumbing the system, returning at least some portion of the sheet flow back onto the landscape.

Figure 1 shows the problem graphically. The pre-anthropogenic situation allowed Lake Okeechobee to overflow its natural levees, delivering sheet flow to the downstream wetlands and eventually to coastal estuaries. Shark Valley's River of Grass is the most notable of the affected waterways. (For a more thorough

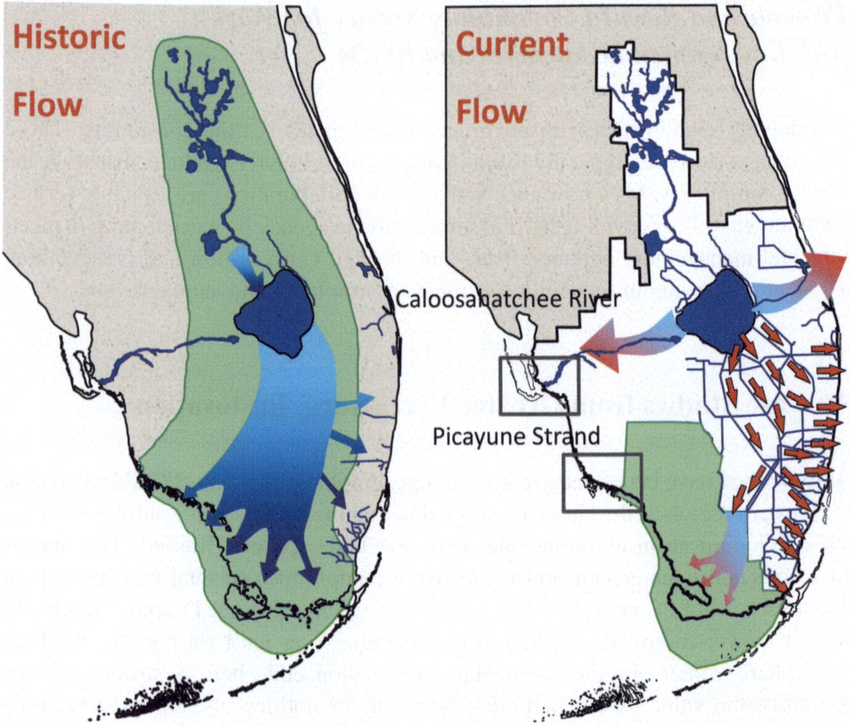

Fig. 1 The two maps show the historic flow patterns through the Greater Everglades prior to human engineering and the current situation, whereby flows entering Lake Okeechobee are diverted. The geographic locations for the two case studies detailed herein (Caloosahatchee River and Picayune Strand) are shown here. Graphics provided by ACOE, Jacksonville District

review of the historic hydrology and the problems and successes of Everglades' restoration, see Grunwald 2006.) In order to protect property and to better serve agriculture south of Lake Okeechobee: (1) canals were constructed to reduce risk to the population center in southeast FL; and (2) the Caloosahatchee River, in the west, and St. Lucie River, in the east, were artificially connected to the lake to serve as shunts for quick removal of freshwater to the estuaries. As a result, sheet flow has become much reduced throughout most of the region, and some estuaries are bombarded with massive freshwater flows when water levels threaten human life and property. Greater Everglades' restoration is tasked with restoring as much of the natural hydrology as possible, while still protecting society from flooding.

The two most critical problems for Southwest Florida are managing flow through the Caloosahatchee River (Case Study 1) and restoring sheet flow through the Picayune Strand (Case Study 2). Both have benefitted, in part, by insights provided by conservation paleobiology.

Case Study 1: Water Management of the Caloosahatchee River

The Caloosahatchee River estuary has suffered from the voluminous releases of freshwater from Lake Okeechobee. Two severe impacts have been documented. One, the freshets stress benthic ecosystems, particularly oyster reefs and seagrass beds, that historically flourished in the estuary. Two, the flushing of freshwater, accompanied by enriched nutrients, causes harmful algal blooms which trigger hypoxia events and fin- and shellfish kills throughout the estuary and well out into the Gulf of Mexico (Anderson et al. 2008; Vargo et al. 2008). The South Florida Water Management District (SFWMD) wanted to document the effects on the benthos and then use this information to better regulate the freshwater releases from the Lake.

An FGCU graduate student, Jorge Agobian, and I proposed a comparative taphonomic study of mollusk life and death assemblages to document the effects of seasonal salinity changes, influenced most dramatically by freshwater releases on the infaunal and epifaunal mollusks residing in the estuary. The SFWMD was aware of the suspected influence on oyster productivity and upon the distribution of the seagrass *Vallisneria americana* (tape grass), but did not have a clearly documented cause and effect relationship between patterns of salinity change and estuarine health (Doering et al. 2002; Barnes et al. 2007). Following a methodological and conceptual approach developed by Susan Kidwell (2007, 2009), mollusk assemblages were collected and analyzed quarterly from six locations along the estuary's salinity gradient (Fig. 2). Estimates of the time averaging of mollusk death assemblages from the taphonomically active zone, acquired through radiocarbon dating, demonstrated that death assemblages are taphonomically inert and represent 600 years of accumulation with all samples dating from prior to the middle 1900s. They, therefore, clearly reflect the recent pre-anthropogenic environmental

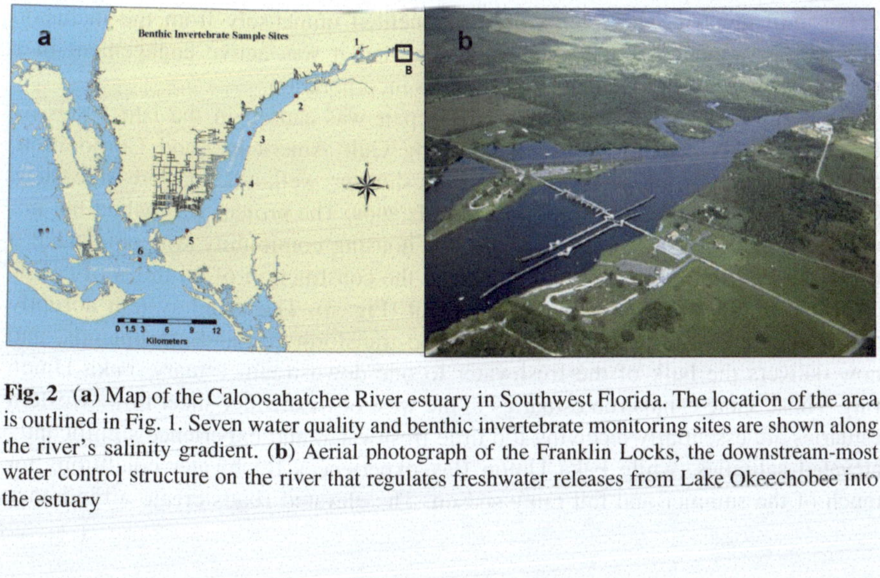

Fig. 2 (**a**) Map of the Caloosahatchee River estuary in Southwest Florida. The location of the area is outlined in Fig. 1. Seven water quality and benthic invertebrate monitoring sites are shown along the river's salinity gradient. (**b**) Aerial photograph of the Franklin Locks, the downstream-most water control structure on the river that regulates freshwater releases from Lake Okeechobee into the estuary

condition and the salinity pattern prior to the engineered connection of the river to Lake Okeechobee. When the compositions of the life and death assemblages were compared using Pearson product-moment and Spearman rank-order correlations, and a detrended correspondence analysis, the influence of lowered salinities was evident within the life assemblages, characterized by faunas tolerant of significantly lower salinity.

The study was made possible because of an established relationship we had with the District. One of the District's lead scientists at the time was an estuarine ecologist specializing in malacology and someone with whom we had worked previously. Though the District did not support the research with funding, it did make critical water quality and benthic monitoring data available, and assisted in the study's design and implementation. The study, though never published in a peer-reviewed journal (see thesis: Agobian 2010), was submitted in technical report form to the District, a final presentation was made to their staff, and the District scientist served as an ex-officio member of Agobian's thesis committee. The ultimate measure of success though is how the results were employed. The District used these results, coupled with the effects of freshets on oysters and seagrasses, to alter their water release practices through the locks on the Caloosahatchee River. Water is now released more modestly with prescribed flows to generate tolerable salinities, and excessive releases, if necessary, are reserved for non-reproductive months for benthic mollusks.

Case Study 2: Picayune Strand Restoration Project

Picayune Strand is the most ambitious restoration project funded through the Comprehensive Everglades Restoration Plan (CERP), estimated, upon completion, to cost $350 M (at time of publication of the Project Implementation Report; U.S. Army Corps of Engineers 2004). The restoration plan that was adopted is currently being implemented. That plan's design benefited immensely from the inclusion of a conservation paleobiological approach, and it was active engagement in a management collaborative that resulted in its incorporation.

The environmental problem requiring repair was caused in the late 1960s by a housing development project financed by Gulf American Land Corporation, South Golden Gate Estates, in former freshwater wetlands located in eastern Collier County north of the Ten Thousand Islands. The project was halted but not until after the infrastructure to support the housing community was emplaced: a network of 279 miles of elevated roads and the construction of 48 miles of canals affecting 70,000 hectares of wetland habitat (Fig. 3). The canal system abruptly and significantly lowered the water table to transform wetlands to uplands, and now delivers the bulk of the freshwater to one downstream estuary, Faka Union Bay. Those same canals rob estuaries to the west of freshwater sheet flow. Western estuaries are essentially receiving too little freshwater, and experience significantly elevated salinities, while Faka Union Bay experiences freshwater conditions for much of the summer and fall rainy season. The elevated roads create a hindrance

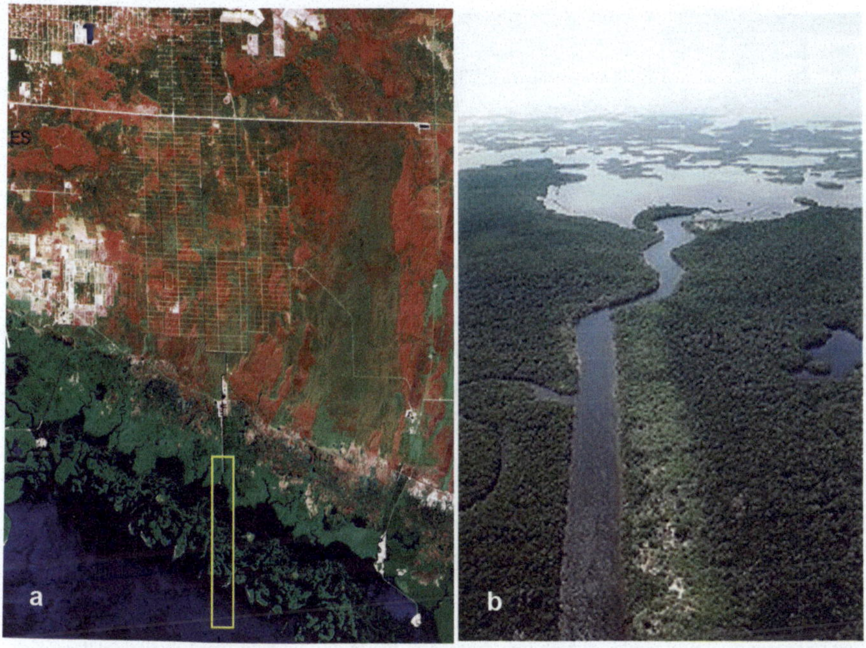

Fig. 3 Images of Picayune Strand. (**a**) The satellite image shows: the system of roads and canals that effectively drained the wetlands, and the trunk canal that delivers the freshwater to Faka Union Bay in the Ten Thousand Islands. (**b**) Aerial photo of region marked by yellow box in A showing the mouth of the main canal as it enters Faka Union Bay

to the restoration of sheet flow because of their damming effect. Any restoration design, and the one ultimately chosen, necessitates the lowering of road grade and an interruption of canal flow. Because this restoration project involved both federal and state monies, the lead agencies overseeing the project's design and implementation are ACOE and the SFMWD. Their process requires all stakeholders to be involved, and, for that reason, a management collaborative was employed; the SFWMD and ACOE nomenclature calls these "project development teams" (PDT; U.S. Army Corps of Engineers 2012).

When the environmental problem was first identified and prioritized, it was conceived as a freshwater wetland project. At the time, because of my former involvement with the SFWMD, I was invited as an FGCU representative for the PDT. The group's early conversations, in part because of my participation, resulted in the extension of the project's scope to the downstream estuaries. It was for this aspect of the project that conservation paleobiology made a significant contribution. Without this participation in the management collaborative, conservation paleobiology would have been a completely unknown discipline among the environmental scientists and managers. The tenor of the group and the project's ultimate design was transformed.

Fig. 4 Photograph of an oyster reef exposed at low tide. When oyster reefs become emergent, they trap red mangrove propagules, which may transform the reef to a mangrove-forested island

In order to restore the hydrology, salinity targets were needed for the affected downstream estuaries. Unfortunately, prior to the emplacement of the canals and roadways in the 1960s, no one had the foresight to monitor the estuarine water quality; no record of pre-anthropogenic salinity variability existed. The estuaries of the Ten Thousand Islands are ideal habitat for oyster productivity; oyster reefs are prolific (Savarese et al. 2003; Savarese and Volety 2008). In fact, oyster reef development throughout the last 3200 years of the late Holocene created the Ten Thousand Island geomorphology; because oyster reef accretion exceeded the rate of sea-level rise through this interval of time, the coast prograded and the oyster reefs succeeded into mangrove-forested islands (Parkinson 1989), thereby generating the Ten Thousand Island estuarine-scape (Figs. 4 and 5). The eastern oyster, *Crassostrea virginica*, tolerates a wide range of salinities from near fresh to fully marine, but its maximum productivity sits within the brackish range, from 15 to 25 ppt (Tolley et al. 2003; Volety et al. 2009). The conservation paleobiological approach was predicated on the assumption that oyster reef development would have been most prolific within this salinity range. The loci of reef development during pre-anthropogenic history, when water management practices had not yet influenced salinity, would be the geographic targets for those brackish water salinities. Mapping of oyster reefs today, integrated with an understanding of reef history, through stratigraphy, was the methodology proposed. Hydrologic modelers could then model various freshwater flow scenarios through the wetlands to deliver the appropriate salinities throughout the estuarine-scape.

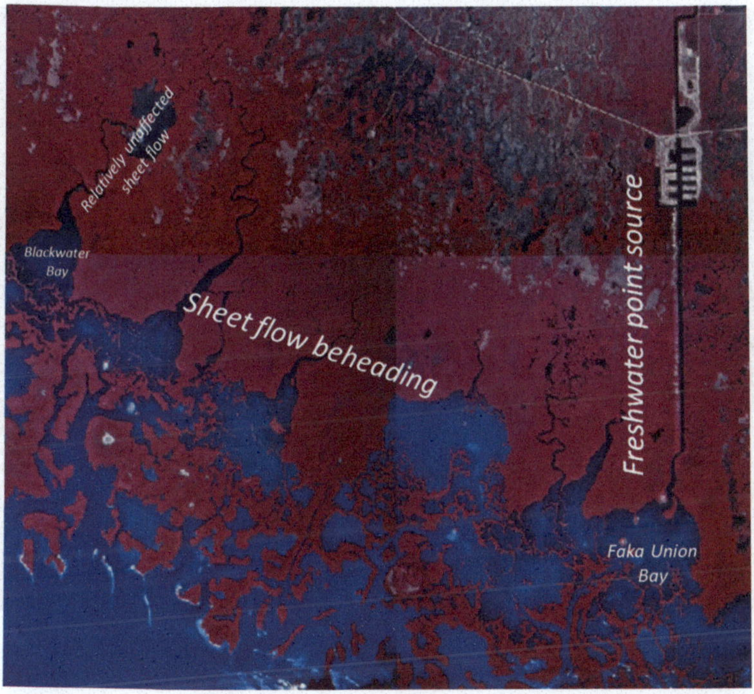

Fig. 5 Google Earth image of the Ten Thousand Islands. Faka Union Bay, the recipient of the canal system's freshwater outfall, has become a freshwater point source during the rainy season; the estuaries to the west have had their freshwater sheet flow beheaded, creating unnatural high salinities; further west, at Blackwater Bay, sheet flow has been relatively undisturbed

A first step, involving both conservation paleobiology and estuarine ecology, concerned the mapping of existing oyster reefs and an assessment of oyster health and productivity on those oyster reefs. The reef-distribution maps comparing an estuary far enough west to have avoided effects of canalization, Blackwater Bay, with Faka Union Bay, the estuary receiving excessive freshwater flow, show that modern reefs with the greatest productivity are spatially skewed. In Blackwater Bay, greatest reef number and greatest oyster productivity resides in the central region of the estuarine track, in Homologue 3 (Fig. 6). (Regions of the same homologue in neighboring estuaries exhibit similar geomorphology and experienced similar hydrology prior to anthropogenic influence.) This is where the estuarine geomorphology is richest in oyster-reef produced landforms—the now mangrove-forested islands. Faka Union Bay, on the other hand, shows greatest oyster productivity in Homologue 5, while relict reefs and reef-produced landforms are most extensive in Homologue 3 (Fig. 7). Conditions for best growth and reproduction are displaced downstream in Faka Union relative to Blackwater Bay, a condition one would predict because of the volume of freshwater output into this estuary.

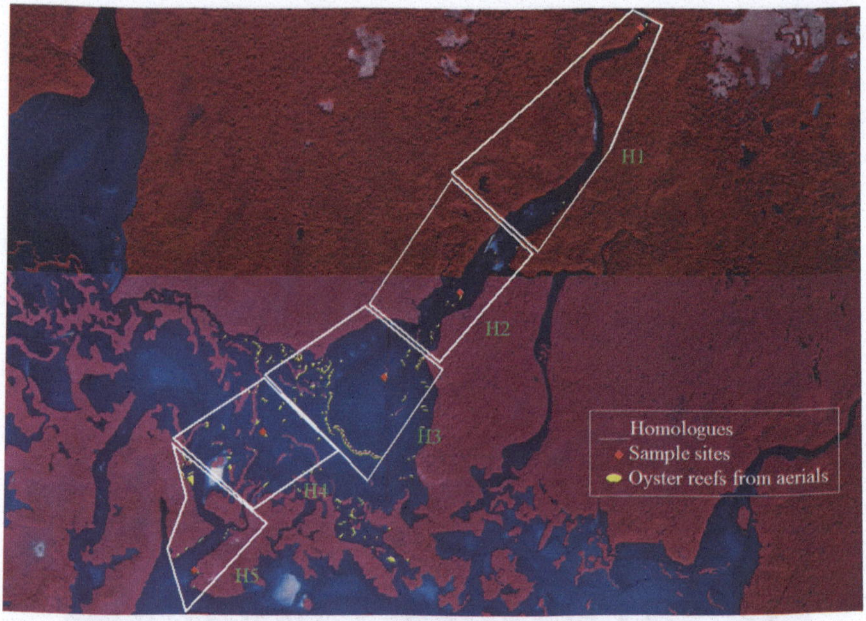

Fig. 6 Satellite image showing the distribution of oyster reefs within Blackwater Bay estuary (in yellow), located far enough west in the Ten Thousand Islands to be unaffected by sheet flow beheading. Homologues 1–5 (H1, H2, etc.) are regions in the estuary that have specific and comparable geomorphology and pre-anthropogenic salinities to neighboring estuaries. Consequently, similar homologues should have comparable states of reef development. In Blackwater Bay, oyster reefs are most numerous and most productive in H3

The second conservation paleobiologic step was to interpret the history of reef development in both estuaries in Homologue 3. Both Blackwater and Faka Union Bay oyster reefs in Homologue 3 have an extensive history, dating back at least 2000 ybp, with high oyster productivity. This indicates Homologue 3 throughout the Ten Thousand Islands maintained the ideal salinity for maximum oyster health in the pre-anthropogenic condition. Delivering brackish water salinities to this medial region through sheet flow restoration was justified, and the hydrologic modelers targeted the needed salinities on this region of the estuarine-scape.

The ultimate restoration design chosen (Fig. 8) involves a number of engineered features, included the lowering of road grade, the filling or plugging of canals, and the emplacement of pumps and spreader canals. The pumps and spreader canals permit quick transferal of freshwater and sheet flow initiation from the region to the north, the existing housing development of North Golden Gate Estates, to the Picayune Strand during times of excessive rainfall. The restoration design is adaptive in that freshwater flow through the pump stations can be regulated to ensure the appropriate salinities are delivered to the estuaries.

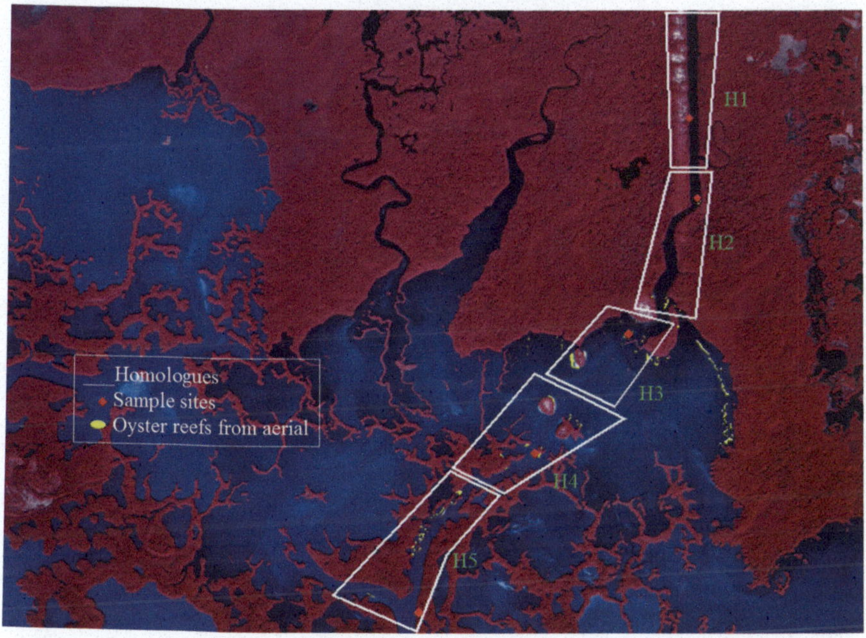

Fig. 7 Similar satellite image as in Fig. 6 but for Faka Union Bay, the estuary receiving excessive freshwater. The greatest oyster reef productivity is shifted downstream and resides in H5

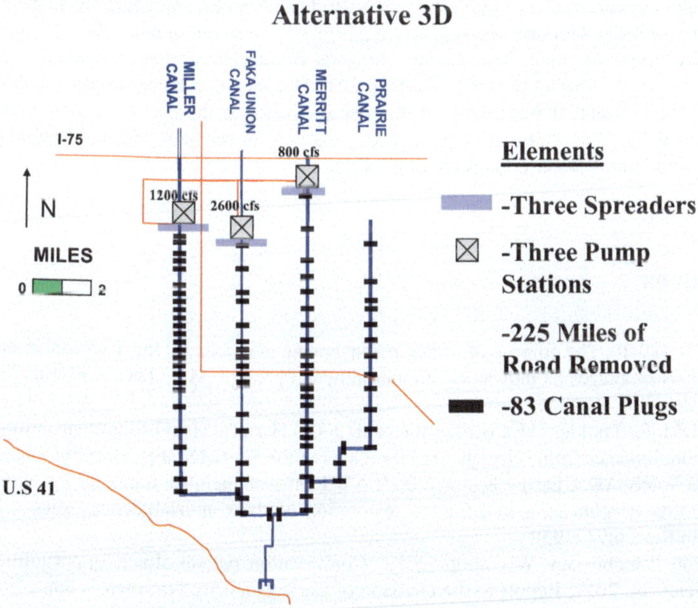

Fig. 8 Restoration design "Alternative 3D," the engineering plan that was ultimately chosen for Picayune Strand restoration. The solution involves the removal of roads, the filling or plugging of canals, and the construction of three pump stations and associated spreader canals

5 Conclusions

Paleontologists have a wealth of unrealized experience and knowledge to share with the environmental management and restoration world. Those skills are often unappreciated by the environmental management community for two fundamental reasons: the environmental profession is filled predominately by professionals that lack training in geoscience, and therefore lacks an understanding of our abilities to interpret environmental history; and the paleobiologic community does not effectively communicate or collaborate with the environmental field. The keys to our productive involvement in environmental science and management include self-promotion, marketing, and relationship building, over the short term, and a transformation in how we teach and to whom we teach paleontology, over the long term. Conservation paleobiology is societally relevant and necessary. It provides a new purpose for our science and, ultimately, greater opportunities for employment of our students.

Acknowledgments Many agencies over the years have generously supported conservation paleobiologic research for Greater Everglades' management and restoration. The granting agencies include: South Florida Water Management District, Florida Department of Environmental Protection, City of Naples, Charlotte Harbor National Estuary Program, National Science Foundation, National Oceanic and Atmospheric Administration, Environmental Protection Agency, and the U.S. Department of Education (for support of student involvement). A number of individuals from the environmental management world are owed thanks for their appreciation and advocacy for conservation paleobiology: Michael Bauer, Peter Doering, Kim Dryden, Michael Duever, Kevin Godsea, Patty Goodman, Gary Lytton, Ananta Nath, Janet Starnes, and Clarence Tears. Thanks to my graduate students who over the years have advanced conservation paleobiologic applications. They include Jorge Agobian, Kim Andres, Amanda Booth, Nicole Fronczkowski, Brian Hoye, and Sasha Wohlpart. Special thanks to the Fall, 2013 Conservation Paleobiology graduate course at FGCU. The conceptual framework of this paper developed through the course's curriculum and conversations. This manuscript benefitted greatly from the thoughtful reviews by Rowan Lockwood and Lynn Wingard; thanks to both of them.

References

Agobian JN (2010) The impact of water management practices in the Caloosahatchee River: mollusk assemblages as indicators of environmental change. M.S. Thesis, Florida Gulf Coast University

Anderson DM, Burkholder JM, Cochlan WP et al (2008) Harmful algal blooms and eutrophication: examining linkages from selected coastal regions of the United States. Harmful Algae 8:39–53

Barnes TK, Volety AK, Chartier K et al (2007) A habitat sustainability index for the eastern oyster (*Crassostrea virginica*), a tool for the restoration of the Caloosahatchee estuary, Florida. J Shellfish Res 26:949–959

Conservation Paleobiology Workshop (2012) Conservation paleobiology: opportunities for the Earth sciences. 2011. Report to the Division of Earth Sciences, National Science Foundation. Paleontological Research Institution, Ithaca

Cronon W (1996) The trouble with wilderness: or, getting back to the wrong nature. Environ Hist 1:7–28

Denevan WM (1992) The pristine myth: the landscape of the Americas in 1492. Ann Assoc Am Geogr 82:369–385
Dietl GP, Flessa KW (2011) Conservation paleobiology: putting the dead to work. Trends Ecol Evol 26:30–37
Doering P, Chamberlain RH, Haunert DE (2002) Using submerged aquatic vegetation to establish minimum and maximum freshwater inflows to the Caloosahatchee estuary, Florida. Estuaries 25:1343–1354
Grunwald M (2006) The swamp: the Everglades, Florida, and the politics of paradise. Simon and Schuster, New York
Jackson JBC, Kirby MX, Berger WH et al (2001) Historical overfishing and the recent collapse of coastal ecosystems. Science 293:629–638
Kidwell SM (2007) Discordance between living and death assemblages as evidence for anthropogenic ecological change. Proc Natl Acad Sci 104:17701–17706
Kidwell SM (2009) Evaluating human modification of shallow marine ecosystems: mismatch in composition of molluscan living and time-averaged death assemblages. In: Dietl GP, Flessa KW (eds) Conservation paleobiology: using the past to manage for the future. The Paleontological Society papers, vol 15. pp 113–139
Knowlton N, Jackson JB (2008) Shifting baselines, local impacts, and global change on coral reefs. PLoS Biol 6(2):e54. https://doi.org/10.1371/journal.pbio.0060054
Parkinson RW (1989) Decelerating Holocene sea-level rise and its influence on Southwest Florida coastal evolution: a transgressive/regressive stratigraphy. J Sediment Petrol 59:960–972
Saltmarsh J, Giles DE Jr, Ward E et al (2009) Rewarding community-engaged scholarship. New Dir High Educ (147):25–35
Savarese M, Volety A (2008) Oyster reef health in Pumpkin and Fakahatchee Estuaries: baseline monitoring for Ten Thousand Islands Restoration. South Florida Water Management District, Technical Report
Savarese M, Volety A, Tolley G (2003) Influence of watershed alteration on oyster health and oyster-reef habitat: management implications for the Faka-Union and Estero Bays. South Florida Water Management District, Technical Report
Sklar FH, Chimney MJ, Newman S et al (2005) The ecological-societal underpinnings of Everglades restoration. Front Ecol Environ 3:161–169
Tolley G, Volety A, Savarese M (2003) Shellfish research and adaptive resource management in Southwest Florida: oysters as sentinels of ecosystem health. World Aquaculture 34:64–66
U.S. Army Corps of Engineers (2004) Project documents: Picayune Strand Restoration Integrated Project Implementation Report (PIR)/Environmental Impact Statement (EIS). EvergladesRestoration.gov. URL: http://141.232.10.32/pm/projects/docs_30_sgge_pir_final.aspx.
U.S. Army Corps of Engineers (2012) SMART restoration planning guide: the project delivery team. EvergladesRestoration.gov. http://planning.usace.army.mil/toolbox/smart.cfm?Section=2&Part=1.
U.S. Department of the Interior Office of Everglades Restoration Initiatives (2016) EvergladesRestoration.gov: Restoring America's Everglades. http://www.evergladesrestoration.gov/index.htm.
Vargo GA, Heil CA, Fanning KA et al (2008) Nutrient availability in support of *Karenia brevis* blooms on the central West Florida Shelf: what keeps *Karenia* blooming? Cont Shelf Res 28:73–98
Volety AK, Savarese M, Tolley SG et al (2009) Eastern oysters (*Crassostrea virginica*) as an indicator for restoration of Everglades' ecosystems. Ecol Indic 9(Suppl 6):S120–S136
Whitmer A, Ogden L, Lawton J et al (2010) The engaged university: providing a platform for research that transforms society. Front Ecol Environ 8:314–321

Using the Fossil Record to Establish a Baseline and Recommendations for Oyster Mitigation in the Mid-Atlantic U.S.

Kristopher M. Kusnerik, Rowan Lockwood, and Amanda N. Grant

Abstract Eastern oyster populations throughout the Mid-Atlantic region of the USA have been in decline for centuries due to overharvesting, disease, increased sediment pollution, and habitat destruction. By studying Pleistocene fossil oyster assemblages, it is possible to reconstruct baseline conditions and develop recommendations for oyster mitigation. Fossil assemblages were studied from five Pleistocene sites located in Maryland, Virginia, and North Carolina. Reconstructions of paleosalinity and temperature were used to identify modern and colonial sites with similar environmental parameters for comparison. Shell height and life span in Chesapeake Bay oysters declined significantly from the Pleistocene to today, at the same time that ontogenetic growth rates have increased. This pattern is driven by age truncation, in which both harvesting and disease preferentially remove the larger, reproductively more active and primarily female members of the population. By contrast, Pleistocene oysters from North Carolina did not differ significantly, in shell height, life span, or growth rates, from modern oysters.

Although oyster management in the Mid-Atlantic States has focused historically on protecting and supplementing early life stages, this study recommends three potential management solutions to the age truncation revealed by comparison with Pleistocene oysters. Possible solutions include (1) implementation of a maximum size or slot limit on the fishery, (2) establishment of marine protected areas (MPA), or (3) significant lowering of exploitation rates.

K. M. Kusnerik (✉)
Division of Invertebrate Paleontology, Florida Museum of Natural History, Gainesville, FL, USA
e-mail: kmkusnerik@ufl.edu

R. Lockwood
Department of Geology, The College of William and Mary, Williamsburg, VA, USA
e-mail: rxlock@wm.edu

A. N. Grant
School of Earth Sciences and Environmental Sustainability, Northern Arizona University, Flagstaff, AZ, USA
e-mail: angrant@email.wm.edu

C. L. Tyler, C. L. Schneider (eds.), *Marine Conservation Paleobiology*, Topics in Geobiology 47, https://doi.org/10.1007/978-3-319-73795-9_5

Keywords *Crassostrea virginica* · Pleistocene · Growth rates · Chesapeake Bay · Fishery · Aquaculture · Aquatic resource management · Oyster reef · Restoration · Virginia · Maryland · North Carolina

1 Introduction

The Eastern Oyster (*Crassostrea virginica*) plays a vital role in the ecosystem of the Chesapeake Bay and Mid-Atlantic regions (Mann et al. 2009a). Oysters are ecosystem engineers that build habitat for fish and other invertebrate species, boost water quality by filtering bacteria and contaminants, and represent an important component of the food web (Haven and Morales-Alamo 1970; Meyer and Townsend 2000; Cressmann et al. 2003; Hoellein et al. 2015). Historically, oyster harvests have provided a key economic resource for the region (Paolisso and Dery 2010; Rick and Lockwood 2013).

The Pleistocene record provides evidence of widespread, thriving oyster reefs predating human settlement in the region, during the formation of the proto-Chesapeake Bay (Hargis and Haven 1995; USGS 1998; Reshetiloff 2004; Rick and Lockwood 2013). The earliest record of human harvest of Mid-Atlantic oysters dates back to the Late Archaic (ca 2500–2000 cal yr. BC) in the form of oyster middens, or archaeological deposits of kitchen waste material (Waselkov 1982; Custer 1989; Thompson and Worth 2011). Native American harvesting occurred for thousands of years, across the bay region, and is assumed to have involved harvesting of small clusters of shells from easily accessible reefs (Rick et al. 2014, 2016).

European settlement of the region began with the Jamestown Colony, which was established along the James River (Virginia) in 1607 (Rountree et al. 2007; Horn 2008). English settlers relied heavily on oysters as a food source, leaving a record of oyster harvesting in the form of dozens of shells recovered from an abandoned well within the settlement (Kelso 2004; Harding et al. 2008, 2010a). Although oyster harvesting has been a key component of the regional economy for thousands of years, these harvests have declined massively over the last 150–200 years and now represent a mere 1% of peak productivity (Rothschild et al. 1994; Harding et al. 2008; Beck et al. 2011; Wilberg et al. 2011).

By the late 1800s, harvest by industrial dredging caused a massive reduction in reef height through the removal of shell material faster than living oysters could replenish it (Hargis and Haven 1999). Natural oyster reefs have effectively disappeared in the modern bay, due to overfishing, disease, increased sediment input, and habitat destruction, leaving oyster populations depleted and in need of serious mitigation efforts (Rothschild et al. 1994; Mann and Powell 2007).

Oyster populations in the region have also been impacted by two prevalent parasitic diseases that increase oyster mortality rates (Carnegie and Burreson 2009). Dermo disease is caused by the parasite *Perkinsus marinus*, is prevalent in intermediate salinity (12–15 ppt) warmer waters, and was first documented in the Chesapeake Bay in 1949 (Andrews 1996; Burreson and Ragone Calvo 1996; CTDOAG 2016a). MSX (Multinucleated Sphere Unknown) is caused by the spore-forming protozoan *Haplosporidium nelsoni*, prefers more saline conditions (>15 ppt) and was first documented in the Mid-Atlantic region in 1957 (CTDOAG 2016b) as a result of the intentional introduction of the Japanese oyster (*Crassostrea gigas*) to Delaware Bay (Andrews and Wood 1967). Increased sediment influx from land clearance has further complicated the situation by hastening habitat destruction in areas of optimal oyster growth (Hargis and Haven 1999).

Early management approaches to Mid-Atlantic oysters focused on helping the fishery recover from natural and anthropogenic problems primarily for the benefit of the local economy. These approaches included the genetic enhancement of broodstock, the release of spat raised through aquaculture, and planting of dead shell (cultch) to act as substrate for settlement (Bartol and Mann 1997, 1999a, 1999b; Mann and Evans 1998; Southworth and Mann 1998; Wesson et al. 1999; Luckenbach et al. 1999; Mann 2000; Southworth et al. 2000). More recent efforts have embraced a broader strategy, prioritizing the mitigation of ecosystem services via protected areas, reduction of harvesting, and large-scale three-dimensional reef restoration (Luckenbach et al. 2005, Coen et al. 2006, 2007; North et al. 2010; Beck et al. 2011; Grabowski et al. 2012).

Pleistocene fossil reefs may provide insight into how oyster ecosystems functioned before human intervention in the Mid-Atlantic U.S. Although natural oyster reefs have dwindled to extinction in the modern Chesapeake Bay, Pleistocene oysters can yield information on shell size and growth rates that may prove vital for mitigation efforts. The goal of this study is to quantify oyster size and growth rates in the Mid-Atlantic region, across three timescales—Pleistocene, colonial, and modern—to assess the implications of this conservation paleobiological approach for oyster mitigation.

2 Methods

Pleistocene Localities

Samples of Pleistocene oysters were examined from the Virginia Museum of Natural History (VMNH) and field collections across five localities, distributed from southern Maryland to North Carolina (Fig. 1; Table 1). The northernmost of the Chesapeake Bay sites, Wailes Bluff (WB), is located at the mouth of the Potomac River in St. Mary's County, Maryland. This site, which has since been covered by a seawall, yielded molluscan material collected by L. W. Ward in 1971 (VMNH 71LW93). Bulk samples were collected from *C. virginica* shell layers distributed in a sandy silt matrix, thought to represent the late Pleistocene Tabb Formation (Fig. 2; Thompson 1972; Belknap 1979; Cronin 1979; Wehmiller and Belknap 1982; Rader and Evans 1993). Many of the molluscan species identified at Wailes Bluff are now

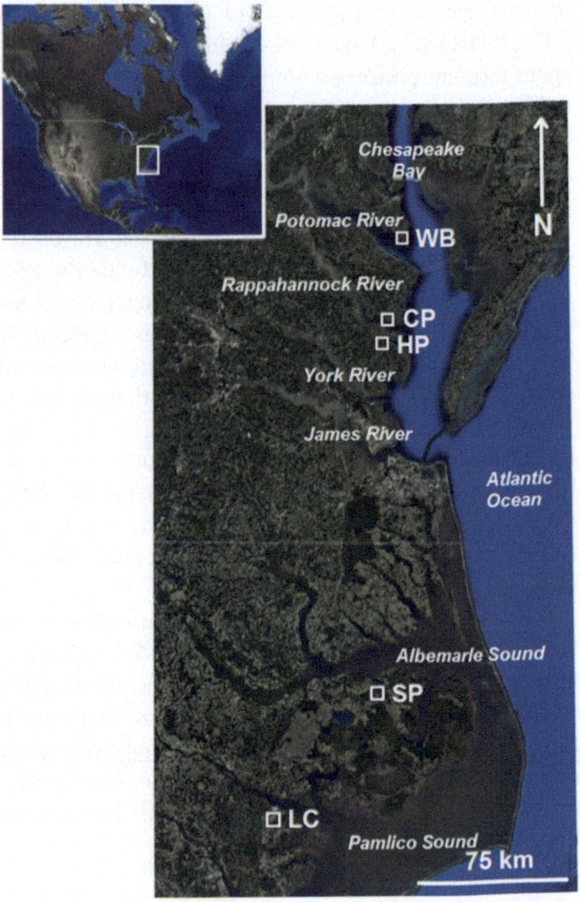

Fig. 1 Map of five localities in Maryland, Virginia, and North Carolina, from which Pleistocene fossil oysters were sampled (*WB* Wailes Bluff, *CP* Cherry Point, *HP* Holland Point, *SP* Stetson Pit, *LC* Lee Creek)

restricted to more southerly latitudes, suggesting a paleotemperature warmer than today (Blake 1953). Several brackish water taxa have been documented at this site, supporting a paleosalinity of approximately 15–30 ppt (Blake 1953; Cronin 1979).

Cherry Point (CP, also known as Norris Bridge) is a middle Pleistocene site located in Lancaster County, Virginia, near the mouth of the Rappahannock River (Fig. 1; Table 1). Although the site is no longer available for collecting due to development, B. W. Blackwelder and T. M. Cronin collected oyster specimens in 1978 (VMNH 78BB79A, B; VMNH T8TC56) from a fossiliferous sandy silt unit thought to represent the Shirley Formation (Fig. 2; Mirecki 1990; Mirecki et al. 1995). Ostracode assemblages at this site indicate a Pleistocene bottom temperature between 12.5 and 15 °C during winter and 27.5 °C during summer months (Cronin 1979), both of which are warmer than modern conditions at the site by at least 1–2 °C (Massmann et al. 1952). Paleosalinity is thought to have ranged between open sound (15–35 ppt) and estuarine (2–15 ppt) conditions (Cronin 1979).

Table 1 Location, stratigraphic unit, and geologic age of the five localities sampled for Pleistocene oysters

	State	Sample size (≥35 mm)	Latitude/longitude	Stratigraphic unit	Pleistocene interval
Wailes Bluff (WB)	MD	36 (36)	38.065560/76.365280°	Tabb	Late
Cherry Point (CP)	VA	36 (36)	37.634184/76.412830°	Shirley	Mid
Holland Point (HP)	VA	865 (611)	37.512088/76.432121°	Shirley	Mid
Stetson Pit (SP)	NC	225 (225)	35.866291/76.293768°	Undetermined	Late
Upper Lee Creek (ULC)	NC	85 (85)	35.324287/76.800213°	James City	Early
Lower Lee Creek (LLC)	NC	21 (21)	35.324287/76.800213°	Flanner Beach	Mid

The Lee Creek locality is divided into two sections: Upper Lee Creek (ULC) and Lower Lee Creek (LLC), respectively

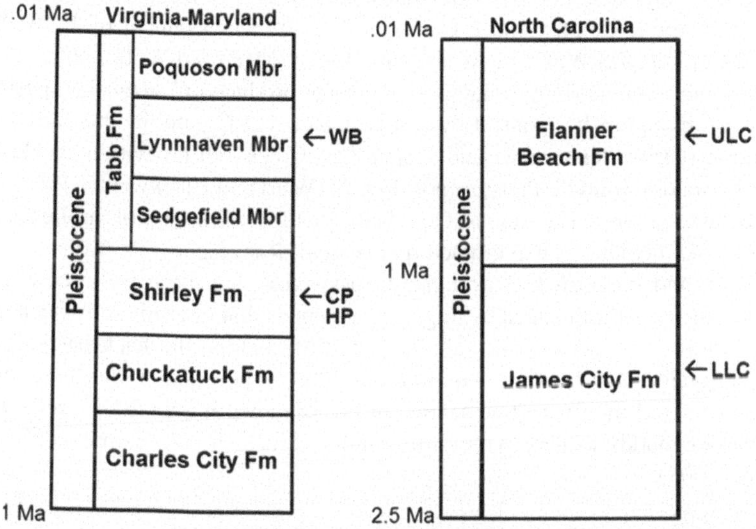

Fig. 2 Stratigraphic framework for four of the Pleistocene localities (*WB* Wailes Bluff, *HP* Holland Point, *CP* Cherry Point, *ULC* Upper Lee Creek, *LLC* Lower Lee Creek). The stratigraphic unit for Stetson Pit (SP) is undetermined

Holland Point (HP), the southernmost of the Chesapeake Bay sites, is located on the Piankatank River (Fig. 1; Table 1). Amino acid racemization dating of *C. virginica* and *Mercenaria* specimens suggests an age of approximately 195–243Ka (MIS 7 or 9, J. Wehmiller, personal communication 2016). The unit in which the oyster deposit occurs is thought to represent the Shirley Formation, which is middle Pleistocene in age (Fig. 2; C. R. Berquist, personal communication 2016). The exposed oyster deposit at Holland Point is laterally extensive (up to 25 m) and thick (up to 3 m), containing thousands of oysters, many of which are articulated and preserved in life position within a fine sandy, clayey silt matrix. Reconstructions of both paleotemperature and salinity were accomplished as part of the current study.

The northernmost site in North Carolina is Stetson Pit (SP, Fig. 1, Table 1), located at the mouth of the Albemarle Sound in Dare County, North Carolina, and subsequently covered by landfill material. Bulk samples, collected in 1979 by B. W. Blackwelder (VMNH 79BB32(D)), yielded several *C. virginica*. This shelly, sandy mud unit that was sampled has never been attributed definitively to a specific stratigraphic unit (Miller 1982), but its position above a U-series-dated coral suggests that it is late Pleistocene (late MIS 5a, J. Wehmiller personal communication 2016) in age. Using assemblages of temperature-sensitive ostracodes, York et al. (1989) identified a high proportion of cryptophilic species, suggesting relatively cooler temperatures compared to present-day Cape Hatteras, North Carolina. These ostracode assemblages, and the presence of molluscan taxa including *Rangia cuneata*, indicate a brackish paleosalinity (York et al. 1989).

The final site, Lee Creek (PCSD Phosphate Mine), is located on the Pamlico River in Beaufort County, North Carolina. The site was sampled by L. W. Ward in 1972 (VMNH 72LW8C, 72LW1B) and 1992 (VMNH 92LW60a) although it is currently inaccessible (Fig. 1; Table 1). This site produced *C. virginica* specimens at two stratigraphic horizons: Lower Lee Creek (LLC) within the James City Formation (early Pleistocene) and Upper Lee Creek (ULC) within the Flanner Beach Formation (middle Pleistocene) (Fig. 2; (Ward and Blackwelder 1987; Ward and Bohaska 2008). The James City beds yielded a variety of mollusks in a fossiliferous, medium-coarse grained quartz sand. Rare records of freshwater (i.e., *Corbicula*) and brackish (i.e., *Rangia*) taxa suggest that these beds represent an offshore barrier bar influenced by migrating channels that intermittently opened and closed, with more open-marine salinity conditions behind the back barrier (Ward and Blackwelder 1987; Ward and Bohaska 2008). The Flanner Beach sediments are characterized by a very fine sandy silt (Ward and Bohaska 2008), reflecting a back barrier muddy estuary with corresponding brackish paleosalinity (Ward and Bohaska 2008).

Field and Museum Sampling

Of the five sites studied, Holland Point is the only one still accessible for field sampling. We sampled the exposed oyster deposit at Holland Point in July 2011, in addition to describing the sedimentology and measuring the stratigraphic section. Samples were collected in five columns spaced approximately 3 m apart along the lateral extent of the deposit. We collected three bulk samples (spaced evenly apart according to deposit thickness) from each column (n = 15 samples total), using a 0.5 m^2 quadrat (Fig. 3).

Fig. 3 Sampling transects of the oyster deposit at Holland Point. Samples were collected in five columns spaced approximately 3 m apart along the lateral extent of the deposit. Three bulk samples (spaced evenly apart according to deposit thickness) were collected from each column ($n = 15$ samples total), using a 0.5 m^2 quadrat

Bulk samples of *C. virginica* from the Virginia Museum of Natural History in Martinsville, Virginia, were examined from Wailes Bluff (VMNH 71LW93), Cherry Point (VMNH 78BB79A, B; 78TC56), Stetson Pit (VMNH 79BB32), and Lee Creek (VMNH 72LW1B, 72LW8C, 92LW60A).

Oyster Size and Abundance Data

Samples from Holland Point were sieved using a 4 mm mesh size, sorted, and molluscan whole specimens and hinge fragments were identified to the lowest taxonomic level possible (Abbott 1974; Spencer and Campbell 1987).

For each museum and field sample, we used digital calipers to measure shell height for all whole left valves of *C. virginica.* Shell height was measured as the distance from the umbo to the ventral most edge of the shell. Although this distance is commonly referred to as "shell length," it is more accurate to use the term "shell height" (Galtsoff 1964). The field site at Holland Point yielded 865 whole left valves of *C. virginica* for measurement. The number of museum specimens available for each Pleistocene site varied from 21 to 225 whole left valves.

A subset of these left valves was randomly selected to be sawed in half, using a diamond-tipped tile saw. The resulting bisected hinges were used to count shell bands in cross-section. We recorded the number of thick, dark gray shell bands that were continuous from the hinge to the outer shell layer (Fig. 4), to provide a proxy for biological age (e.g., life span) of each specimen, following Harding and Mann (2006) and Zimmt et al. 2016).

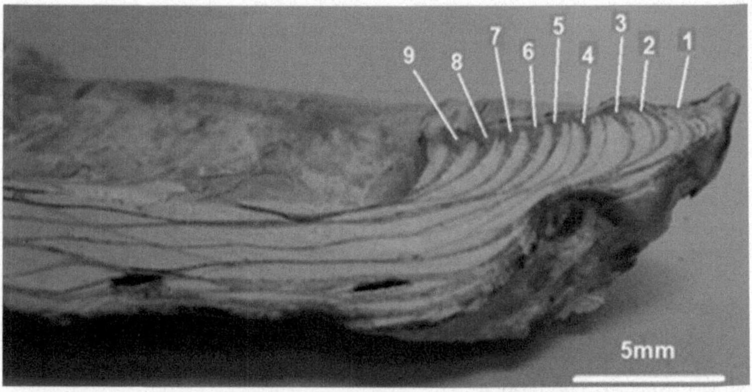

Fig. 4 Cross-section of bisected hinge from Holland Point, showing gray and white growth lines; *1–9* represent thick, gray shell bands that were continuous from the hinge to the outer shell layer and were counted to provide an approximate biological age (i.e., life span) for oyster specimens

Reconstructing Paleotemperature and Salinity

In order to standardize for paleoenvironment, we compiled information on paleosalinity and paleotemperature for each Pleistocene locality, except for Holland Point, from the literature.

To estimate paleosalinity at Holland Point, the raw abundance of all other molluscan specimens identified to the species level was assessed. For bivalves, whole shells and hinge fragments were sorted into left versus right valves and the larger of the two values was used to represent raw abundance for each sample. For gastropods, whole and fragmented specimens were sorted into apertures and apices and the larger of the two values was used to represent abundance for each sample. We compiled modern salinity tolerance ranges from the literature for those species with more than one occurrence at Holland Point (Federighi 1931; Andrews 1953; Menzel et al. 1966; Castagna and Chanley 1973; Buroker 1983; Zimmerman and Pechenik 1991; Grabe et al. 1993; Griffin 2001; Hill 2004, Zachary and Haven 2004; Wilson et al. 2005; Harding et al. 2010b; Cohen 2011).

To assess paleotemperature at Holland Point, one specimen of *C. virginica* was assessed using clumped isotope analysis by G. A. Henkes (Johns Hopkins University). A sub-sample of shell material was collected from the hinge area using a low-speed Dremel drill and analyzed following the methods of Henkes et al. (2013).

Additionally, two articulated *C. virginica* and one articulated *M. mercenaria,* all preserved in life position from Holland Point, were assessed for paleotemperature using sclerochronology under the supervision of G.S. Herbert (University of South Florida). Sub-samples (one sample per every 0.78–1.75 mm) were collected from each annual growth band of sectioned valves using a Dremel drill. Powdered material was dissolved in 100% H_3PO_4 at 25 °C for 24 h. The resulting CO_2 was separated, focused, and analyzed on a Thermo Finnigan Delta V Advantage IRMS in continuous flow mode coupled to a Gasbench II preparation device (Harke et al. 2015). Growing season paleotemperatures were calculated using the Craig (1965) calcite-water equation for *C. virginica* and the Grossman and Ku (1981) aragonite-water equation for *M. mercenaria*. The salinity range estimated from HP molluscan occurrences was used to determine $\delta^{18}O_{seawater}$ for these equations.

Modern and Colonial Data

Reconstructions of paleosalinity from the Pleistocene sites were used to identify modern and colonial sites with *C. virginica* living in similar salinity conditions. Data on shell height and growth rates in colonial and modern oysters were compiled from the published literature and management agencies as described below.

In Maryland, colonial data on mean shell height and growth rates were compiled from Miller (1986) and Catts et al. (1998) for four sites from similar salinity regimes (15–25 ppt), including St. Mary's City (sample size not published) and Ashcomb's Quarter ($n = 99$ shells). Data on modern shell height and growth rates were provided by M. Tarnowski (Maryland Department of Natural Resources) for six sites ($n = 1176$ shells) from Pocomoke and Tangier Sounds (15–25 ppt salinity zone) sampled from 2013 to 2015.

In Virginia, colonial data on shell heights and growth rates were acquired from Harding et al. (2008, 2010a) from the Jamestown Colony ($n = 363$ shells, salinity 15–30 ppt). Modern shell height and growth rate data were also compiled from published sources (Harding et al. 2008; Sisson et al. 2011) for eight sites ($n = 6916$ shells) in the James River (collected from 2006 to 2008) and Lynnhaven River (collected from 2005 to 2008), from the same salinity range.

We were unable to locate any colonial aged oyster data from North Carolina. Data on shell height and growth rates for modern North Carolina oysters were obtained from Puckett and Eggleston (2012), for six sites sampled from 2006 to 2008 ($n = 5443$ shells) in Pamlico Sound (15–30 ppt salinity zone).

3 Results

Paleoenvironmental Reconstruction of Holland Point

Paleotemperature

Clumped isotopic analysis of a single *C. virginica* valve from the Holland Point fossil deposit produced a $\delta^{13}C_{carb}$ composition of 0.47 ± 0.02‰PDB, $\delta^{18}O_{carb}$ of 2.08 ± 0.01‰PDB, and Δ_{47carb} of 0.725 ± 0.014‰ (Ghosh)(Ghosh et al. 2006; Huntington et al. 2009). This Δ_{47} value corresponds to an oyster growth temperature (i.e., averaged temperature at which the oyster was growing at this particular site, at this particular time) of 14.1 ± 5 °C. Given this paleotemperature and measured oxygen isotope value, solving the Kim et al. (2007) equilibrium oxygen isotope fractionation equation provides a $\delta^{18}O_{water}$ value of −3.44‰ SMOW.

Isotopic sclerochronological analysis of one *M. mercenaria* and two *C. virginica* shells from Holland Point yielded comparable results of $\delta^{18}O_{water} = -3.40$‰

SMOW. Using the Craig (1965) and Grossman and Ku (1981) equations, respectively, growing season paleotemperatures ranged from 5.3 to 20.5 °C for *C. virginica* and 10.9 to 20.9 °C for *M. mercenaria*. This growth temperature is cooler than the modern Piankatank River, which routinely reaches temperatures between 25 and 30 °C during the summer growing months of modern *C. virginica* (Harding et al. 2010b). Sclerochronological analyses of other species would be required to determine whether *C. virginica* stopped growing in either the summer or winter months at this site in the middle Pleistocene.

Paleosalinity

Almost 1500 whole and fragmented shells, representing 21 macroinvertebrate species other than *C. virginica*, were identified in the bulk samples collected from the Holland Point deposit. Raw abundance of these species ranged from 1 to 515 (mean = 55.62) per sample. Six taxa were excluded from the salinity analyses because they could not be identified to the species level (undetermined crab, sponge (likely *Cliona*), barnacle (likely *Balanus*)) or because data on salinity tolerance of the modern representatives were not readily available (*Trittia trivittata, Crepidula convexa,* and *Melanella polita*). When rare taxa, those with only a single occurrence, are excluded, a paleosalinity range of 16–32 ppt encompasses all remaining species (Fig. 5). This salinity is higher than that of the modern Piankatank River, which fluctuates between 6 and 23 ppt (Harding et al. 2010b).

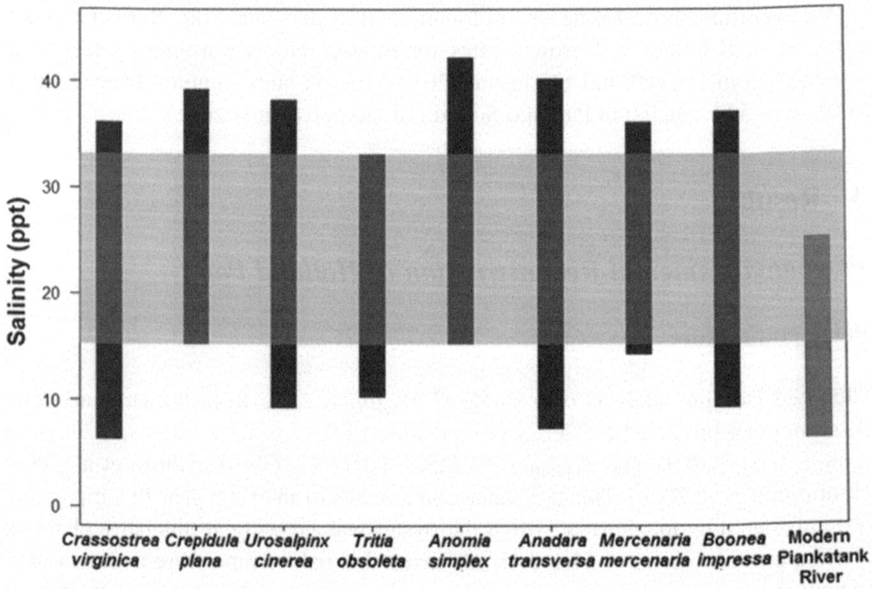

Fig. 5 Modern salinity tolerances of fossil macroinvertebrate species with abundance greater than 1 recorded at Holland Point (Virginia); shaded region (15–32 ppt) encompasses salinity range within which all taxa could co-occur

Shell Height

Shell height was compared across all of the Pleistocene, colonial, and modern localities using size frequency distributions and non-parametric statistical tests. We excluded specimens ≤35 mm in shell height in order to avoid sampling spat (Mann et al. 2009a). Kolmogorov-Smirnov tests for normality indicated that the shell height data were non-normally distributed for most fossil ($KS_{21\text{–}655} = 0.08\text{–}0.22$, $p = 0.20\text{–}0.0001$) and modern ($KS_{24\text{–}5784} = 0.05\text{–}0.17$, $p = 0.20\text{–}0.0001$) samples. We therefore used a non-parametric test (Mann-Whitney U) to test for differences in mean shell height among Pleistocene, colonial, and modern oysters across three geographic regions: (1) upper Chesapeake Bay (Maryland), (2) lower Chesapeake Bay (Virginia), and (3) North Carolina.

Starting with the Maryland portion of the Chesapeake Bay, late Pleistocene oysters tend to be larger than colonial or modern oysters from similar salinity regimes (15–25 ppt, Fig. 6, Table 2). Pleistocene oysters also exhibit more strongly right-skewed size frequency distributions than modern oysters (Fig. 6).

Moving south, into the Virginia portion of the bay, middle Pleistocene (MP) oysters were statistically significantly larger than both colonial and modern oysters from similar salinity regimes (15–30 ppt, Fig. 7, Table 2). Pleistocene oysters reached a maximum size of nearly 260 mm, in comparison to colonial (124 mm) and modern (148 mm) specimens. The right tails of the size frequency distributions for both the modern and colonial oysters appear truncated. These distributions are missing the larger adults that contribute to the strongly right-skewed distribution of the Pleistocene sample (Fig. 7).

In Albemarle and Pamlico Sounds in North Carolina, MP and LP oysters show no significant difference in size or the shape of the size frequency distribution from modern oysters from similar salinity regimes (15–30 ppt, Fig. 8, Table 2).

Growth Rate

Growth rates were plotted for Pleistocene, colonial, and modern oysters by plotting biological age (i.e., life span) of each specimen versus shell height (Figs. 9 and 10). We excluded specimens ≤35 mm in shell height in order to avoid sampling spat (Mann et al. 2009a). We calculated the slopes of these growth trajectories (i.e., growth rates) and compared them using an analysis of covariance (ANCOVA, modelled as linear). Growth rates were compared among Pleistocene, colonial, and modern oysters across three geographic regions: (1) upper Chesapeake Bay (Maryland), (2) lower Chesapeake Bay (Virginia), and (3) North Carolina.

In the upper Chesapeake Bay, growth trajectories in Pleistocene oysters extend beyond 12 years (Fig. 9). In contrast, growth trajectories for colonial and modern oysters rarely extend beyond 5 years (Fig. 9). Using ANCOVA to compare growth rates in Maryland oysters between 0 and 5 years of age suggests that modern oysters are growing significantly faster than Pleistocene or colonial oysters from similar salinity regimes (15–30 ppt, Fig. 9, Table 3). Whereas modern oysters record

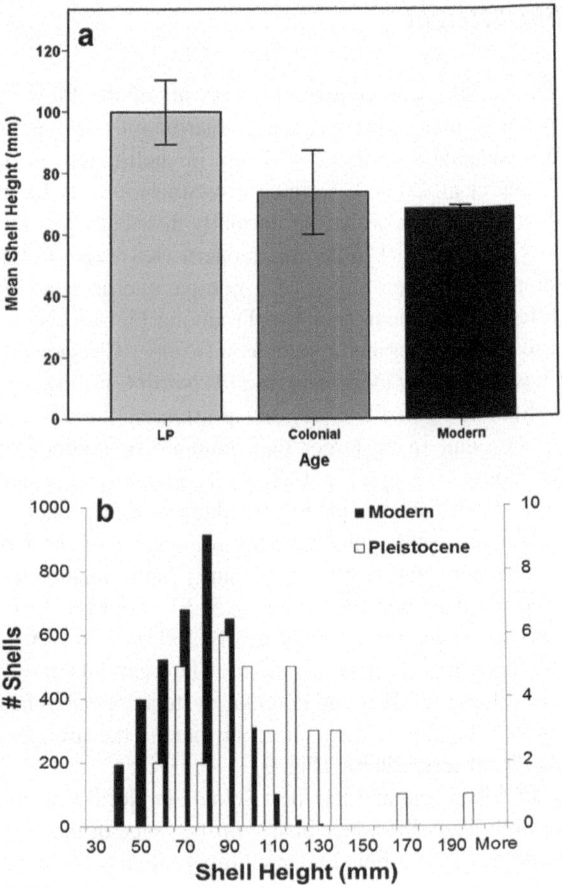

Fig. 6 Shell height (mm) trends in late Pleistocene (LP), colonial and modern oysters from sites in the upper Chesapeake Bay region (Maryland) with salinities ranging from 15 to 25 ppt. (**a**) Mean shell height ± S.E.; (**b**) Size frequency distributions for Pleistocene and modern oysters

Table 2 Mann-Whitney U tests comparing shell height (mm) in Pleistocene, colonial, and modern oysters from Maryland, Virginia, and North Carolina

	Maryland	Virginia	North Carolina
MP vs. LP	–	–	$\mathbf{Z_{225,21} = -2.93, p = 0.003}$
MP vs. colonial	–	$\mathbf{Z_{647,364} = -4.48, p < 0.0001}$	–
MP vs. modern	–	$\mathbf{Z_{647,6916} = -15.15, p < 0.0001}$	$Z_{21,5443} = -1.05, p < 0.29$
LP vs. colonial	$Z_{36,3} = -1.69, p = 0.09$	–	–
LP vs. modern	$\mathbf{Z_{36,1176} = -6.44, p < 0.0001}$	–	$\mathbf{Z_{225,5443} = -3.76, p < 0.0001}$
Colonial vs. modern	$Z_{3,1176} = -0.69, p = 0.49$	$\mathbf{Z_{364,6916} = -9.56, p < 0.0001}$	

Statistically significant differences ($p \leq 0.05$) highlighted in bold; all oyster comparisons from similar salinity regimes; *MP* middle Pleistocene, *LP* late Pleistocene

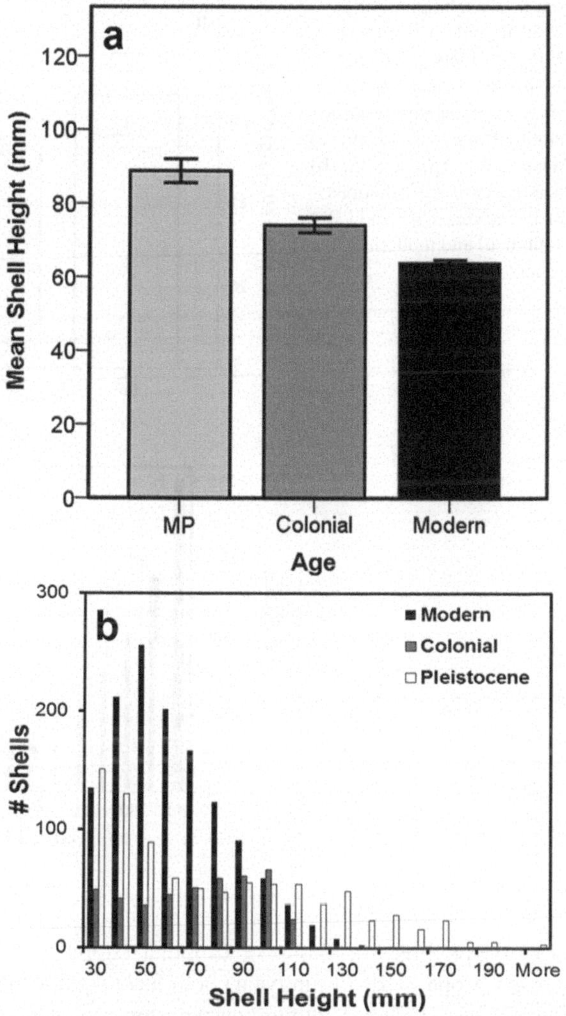

Fig. 7 Shell height (mm) trends in middle Pleistocene (MP), colonial, and modern oysters from sites in the lower Chesapeake Bay region (Virginia) with salinities ranging from 15 to 30 ppt; (**a**) Mean shell height ± S.E.; (**b**) Size frequency distributions for Pleistocene (MP), colonial, and modern oysters

average growth rates of 14 mm/year (slope), colonial (slope = 11.78 mm/year), and Pleistocene oysters grow significantly more slowly (slope = 8.51 mm/year).

Comparisons of growth trajectories among Pleistocene, colonial, and modern oysters in the lower Chesapeake Bay yield similar results. Growth trajectories in Pleistocene oysters extend beyond 20 years, while colonial and modern oysters rarely live longer than 5 years (Fig. 10). Mean biological age (i.e., life span, x axis of Fig. 10) ranges from 8.6 years at the Holland Point fossil locality to 2.38 years at the Jamestown colonial site to 2.59 years in modern bay sites with salinities ranging from 15 to 30 ppt. Comparison of growth rates in Virginia via ANCOVA (oysters aged 0–5 years) reveals that Pleistocene oysters have slower growth rates than either colonial or modern oysters from similar salinity regimes

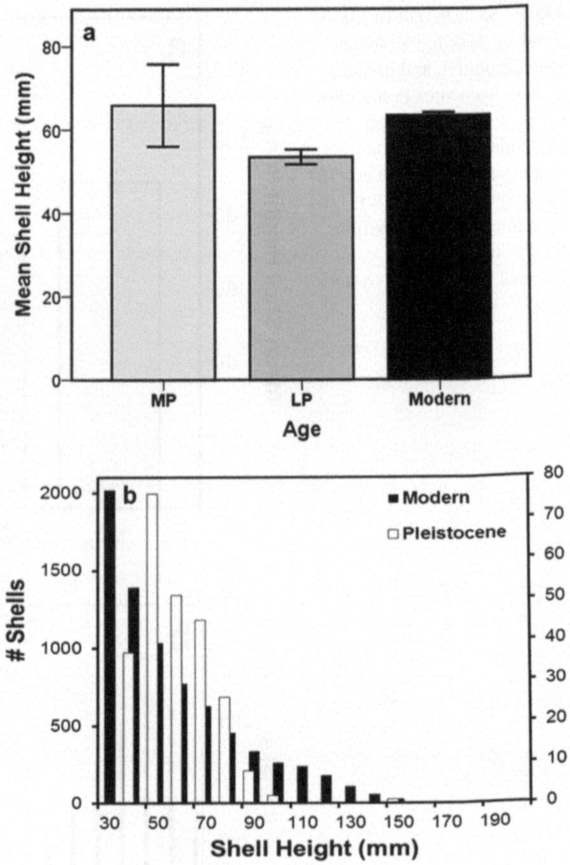

Fig. 8 Shell height (mm) trends in middle Pleistocene (MP), late Pleistocene (LP), and modern oysters from North Carolina with salinities ranging from 15 to 30 ppt; (**a**) Mean shell height ± S.E.; (**b**) Size frequency distributions for Pleistocene (MP and LP combined) and modern oysters

(15–30 ppt, Fig. 10, Table 3). Colonial (slope = 28.94 mm/year) and modern oysters (slope = 19.73 mm/year) both have significantly greater growth rates than Pleistocene (slope = 8.24 mm/year) oysters.

Interestingly, growth trajectories in North Carolina oysters do not extend beyond 4 years of age, for either Pleistocene or modern oysters. ANCOVA reveals that growth rates in oysters $\leq$ 5 years are significantly steeper in modern, in comparison to Pleistocene oysters, when salinity is controlled for (15–30 ppt, Fig. 11, Table 3). The slope of modern oysters (23.77 mm/year) is almost three times that of Pleistocene oysters (8.87 mm/year).

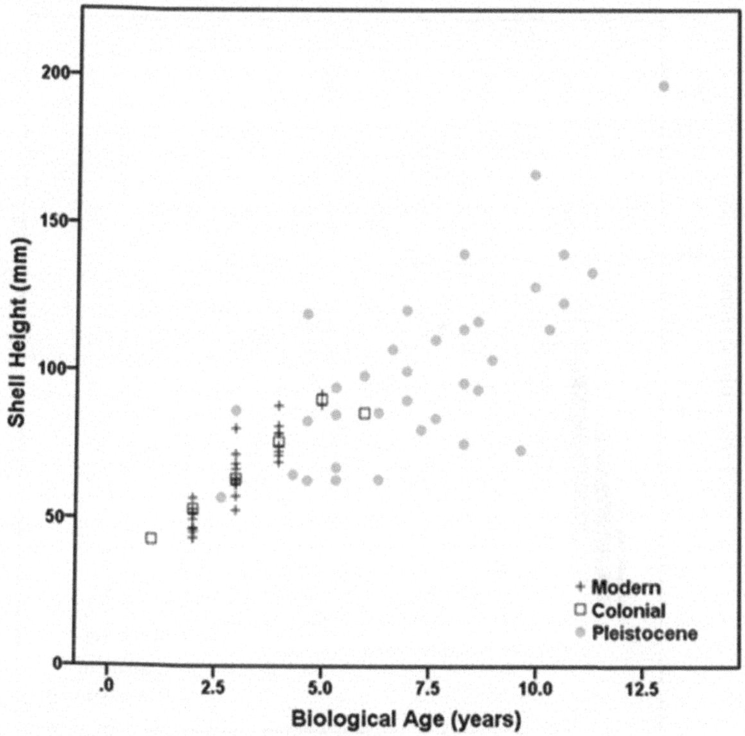

Fig. 9 Growth rate trends in Pleistocene and modern oysters from sites in the upper Chesapeake Bay region (Maryland) with salinities ranging from 15 to 30 ppt

Table 3 Results for ANCOVA comparing growth rates (shell height (mm) vs. biological age (years)) in Pleistocene, colonial, and modern oysters from Maryland, Virginia, and North Carolina

	Maryland	Virginia	North Carolina
Pleistocene vs. colonial	$F_{2,11} = 3.64, p = 0.08$ Not significant	**$F_{2,33} = 9.93, p < 0001$ Colonial > Pleistocene**	–
Pleistocene vs. modern	**$F_{3,43} = 29.52,$ $p < 0001$ Modern > Pleistocene**	**$F_{2,1067} = 570, p < 0001$ Modern > Pleistocene**	**$F_{2,875} = 472.53,$ $p < 0001$ Modern > Pleistocene**
Colonial vs. modern	**$F_{2,37} = 104.18,$ $p < 0001$ Modern > Pleistocene**	**$F_{2,1057} = 571.02,$ $p < 0001$ Modern > Colonial**	–

Growth rates modelled as linear for oysters ≤5 years; statistically significant differences ($p \leq 0.05$) highlighted in bold, all comparisons are controlled for salinity regime

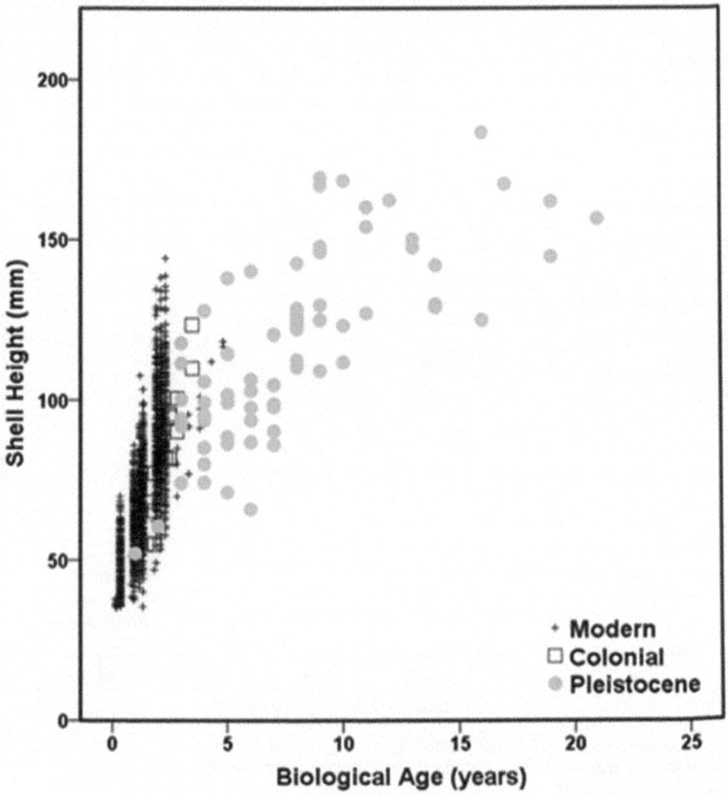

Fig. 10 Growth rate trends in Pleistocene (middle and late Pleistocene combined), colonial, and modern oysters from sites in the lower Chesapeake Bay region (Virginia) with salinities ranging from 15 to 30 ppt

4 Discussion

Comparing Pleistocene to Modern Oysters

Fossil oysters from the Chesapeake Bay are 1.3–1.6 times larger than oysters from either colonial or modern times (depending on location). This difference is not the result of slower growth rates. In fact, colonial and modern oysters grew approximately 3–4 times faster than Pleistocene oysters in the bay, even when growth rates are only calculated for younger, faster-growing individuals (0–5 years). The size difference is driven by the presence of significantly longer-lived (up to 3.6 times longer) adults in the Pleistocene assemblages. Growth trajectories in fossil Chesapeake Bay oysters continued beyond 5 years of age (Figs. 9 and 10), but colonial and modern bay populations rarely lived that long.

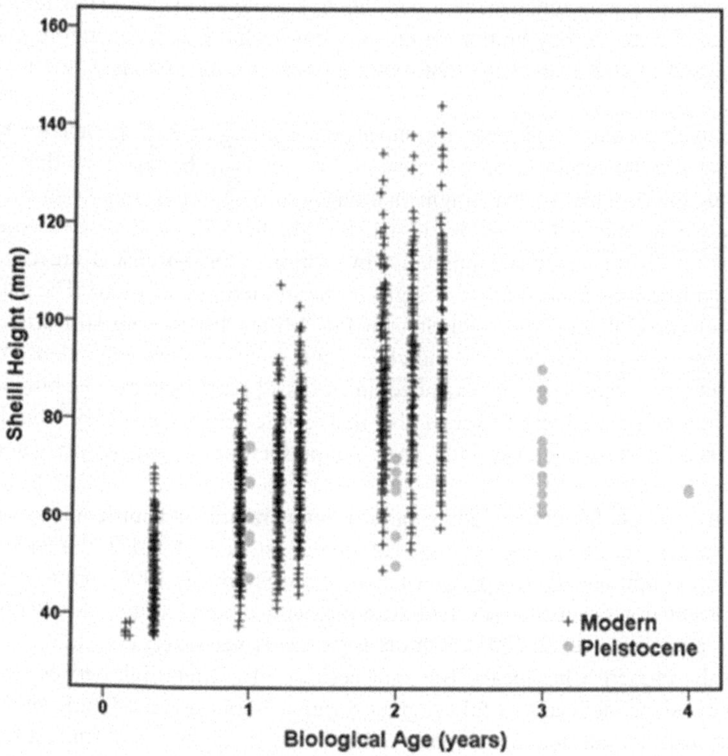

Fig. 11 Growth rate trends in Pleistocene (middle and late Pleistocene combined) and modern oysters from sites in the Pamlico Sound region of North Carolina with salinities ranging from 15 to 30 ppt

The same pattern does not hold for North Carolina. While there is no significant difference in size between middle Pleistocene and modern oysters, late Pleistocene oysters are significantly smaller than modern ones. Although growth rates in modern oysters in Pamlico Sound are 6–7 times faster than fossil oysters, the short life spans of both fossil and modern oysters result in little to no difference in these population parameters through time.

Environmental Controls on Oyster Size

A number of factors could be responsible for this significant decrease in oyster size in the Chesapeake Bay from the Pleistocene to the modern. Sampling, for both modern and fossil localities, was accomplished via bulk sampling, either by hand (Pleistocene) or with the use of hydraulic patent tongs (modern). The open dimensions of patent tongs make it possible to sample all of the oysters inhabiting one square meter of bay bottom. Both sampling techniques are therefore random with respect to size, suggesting that sampling bias is unlikely to be affecting these trends.

The likelihood of fossil preservation varies according to shell size and thickness, such that smaller mollusks are, in general, less likely to be preserved than larger ones, due to processes such as fragmentation and dissolution (Cummins et al. 1987; see reviews in Martin 1999; Kidwell 2013). In the field, these Pleistocene sites are extremely well-preserved and qualitative inspection of the oyster shells reveals little taphonomic damage (including few signs of dissolution or corrosion). The majority of oysters at Holland Point and Stetson Pit (Miller 1982) were still articulated and oriented in life position and the clayey-silty matrix shows no evidence of the molds and casts that would be expected under dissolution regimes. The bimineralic and relatively thick shell of oysters also makes it less likely that they would have experienced dissolution, especially given the preservation of aragonitic material at the same localities.

Environmental factors, including salinity, temperature, and nutrients, are closely tied to oyster size in modern settings. Although oysters are tolerant of a wide range of salinity conditions (0–40 ppt; Quast et al. 1988; Shumway 1996), optimal growth and reproduction occur between 10 and 28 ppt, with larval recruitment only possible above 6 ppt (Wilson et al. 2005). Salinity is generally negatively correlated to oyster size in the modern Chesapeake Bay, not because of a direct link between salinity and oyster size, but because of the high occurrence of disease, predators, and boring sponges in fully marine conditions (Galtsoff 1964; Paynter and Burreson 1991; VOSARA 2016). By explicitly limiting modern comparisons to those with similar salinity regimes, we have controlled for the complicating effects of salinity on these data.

Growth studies in the Chesapeake Bay found that modern *Crassostrea virginica* grow between July and October, with growth stopping when mean temperatures dropped below 10 °C (Paynter and Dimichele 1990). Though environmental conditions are similar during some spring months, the oysters exhibit no growth during these periods (Paynter and Dimichele 1990). The Pleistocene sites sampled in this study all represent interglacial intervals. Bottom water temperatures at these sites in the Pleistocene range from substantially warmer (+10 °C at Cherry Point, Virginia) to substantially colder (−10 °C at Holland Point, Virginia) than the same locations today (NDBC 2016). The colder paleotemperatures for Holland Point (Virginia) are similar to those recorded today between Ocean City, Maryland and Lewes, Delaware (NOAA NCEI 2016). The warmer paleotemperatures documented at Cherry Point (Virginia) correspond to temperatures observed today near Wilmington, North Carolina. If the changes in shell height were driven by temperature, we would expect oysters from Pleistocene bay localities with warmer temperatures (i.e., Cherry Point and Wailes Bluff, Maryland) to differ from those from localities with cooler temperatures (i.e., Holland Point, Virginia). In reality, there is no significant difference in average or maximum shell height at these three localities.

Like most marine invertebrates, oysters grow larger under higher nutrient conditions (Berg and Newell 1986; Rice and Rheault 1996). Although the complex history of nutrient pollution in the Chesapeake Bay has been reconstructed in detail for the last millennium (Cooper and Brush 1991; Cooper 1995; Zimmerman and Canuel 2002; Kemp et al. 2005), considerably less is known about nutrients along the Mid-Atlantic Coastal Plain during the Pleistocene. Krantz (1990) sampled growth bands of middle Pliocene (Yorktown Formation) to early Pleistocene (James City Formation) scallops from Virginia and found evidence of seasonal increases in productivity (interpreted as spring phytoplankton blooms), but no record of upwelling. This result suggests that local nutrient levels were most likely lower in the Pleistocene than during historic and modern times. This increase in nutrients may be driving the increase in oyster growth rates observed in the historic and modern Chesapeake Bay (Kirby and Miller 2005; Harding et al. 2008; Mann et al. 2009b), but it cannot explain the smaller shell sizes and shorter longevities. Until proxy data for local productivity are available throughout the Pleistocene, it will be difficult to quantify the effects of primary productivity on oyster growth in the Mid-Atlantic region through time.

Human Factors Influencing Oyster Size

In addition to environmental factors, two anthropogenically driven factors are affecting historical and modern oyster sizes: disease and overharvesting. Two diseases, Dermo and MSX, exert a massive influence on oyster abundance in the bay today. Mortality is higher with MSX although the virulence of Dermo appears to have increased rapidly after the introduction of MSX (Burreson and Ragone Calvo 1996; Carnegie and Burreson 2009). Today, MSX is thought to kill the majority of oysters larger than 51 mm in high salinity regions of the bay. Disease-resistant strains of *C. virginica* exist (Brown et al. 2005; Encomio et al. 2005; Carnegie and Burreson 2011), but the majority of oyster growers prefer to grow triploid (non-reproductively active) oysters. These oysters, which have three sets of chromosomes, are just as susceptible to disease but tend to grow faster and reach market size before dying.

Studies of the sustainability of Native American and colonial harvesting are just beginning, but a compilation of oyster size across 28 archaeological sites spanning 3500 years of Native American harvest suggests that shell height remained relatively stable throughout this interval (Rick et al. 2016). Although Native Americans harvested oysters for thousands of years prior to European colonization, early English settlers reported massive oyster reefs covering the bay and its tributaries (Wharton 1957; Hargis and Haven 1999; Mann et al. 2009b; Rick et al. 2014).

Measures of Colonial Era oyster shell height and growth rates, using similar techniques to those used here, suggest that Jamestown Colony oysters were intermediate in size between Pleistocene and modern oysters (Harding et al. 2008, 2010a). Their growth rates were elevated, relative to both modern and Pleistocene growth rates (Harding et al. 2008). Kirby and Miller (2005) observed the same pattern in colonial oysters from the St. Mary's and Patuxent Rivers (Maryland) and argued that this pattern was the result of increased nutrient availability due to land clearance. By 1860, oyster growth rates began to decrease, with the initiation of harvesting by dredging, combined with hypoxia and harmful algal blooms (Kirby and Miller 2005).

Today, oyster size and population density are at historic lows, in part due to the culling of the larger tail of the size frequency distribution that is highlighted by comparisons between Pleistocene and modern oyster sizes. Like disease, harvesting disproportionately affects the larger, more reproductively active adult oysters (Hutchings and Reynolds 2004). In fact, the average mortality of these larger oysters exceeds 60% on natural oyster bars in the Piankatank (Harding et al. 2010b), James (Mann et al. 2009a), and Great Wicomico (Southworth et al. 2010) Rivers.

Implications for Restoration

This study has documented a substantial decrease in oyster shell size in the Chesapeake Bay from the Pleistocene to today. This difference in shell size is not due to a decrease in growth rates or culling of smaller oysters. Instead, it is driven by the culling of oysters at the larger end of the size frequency distribution in the modern bay, due to both overharvesting and disease. The elimination of large adults from the population is a common sign of overharvesting in many marine species (Hutchings and Reynolds 2004; Berkeley et al. 2004; Birkeland and Dayton 2005; Hsieh et al. 2006).

With that in mind, it is important to note that conservation efforts for Chesapeake Bay oysters focus almost exclusively on early life stages—in particular larvae and spat. These approaches include: (1) distribution of cultch on the bay floor to increase the likelihood of larval settlement, (2) rearing and release of oyster larvae, and (3) minimum size restriction on oysters harvested from both aquaculture (2 in.) and natural beds (3 in.)(Kennedy 1989; Mann and Powell 2007; Kennedy et al. 2011; Wilberg et al. 2011; Md. Code Regs. § 08.02.04.11; 4VA Admin Code 20-260-30). Millions of dollars are spent on these approaches each year (Luckenbach et al. 1999; Mann and Powell 2007; Beck et al. 2011; zu Ermgassen et al. 2012) but, from a statistical standpoint, very few of these oysters will actually settle and grow to market size.

Efforts devoted to preserving adult oysters are minimal by comparison. Only a small percentage of oyster habitat is protected from harvesting for the long term (= sanctuaries, 9–25% or 9000 acres in Maryland, <2% or 200 acres in Virginia; Schulte et al. 2009; MDDNR 2016; VMRC 2016a), which, in turn, protects both the early and late stage oysters growing there. The majority of protected areas in Virginia are closed to harvesting for only 1 to 3 years at a time (VMRC 2016a). This approach means that adult oysters are afforded little to no protection in the bay, despite the fact that they are the most reproductively important members of the population (Mann et al. 2009a, b). Because older oysters put exponentially more energy into reproduction and less into shell growth, lack of protection of adults has a catastrophic effect on oyster population growth. Similar patterns have been documented across a wide range of marine and freshwater fisheries (Berkeley et al. 2004; Birkeland and Dayton 2005; Hsieh et al. 2006; Venturelli et al. 2009; Arlinghaus et al. 2010). In the Chesapeake Bay, the overfishing of larger specimens and culturing of triploid specimens also means that disease resistance is evolving exceptionally slowly in oysters (Encomio et al. 2005). These problems are further compounded by the fact that oysters are sequential protandric hermaphrodites and that the ratio of females in a population increases with increasing age/size (Kennedy 1983; Heffernan et al. 1989; Harding et al. 2013). Oysters start their lives as male, and, in the Chesapeake Bay, do not transition to female until they are at least 60 mm long (approximately 1.6 years old; Harding et al. 2013). Thus, the culling of the larger sizes of oysters also preferentially removes females from the population.

The exception to preferential removal is areas of the bay that have historically avoided long-term harvesting because they are privately owned or excessively polluted. For example, larger, more disease-resistant oysters have been documented recently in locations such as Tangier Sound (Blankenship 1997; Encomio et al. 2005) and the Elizabeth River (Schulte et al. 2009; CBF 2016). This finding suggests that Chesapeake Bay oysters have the potential, if not the opportunity, to evolve disease resistance and grow to larger sizes that approach those seen in the Pleistocene record.

The Pleistocene record of Chesapeake Bay oysters emphasizes the significant extent to which colonial and modern populations have experienced age and size truncation. Management solutions to age truncation in marine fisheries include: (1) implementation of a maximum size or slot limit, (2) the establishment of marine protected areas (MPA), or (3) significant lowering of exploitation rates (Berkeley et al. 2004; Venturelli et al. 2009; Hixon et al. 2014). In many freshwater and some marine species, fishery size restrictions include both a maximum and minimum (slot limit), or a restriction based on reproductive stage. For example, management of the blue crab (*Callinectes sapidus*) in the Chesapeake Bay includes: (1) an MPA (Lipcius et al. 2003) and (2) fishing restrictions based on egg mass and molting stage, in addition to overall size (MDDNR 2016; VMRC 2016b). The reproductive stage of oysters would be difficult for waterpeople to determine because they are broadcast spawners (Kennedy 1983; Heffernan et al. 1989). Maximum size restrictions could, however, be enforced because minimum size restrictions are already in effect for oysters collected in the bay.

MPAs are areas of oceans or lakes that are protected from human activity to conserve natural or cultural resources. MPA approaches have proven useful across a wide variety of fishery species (Russ 2002; Halpern 2003; Pelletier et al. 2005), but effectiveness varies according to the size of the MPA and its duration (Claudet et al. 2008; Edgar et al. 2014). Harvest moratoria are controversial but have been suggested in the past by both ecological and economic managers to preserve oyster resources in the bay (Schulte et al. 2009; Kasperski and Wieland 2010; Wilberg et al. 2011).

If conservation funding were to shift toward protection of older, reproductively more active oysters, it would represent a more efficient approach, as each adult female produces an average of 2–115 million eggs per year, increasing with age (Brooks 1996). This approach is not without its challenges, especially given the importance of sustaining a culture supporting the local waterpeople (Paolisso 2007; Paolisso and Dery 2010). But the benefits of shifting funding priorities to preserving large, disease-resistant oysters make it worthwhile. The importance of prioritizing adult over early life stages has been recognized for several other aquatic species (Berkeley et al. 2004; Birkeland and Dayton 2005; Hsieh et al. 2006; Venturelli et al. 2009).

A Role for Conservation Paleobiology

Oysters from the Mid-Atlantic states represent an excellent example of how fossil data can inform conservation issues. Because humans have inhabited this region for at least 13,000 years (Dent 1995; Lowery et al. 2010), and harvests were not tracked until the 1870s (Lotze 2010; zu Ermgassen et al. 2012), a sizeable gap exists in our understanding of how these ecosystems have changed over long timescales. By the time marine scientists established monitoring of bay oysters in the 1940s, oyster populations were already decimated by 75 years of dredging (Haven et al. 1978; Rothschild et al. 1994; Rountree et al. 2007; zu Ermgassen et al. 2012). Given the effects of introduced disease and increased sediment influx, the oyster beds that we study today tell us very little about how healthy oyster reefs function, either in the past or present. One could argue that bay managers have never seen a healthy oyster reef.

Pleistocene fossil assemblages can serve as a baseline for Chesapeake Bay oyster mitigation. These assemblages allow us to quantify body size, growth rates, and other factors that can be used to shape modern mitigation efforts.

Combining historical (zu Ermgassen et al. 2012), archaeological (Rick and Lockwood 2013; Rick et al. 2016), and paleobiological (this volume) approaches makes it possible to bridge these timescales and to assess how oysters have responded to various pressures, including harvesting, climate, and sea level change in the past. Conservation paleobiology plays a particularly important role, as the only approach to yield information on ecosystems before human settlement. In the end, the fossil record provides us with a crucial baseline for mitigation, a glimpse into the world of Mid-Atlantic oysters before humans.

5 Conclusion

In conclusion, Pleistocene oysters from the Chesapeake Bay region are larger, and longer-lived than either colonial or modern oysters. This pattern is not the result of environmental shifts in salinity or temperature. Instead, it is driven by culling of the larger tail of the shell size frequency distribution in the modern bay, suggesting that both human harvesting and disease are eliminating the larger, reproductively more active female members of the population. Solutions to this size and age-skewed population structure include: (1) decreasing harvest pressure, (2) establishing a maximum size limit, and (3) designating MPAs specifically for oysters in the Chesapeake Bay. The conservation paleobiologic approach applied in this study emphasizes the impact that human activities have had on these ecosystem, at the same time providing a baseline for future mitigation.

Acknowledgments We would like to thank Admiral Pete Roane and his family for access to the study site at Holland Point, Virginia. Additionally, we would like to thank Dr. Roger Mann (Virginia Institute of Marine Science) for information on the Holland Point site; Dr. Buck Ward (Virginia Museum of Natural History) for field assistance and access to museum specimens; students Chris Young, Sam Bonanni, and Eric Dale (The College of William and Mary); Mitch Tarnowski (Maryland Department of Natural Resources) for Maryland oyster data; Dr. Gregory Henkes (Johns Hopkins University) for clumped isotope analyses; Dr. Gregory Herbert (University of South Florida) for help with sclerochronology lab work; Dr. Henry Miller (Historic St. Mary's City), Keith Doms (Newlin Grist Mill), and Drs. Torben Rick and Leslie Reeder-Myers (Smithsonian Institution) for Maryland archaeological data; Dr. John Wehmiller (University of Delaware) for amino acid racemization analyses; Rick Berquist (Virginia Division of Geology and Mineral Resources) for designation of stratigraphic units; and Dr. Doug Jones (Florida Museum of Natural History) for help with data interpretation. Comments from two reviewers, Drs. Patricia Kelley (University of North Carolina at Wilmington) and Nicole Bonuso (California State University at Fullerton) greatly improved the quality of the manuscript. Lastly, we would like to thank the Ellen Stofan Fund from the College of William and Mary for financial support of this research project.

References

Abbott RT (1974) American seashells. Litton Educational Publishing Inc., New York

Andrews JD (1953) Fouling organisms of the Chesapeake Bay. The Chesapeake Bay Institute of the Johns Hopkins University Inshore Survey Program Interim Report XVII, 53-3

Andrews JD (1996) History of *Perkinsus marinus*, a pathogen of oysters in Chesapeake Bay 1950-1984. J Shellfish Res 15:13–16

Andrews JD, Wood JL (1967) Oyster mortality studies in Virginia. VI. History and distribution of *Minchinia nelsoni*, a pathogen of oysters, in Virginia. Chesap Sci 8(1):1–13

Arlinghaus R, Matsumura S, Dieckmann U (2010) The conservation and fishery benefits of protecting large pike (*Esox lucius* L.) by harvest regulations in recreational fishing. Biol Conserv 143(6):1444–1459

Bartol IK, Mann R (1997) Small-scale settlement patterns of the oyster *Crassostrea virginica* on a constructed intertidal reef. Bull Mar Sci 61:881–897

Bartol IK, Mann R (1999a) Small-scale patterns of recruitment on a constructed intertidal reef: the role of spatial refugia. In: Luckenbach MW, Mann R, Wesson JA (eds) Oyster reef habitat restoration: a synopsis of approaches. Virginia Institute of Marine Science, Gloucester Point, pp 159–170

Bartol IK, Mann R (1999b) Growth and mortality of oysters (*Crassostrea virginica*) on constructed intertidal reefs: effects of tidal height and substrate level. J Exp Mar Biol Ecol 237:157–184

Beck MW, Brumbaugh RD, Airoldi L et al (2011) Oyster reefs at risk and recommendations for conservation, restoration, and management. Bioscience 61(2):107–116

Belknap DF (1979) Application of amino acid geochronology to stratigraphy of late Cenozoic marine units of the Atlantic coastal plain. Dissertation, University of Delaware

Berg JA, Newell RI (1986) Temporal and spatial variations in the composition of seston available to the suspension feeder *Crassostrea virginica*. Estuar Coast Shelf Sci 23(3):375–386

Berkeley SA, Hixon MA, Larson RJ et al (2004) Fisheries sustainability via protection of age structure and spatial distribution of fish populations. Fisheries 29(8):23–32

Birkeland C, Dayton PK (2005) The importance in fishery management of leaving the big ones. Trends Ecol Evol 20(7):356–358

Blake SF (1953) The Pleistocene fauna of Wailes Bluff and Langleys Bluff, Maryland. Smithsonian Misc Coll 121:1–32

Blankenship K (1997) Oysters rescued from VA harvest may return favor. Chesapeake Bay J 7:9

Brooks WK (1996) The oyster. JHU Press, Baltimore

Brown BL, Butt AJ, Meritt D et al (2005) Evaluation of resistance to Dermo in eastern oyster strains tested in Chesapeake Bay. Aquac Res 36(15):1544–1554

Buroker NE (1983) Population genetics of the American oyster *Crassostrea virginica* along the Atlantic coast and Gulf of Mexico. Mar Biol 75:99–112

Burreson EM, Ragone Calvo LM (1996) Epizootiology of *Perkinsus marinus* disease of oysters in Chesapeake Bay, with emphasis on data since 1985. Oceanogr Lit Rev 12(43):1265

Carnegie R, Burreson EM (2009) Status of the major oyster diseases in Virginia 2006–2008: a summary of the annual oyster disease monitoring program. Virginia Institute of Marine Science Report, Gloucester Point; p 20

Carnegie R, Burreson EM (2011) Declining impact of an introduced pathogen: *Haplosporidium nelsoni* in the oyster *Crassostrea virginica* in Chesapeake Bay. Mar Ecol Prog Ser 432:1–15

Castagna M, Chanley P (1973) Salinity tolerance of some marine bivalves from inshore and estuarine environments in Virginia waters on the western mid-Atlantic coast. Malacologia 12:47–96

Catts WP, Fiedel S, Kellogg DC et al. (1998) Phase III data recovery investigations at 18CV362, Ashcomb Quarter (historic component) and Awapantop (prehistoric component), Solomons Naval Recreation Center, Solomons, Calvert County, Maryland. Calvert soil Conservation District (on file with Maryland Historical Trust, Crownsville, Maryland)

CBF (2016) Chesapeake Bay Foundation. http://www.cbf.org/document.doc?id=656. Accessed 2 Mar 2016

Claudet J, Osenberg CW, Benedetti-Cecchi L et al (2008) Marine reserves: size and age do matter. Ecol Lett 11:481–489

Coen LD, Bolton-Warberg M, Stephen JA (2006) An examination of oyster reefs as a biologically-critical estuarine ecosystems. Final report, grant R/ER-10, South Carolina Sea Grant Consortium, p. 214

Coen LD, Brumbaugh RD, Bushek D et al (2007) As we see it: ecosystem services related to oyster restoration. Mar Ecol Prog Ser 341:303–307

Cohen AN (2011) The exotics guide: non-native marine species of the North American Pacific coast. Center for Research on Aquatic Bioinvasions, Richmond, California, and San Francisco Estuary Institute, Oakland. http://www.exoticsguide.org/. Accessed 2 Mar 2016

Cooper SR (1995) Chesapeake Bay watershed historical land use: impact on water quality and diatom communities. Ecol Appl 5:703–723

Cooper SR, Brush GS (1991) Long-term history of Chesapeake Bay anoxia. Science (Washington) 254(5034):992–996

Craig H (1965) Measurement of oxygen isotope paleotemperatures. In: Tongiorgi E (ed) Stable isotopes in oceanographic studies and paleotemperatures. CNR Lab. Geol. Nucl., Pisa, pp 161–182

Cressmann KA, Posey MH, Mallin MA et al (2003) Effects of oyster reefs on water quality in a tidal creek estuary. J Shellfish Res 22:753–762

Cronin TM (1979) Late Pleistocene marginal marine ostracodes from the southeastern Atlantic coastal plain and their paleoenvironmental implications. Géog Phys Quatern 33(2):121–173

CTDOAG (2016a) Dermo disease, Connecticut Department of Agriculture. http://wwwctgov/doag/lib/doag/aquaculture/dermopdf. Accessed 17 Mar 2016

CTDOAG (2016b) MSX disease, Connecticut Department of Agriculture. http://wwwctgov/doag/lib/doag/aquaculture/msxpdf. Accessed 17 Mar 2016

Cummins H, Powell EN, Stanton RJ Jr et al (1987) The size-frequency distribution in palaeoecology: effects of taphonomic processes during formation of molluscan death assemblages in Texas bays. Palaeontology 29:495–518

Custer JF (1989) Prehistoric cultures of the Delmarva Peninsula. University of Delaware Press, Newark

Dent RJ (1995) Chesapeake prehistory: old traditions, new directions. Plenum, New York

Edgar GJ, Stuart-Smith RD, Willis TJ et al (2014) Global conservation outcomes depend on marine protected areas with five key features. Nature 506:216–220

Encomio VG, Stickler SM, Allen SK Jr et al (2005) Performance of "natural dermo-resistant" oyster stocks-survival, disease, growth, condition and energy reserves. J Shellfish Res 24(1):143–155

Federighi H (1931) Salinity death-points of the oyster drill snail, *Urosalpinx cinerea* Say. Ecology 12:346–353

Galtsoff PS (1964) The American oyster. Fishery Bull Fish Wildlife Service 64:1–480

Ghosh P, Adkins J, Affek H et al (2006) 13 C–18 O bonds in carbonate minerals: a new kind of paleothermometer. Geochim Cosmochim Acta 70(6):1439–1456

Grabe, SA, Courtney CM, Lin Z et al (1993) Environmental monitoring and assessment program-estuaries west Indian province 1993 sampling, vol. 3, executive summary: a synoptic survey of the benthic macroinvertebrates and demersal fishes of the Tampa Bay estuarine system. Tampa Bay National Estuary Program Technical Publication 95–12

Grabowski JH, Brumbaugh R, Conrad R (2012) Economic valuation of ecosystem services provided by oyster reefs. Bioscience 62:900–909

Griffin T (2001) *Crepidula* cf. *plana*. Smithsonian Marine Station at Fort Pierce. http://www.sms.si.edu/irlspec/Crepid_cfplan.htm. Accessed 3 Mar 2016

Grossman EL, Ku T-L (1981) Aragonite-water isotopic paleotemperature scale based on benthic foraminifer *Hoeglundia elegans*. Geol Soc Am Abstr Prog 13:464

Halpern BS (2003) The impact of marine reserves: do reserves work and does reserve size matter? Ecol Appl 13:S117–S137

Harding JM, Mann R (2006) Age and growth of wild Suminoe (*Crassostrea ariakensis*, Fugita 1913) and Pacific (*C. gigas*, Thunberg 1973) oysters from Laizhou Bay, China. J Shellfish Res 25(1):73–82

Harding JM, Mann R, Southworth MJ (2008) Shell length-at-age relationships in James River, Virginia, oysters (*Crassostrea virginica*) collected four centuries apart. J Shellfish Res 27(5):1109–1115

Harding JM, Spero HJ, Mann R et al (2010a) Reconstructing early 17th century estuarine drought conditions from Jamestown oysters. Proc Natl Acad Sci U S A 107(23):10549–10554

Harding JM, Mann R, Southworth MJ et al (2010b) Management of the Piankatank River, Virginia in support of oyster (*Crassostrea virginica*, Gmelin 1791) fishery repletion. J Shellfish Res 29(4):1–22

Harding JM, Powell EN, Mann R et al (2013) Variations in eastern oyster (*Crassostrea virginica*) sex-ratios from three virginia estuaries: protandry, growth and demographics. J Mar Biol Assoc U K 93(02):519–531

Hargis WJ Jr, Haven DS (1995) The precarious state of the Chesapeake public oyster resource. In: Hill PL, Nelson S (eds) Toward a sustainable coastal watershed: the Chesapeake experiment, Chesapeake Research Consortium publication no. 149. Solomons, Maryland, pp 559–584

Hargis WJ Jr, Haven DS (1999) Chesapeake oyster reefs, their importance, destruction and guidelines for restoring them. In: Luckenbach MW, Mann R, Wesson JA (eds) Oyster reef habitat restoration: a synopsis of approaches. Virginia Institute of Marine Science, Gloucester Point, pp 329–358

Harke RM, Herbert GS, White NM et al (2015) Sclerochronology of *Busycon sinistrum*: late prehistoric seasonality determination at St. Joseph Bay, Florida, USA. J Archaeol Sci 57:98–108

Haven DS, Morales-Alamo R (1970) Filtration of particles from suspension by the American Oyster *Crassostrea virginica*. Biol Bull 139:248–264

Haven DS, Hargis WJ Jr, Kendall PC (1978) The oyster industry of Virginia: its status, problems and promise (A comprehensive study of the oyster industry in Virginia). Special Papers in Marine Science, Virginia Institute of Marine Science, Gloucester Point (4)

Heffernan P, Walker R, Carr J (1989) Gametogeneic cycle of three marine bivalves in Wassaw Sound, Georgia. II. *Crassostrea virginica* (Gmelin, 1791). J Shellfish Res 8:61–70

Henkes GA, Passey BH, Wanamaker AD Jr et al (2013) Carbonate clumped isotope compositions of modern marine mollusk and brachiopod shells. Geochim Cosmochim Acta 106:307–325

Hill K (2004) Mercenaria, Smithsonian Marine Station at Fort Pierce. http://www.sms.si.edu/irlspec/Mercen_mercen.htm. Accessed 3 Mar 2016

Hixon MA, Johnson DW, Sogard SM (2014) BOFFFFs: on the importance of conserving old-growth age structure in fishery populations. ICES J Mar Sci 71(8):2171–2185

Hoellein TJ, Zarnoch CB, Grizzle RE (2015) Eastern oyster (*Crassostrea virginica*) filtration, biodeposition, and sediment nitrogen cycling in two oyster reefs with contrasting water quality in the Great Bay Estuary (New Hampshire, USA). Biogeochemistry 122:113–129

Horn J (2008) A land as God made it: Jamestown and the birth of America. Basic Books, New York

Hsieh CH, Reiss CS, Hunter JR et al (2006) Fishing elevates variability in the abundance of exploited species. Nature 443(7113):859–862

Huntington KW, Eiler JM, Affek HP et al (2009) Methods and limitations of 'clumped' CO_2 isotope (D47) analysis by gas-source isotope ratio mass spectrometry. J Mass Spectrom 44:1318–1329

Hutchings JA, Reynolds JD (2004) Marine fish population collapses: consequences for recovery and extinction risk. Bioscience 54(4):297–309

Kasperski S, Wieland R (2010) When is it optimal to delay harvesting? The role of ecological services in the Northern Chesapeake Bay oyster fisheries. Mar Resour Econ 24:361

Kelso WM (2004) Jamestown rediscovery, vol 8. Association for the preservation of Virginia antiquities, Charlottesville

Kemp WM, Boynton WR, Adolf J et al (2005) Eutrophication of Chesapeake Bay: historical trends and ecological interactions. Mar Ecol Prog Ser 303(21):1–29

Kennedy VS (1983) Sex ratios in oysters, emphasizing *Crassostrea virginica* from Chesapeake Bay, Maryland. Veliger 25:329–338

Kennedy VS (1989) The Chesapeake Bay oyster fishery: traditional management practices. In: Caddy JF (ed) Marine invertebrate fisheries: their assessment and management. Wiley, New York, pp 455–477

Kennedy VS, Breitburg DL, Christman MC et al (2011) Lessons learned from efforts to restore oyster populations in Maryland and Virginia, 1990 to 2007. J Shellfish Res 30(3):719–731

Kidwell SM (2013) Time-averaging and fidelity of modern death assemblages: building a taphonomic foundation for conservation palaeobiology. Palaeontology 56(3):487–522

Kim S, O'Neil JR, Hillaire-Marcel C et al (2007) Oxygen isotope fractionation between synthetic aragonite and water: influence of temperature and Mg2+ concentration. Geochim Cosmochim Acta 71:4704–4715

Kirby MX, Miller HM (2005) Response of a benthic suspension feeder (*Crassostrea virginica* Gmelin) to three centuries of anthropogenic eutrophication in Chesapeake Bay. Estuar Coast Shelf Sci 62(4):679–689

Krantz DE (1990) Mollusk-isotope records of Plio-Pleistocene marine paleoclimate, US middle Atlantic coastal plain. PALAIOS 5:317–335

Lipcius RN, Stockhausen WT, Seitz RD et al (2003) Spatial dynamics and value of a marine protected area and corridor for the blue crab spawning stock in Chesapeake Bay. Bull Mar Sci 72(2):453–469

Lotze HK (2010) Historical reconstruction of human-induced changes in US estuaries. Oceanogr Mar Biol Annu Rev 48:267–338

Lowery DL, O'Neal MA, Wah JS et al (2010) Late Pleistocene upland stratigraphy of the western Delmarva Peninsula, USA. Quat Sci Rev 29(11):1472–1480

Luckenbach MW, Mann R, Wesson JA (1999) Oyster reef habitat restoration: a synopsis and synthesis of approaches; proceedings from the symposium. Williamsburg, Virginia

Luckenbach MW, Coen LD, Ross PG Jr et al (2005) Oyster reef habitat restoration: relationships between oyster abundance and community development based on two studies in Virginia and South Carolina. J Coastal Res 40:64–78

Mann R (2000) Restoring oyster reef communities in the Chesapeake Bay: a commentary. J Shellfish Res 19:335–340

Mann R, Evans DA (1998) Estimation of oyster, *Crassostrea virginica*, standing stock, larval production and advective loss in relation to observed recruitment in the James River, Virginia. J Shellfish Res 17:239–254

Mann R, Powell EN (2007) Why oyster restoration goals in the Chesapeake Bay are not and probably cannot be achieved. J Shellfish Res 26(4):905–917

Mann R, Southworth M, Harding JH et al (2009a) Population studies of the native oyster, *Crassostrea virginica* (Gmelin, 1791), in the James River, Virginia, USA. J Shellfish Res 28:193–220

Mann R, Harding JM, Southworth M (2009b) Reconstructing pre-colonial oyster demographics in the Chesapeake Bay, USA. Estuar Coast Shelf Sci 85:217–222

Martin RE (1999) Taphonomy: a process approach. Cambridge University Press, Cambridge

Massmann WH, Ladd EC, McCutcheon HN (1952) Rappahannock River survey 1951: a biological survey of the Rappahannock River part II. Virginia Fisheries Laboratory Special Scientific Report 6

MDDNR (2016) Maryland Department of Natural Resources. http://dnr2.maryland.gov/fisheries/Pages/oysters/sanctuaries.aspx. Accessed 1 Mar 2016

Menzel RW, Hulings NC, Hathaway RR (1966) Oyster abundance in Apalachicola Bay, Florida in relation to biotic associations influenced by salinity and other factors. Gulf Res Rep 2:73–96

Meyer DL, Townsend EC (2000) Faunal utilization of created intertidal eastern oyster (*Crassostrea virginica*) reefs in the southeastern United States. Estuaries 23:34–45

Miller W (1982) The paleoecologic history of late Pleistocene estuarine and marine fossil deposits in Dare County, North Carolina. Southeast Geol 23:1–13

Miller HM (1986) Transforming a "splendid and delightsome land": colonists and ecological change in the Chesapeake 1607-1820. J Wash Acad Sci 76:173–187

Mirecki J (1990) Aminostratigraphy, geochronology and geochemistry of fossils from late Cenozoic marine units in southeastern Virginia. Dissertation, University of Delaware

Mirecki JE, Wehmiller JF, Skinner AF (1995) Geochronology of quaternary coastal plain deposits, southeastern Virginia, USA. J Coast Res 11:1135–1144

NDBC (2016) National Data Buoy Center. http://www.ndbc.noaa.gov/. Accessed 1 Mar 2016

NOAA NCEI (2016) NOAA National Centers for Environmental Information. http://www.ncei.noaa.gov/. Accessed 1 Mar 2016

North EW, King DM, Xu J et al (2010) Linking optimization and ecological models in a decision support tool for oyster restoration and management. Ecol Appl 20:851–866

Paolisso M (2007) Cultural models and cultural consensus of Chesapeake Bay blue crab and oyster fisheries. NAPA Bullet 28(1):123–135

Paolisso M, Dery N (2010) A cultural model assessment of oyster restoration alternatives for the Chesapeake Bay. Hum Organ 69(2):169–179

Paynter KT, Burreson EM (1991) Effects of *Perkinsus marinus* infection in the eastern oyster, *Crassostrea virginica*: II Disease development and impact on growth rate at different salinities. J Shellfish Res 10(2):425–431
Paynter KT, Dimichele L (1990) Growth of tray-cultured oysters (*Crassostrea virginica* Gmelin) in Chesapeake Bay. Aquaculture 87:289–297
Pelletier D, García-Charton JA, Ferraris J, David G, Thébaud O, Letourneur Y et al (2005) Designing indicators for assessing the effects of marine protected areas on coral reef ecosystems: a multidisciplinary standpoint. Aquat Living Resour 18:15–33
Puckett BJ, Eggleston DB (2012) Oyster demographics in a network of no-take reserves: recruitment, growth, survival, and density dependence. Mar Coast Fish 4(1):605–627
Quast WD, Johns MA, Pitts DE Jr et al (1988) Texas oyster fishery management plan. Fishery management plan series number 1. Texas Parks and Wildlife Department, Coastal Fisheries Branch, Austin, p 178
Rader EK, Evans NH (1993) Geologic map of Virginia: expanded explanation. Virginia Division of Mineral Resources, Charlottesville, p 80
Reshetiloff K (2004) Chesapeake Bay: introduction to an ecosystem: Chesapeake Bay Program, Annapolis, p 1–35
Rice MA, Rheault RB Jr (1996) Food-limited growth and condition index in the eastern oyster, *Crassostrea virginica* (Gmelin, 1791), and the bay scallop, *Argopecten irradians* (Lamarck, 1819). J Shellfish Res 15(2):271–283
Rick TC, Lockwood R (2013) Integrating paleobiology, archeology, and history to inform biological conservation. Conserv Biol 27(1):45–54
Rick TC, Reeder-Myers LA, Cox CJ et al (2014) Shell middens, cultural chronologies, and coastal settlement on the Rhode River sub-estuary of Chesapeake Bay, Maryland, USA. Geoarchaeology 29(5):371–388
Rick TC, Reeder-Myers LA, Hofman CA et al (2016) Millennial-scale sustainability of the Chesapeake Bay native American oyster fishery. Proc Natl Acad Sci USA 13(23):6568–6573
Rothschild BJ, Ault JS, Goulletquer P et al (1994) Decline of the Chesapeake Bay oyster population: a century of habitat destruction and overfishing. Mar Ecol Prog Ser 111(1–2):9–39
Rountree H, Clark W, Mountford K (2007) John Smith's Chesapeake voyages 1607–1609. University of Virginia Press, Charlottesville
Russ GR (2002) Marine reserve as reef fishery management tools: yet another review. In: Sale PF (ed) Coral reef fishes: dynamics and diversity in a complex ecosystem. Academic Press, San Diego, pp 421–443
Schulte DM, Burke RP, Lipcius RN (2009) Unprecedented restoration of a native oyster metapopulation. Science 325(5944):1124–1128
Shumway SE (1996) Natural environmental factors. In: Kennedy VS, Newell RIE, Eble AF (eds) The eastern oyster *Crassostrea virginica*. Maryland Sea Grant College, University of Maryland, College Park, pp 467–513
Sisson GMA, Kellogg ML, Luckenbach ML et al (2011) Assessment of oyster reefs in Lynnhaven River as a Chesapeake Bay TDML best management practice. Virginia Institute of Marine Science. Department of Physical Sciences, Gloucester Point
Southworth M, Mann R (1998) Oyster reef broodstock enhancement in the great Wicomico River, Virginia. J Shellfish Res 17:1101–1114
Southworth M, Harding J, Mann R (2000) Management of oyster broodstock sanctuaries in the great Wicomico and Piankatank rivers, Virginia: optimal use of cultch to maximize settlement. Final report to Virginia Department of Environmental Quality, Coastal Resources Management Program, p 12
Southworth M, Harding JM, Mann R et al (2010) Oyster (*Crassostrea virginica* Gmelin 1791) population dynamics on public reefs in the great Wicomico River, Virginia, USA. J Shellfish Res 29:271–290
Spencer RS, Campbell LD (1987) The fauna and paleoecology of the late Pleistocene marine sediments of southeastern Virginia. Bull Am Paleontol 92(327):1–124

Thompson DE (1972) Paleoecology of the Pamlico formation. Dissertation, Rutgers University, Saint Mary's County
Thompson VD, Worth JE (2011) Dwellers by the sea: native American adaptations along the southern coasts of eastern North America. J Archaeol Res 19:51–101
USGS (1998) Fact sheet, The Chesapeake Bay: geologic product of rising sea level. http://pubsusgsgov/fs/fs102-98/. Accessed 28 Mar 2012
Venturelli PA, Shuter BJ, Murphy CA (2009) Evidence for harvest-induced maternal influences on the reproductive rates of fish populations. Proc R Soc Lond B Biol Sci 276(1658):919–924
VMRC (2016a) Virginia Marine Resources Commission Oyster Sanctuary Areas. http://mrc.virginia.gov/regulations/fr650.shtm. Accessed 1 Mar 2016
VMRC (2016b) Virginia Marine Resources Commission Commercial Fishing. http://mrc.virginia.gov/commercial.shtm. Accessed 1 Mar 2016
VOSARA (2016) Virginia Oyster Stock Assessment and Replenishment Archive. http://cmap.vims.edu/VOSARA/VOSARA_Viewer/VOSARA.html. Accessed 8 Feb 2016
Ward LW, Blackwelder BW (1987) Late Pliocene and early Pleistocene Mollusca from the James City and Chowan River formations at Lee Creek mine. Smithson Contrib Paleobiol 61:113–283
Ward LW, Bohaska DJ (2008) Synthesis of paleontological and stratigraphic investigations at the Lee Creek mine, Aurora, NC (1958–2007). In: Geology and paleontology of the Lee Creek Mine, North Carolina, IV Virginia Museum of Natural History Special Publication 14, Smithsonian Institution Press, Washington, DC, pp 325–436
Waselkov GA (1982) Shellfish gathering and shell midden archaeology, Dissertation, University of North Carolina
Wehmiller JF, Belknap DF (1982) Amino acid age estimates, quaternary Atlantic coastal plain: comparison with U-series dates, biostratigraphy, and paleomagnetic control. Quat Res 18(3):311–336
Wesson JA, Mann R, Luckenbach M (1999) Oyster restoration efforts in Virginia. In: Luckenbach MW, Mann R, Wesson JA (eds) Oyster reef habitat restoration: a synopsis of approaches. Virginia Institute of Marine Science, Gloucester Point, pp 117–130
Wharton J (1957) The bounty of the Chesapeake. Fishing in Colonial Virginia. University of Virginia Press, Charlottesville
Wilberg MJ, Livings ME, Barkman JS et al (2011) Overfishing, disease, habitat loss, and potential extirpation of oysters in upper Chesapeake Bay. Mar Ecol Prog Ser 436:131–144
Wilson C, Scotto L, Scarpa J et al (2005) Survey of water quality, oyster reproduction and oyster health status in the St Lucie Estuary. J Shellfish Res 24:157–165
York LL, Wehmiller JF, Cronin TM et al (1989) Stetson pit, Dare County, North Carolina: an integrated chronologic, faunal, and floral record of subsurface coastal quaternary sediments. Palaeogeogr Palaeoclimatol Palaeoecol 72:115–132
Zachary A, Haven DS (2004) Survival and activity of the oyster drill *Urosalpinx cinerea* under conditions of fluctuating salinity. Mar Biol 22:45–52
Zimmerman AR, Canuel EA (2002) Sediment geochemical records of eutrophication in the mesohaline Chesapeake Bay. Limnol Oceanogr 47(4):1084–1093
Zimmerman KM, Pechenik JA (1991) How do temperature and salinity affect relative rates of growth, morphological differentiation, and time to metamorphic competence in larvae of the marine gastropod *Crepidula plana*? Biol Bull 180:372–386
Zimmt JB, Lockwood R, Andrus CFT, Herbert GS (2016) Revisiting growth increment counting as a method for biologically aging *Crassostrea virginica* from the U.S. Mid-Atlantic. Geol Soc Am Abstr Prog 48:212
zu Ermgassen PS, Spalding MD, Blake B et al (2012) Historical ecology with real numbers: past and present extent and biomass of an imperiled estuarine habitat. Proc R Soc Lond B Biol Sci 279(1742):3393–3400

Coral Reefs in Crisis: The Reliability of Deep-Time Food Web Reconstructions as Analogs for the Present

Peter D. Roopnarine and Ashley A. Dineen

Abstract Ongoing anthropogenic alterations of the biosphere have shifted emphasis in conservation biology from individual species to entire ecosystems. Modern measures of ecosystem change, however, lack the extended temporal scales necessary to forecast future change under increasingly stressful environmental conditions. Accordingly, the assessment and reconstruction of ecosystem dynamics during previous intervals of environmental stress and climate change in deep time has garnered increasing attention. The nature of the fossil record, though, raises questions about the difficulty of reconstructing paleocommunity and paleoecosystem-level dynamics. In this study, we assess the reliability of such reconstructions by simulating the fossilization of a highly threatened and disturbed modern ecosystem, a Caribbean coral reef. Using a high-resolution coral reef food web from Jamaica, we compare system structures of the modern and simulated fossil reefs, including guild richness and evenness, trophic level distribution, predator dietary breadth, food chain lengths, and modularity. Results indicate that despite the loss of species, guilds, and trophospecies interactions, particularly zooplankton and other soft-bodied organisms, the overall guild diversity, structure, and modularity of the reef ecosystem remained intact. These results have important implications for the integrity of fossil food web studies and coral reef conservation, demonstrating that fossil reef communities can be used to understand reef community dynamics during past regimes of environmental change.

1 Introduction

Recent years have witnessed a transition in the emphasis of conservation biology from an initial concern for individual species at risk to habitat preservation and restoration, and most recently to a focus on entire biological communities and

P. D. Roopnarine (✉) · A. A. Dineen
Invertebrate Zoology and Geology, Institute for Biodiversity Science and Sustainability, California Academy of Sciences, San Francisco, CA, USA
e-mail: proopnarine@calacademy.org; adineen@calacademy.org

C. L. Tyler, C. L. Schneider (eds.), *Marine Conservation Paleobiology*, Topics in Geobiology 47, https://doi.org/10.1007/978-3-319-73795-9_6

ecosystems (Franklin 1993; Schwartz et al. 2000; Meffe et al. 2012; Cardinale et al. 2012; Hewitt et al. 2016). The shift reflects two major developments in conservation biology and community ecology, the first being an acknowledgement that species are parts of larger, integrated systems, themselves part of an integrated biosphere, and that those systems provide services vital to human societies (Mace et al. 2012). The nascent field of conservation paleobiology has already incorporated the transition, using multi-taxon approaches as the bases for forecasting models of risk to entire biotas, and paleocommunity models as tools to examine ecosystem dynamics under conditions which lie outside of documented ecological experiences, such as mass extinction (Villeger et al. 2011; Barnosky et al. 2012).

The second development is the growing understanding of communities and ecosystems as complex and rich dynamical structures which often have profound impacts on individual species. Communities have typically been viewed as stable and somewhat static systems (i.e., species interactions are in balanced equilibria) and that stability promotes complexity, but these notions have been replaced by the idea that communities are capable of transitions among multiple, alternative regimes and dynamic equilibria encompassing a range of community parameters (Holling 1973; Scheffer et al. 2001; Knowlton 2004). Most recently, it has been suggested that community persistence itself might act as an agent of long-term selection, with functional structures and interactions appearing repeatedly within ecosystems over geological time (Roopnarine and Angielczyk 2015). Therefore, attempts to understand and conserve modern communities, or the most recent contemporary states of long-lasting communities, must account for the dynamic nature of those communities on multiple timescales and under different environmental circumstances.

The management of modern marine ecosystems undergoing current and future anthropogenically driven change must be informed by how similar systems have responded to environmental variations in the past. This can be fulfilled at the present time with theoretical and experimental studies, in addition to the examination of the historical and geological records. The latter have the added advantages that the range of environmental conditions to which ecosystems have been subjected in the past is far broader than those within the realm of current scientific and societal experience, and that the possibility of observing alternative dynamic ecosystem regimes is increased when longer intervals of time are considered. For example, a study by Pandolfi and Jackson (2006) showed that despite sea level and climate variability, coral reef communities in Barbados were stable in composition for at least 95,000 years during the Pleistocene. Examining and comparing today's reefs with those of the Pleistocene showed that recent human impacts have resulted in a coral reef structure different from anything seen in the last 220,000 years. Without the establishment of baselines for what non-anthropogenically altered reefs look like, we would not know how altered and degraded today's marine ecosystems are in comparison (Knowlton and Jackson 2008).

As the future of oceanic ecosystems is still very much uncertain, deep-time studies provide our best proxy for what we can anticipate in the Earth's near future (Harnik et al. 2012). For example, major questions in marine conservation and global change biology center on how marine ecosystems will respond to environmental stresses and/or large disturbances, and what makes some communities more vulnerable to perturbation than others (Seddon et al. 2014). These concerns are driven by the degraded state of ecosystems in today's oceans, resulting in the decline of species at an alarming rate and unprecedented magnitude (Cardinale et al. 2012). Coral reef ecosystems are of particular interest because one quarter of all marine species may be found in these threatened marine communities (McClanahan 2002; Hoegh-Guldberg et al. 2007; Carpenter et al. 2008; De'ath et al. 2012).

Preserving the Past

One of the foundations of modern paleontology is understanding the impacts of preservation on interpretations of the fossil record, and the ability to use that information to reconstruct the past (Kidwell and Bosence 1991; Kidwell 2001). Current paleontological studies rarely ignore factors of bias, such as selective preservation among taxa, biased preservation of parts of taxa, and outcrop dimensions. Those concerns must be extended to conservation-based paleobiological approaches. A major complication, however, is that in addition to biases of preservation and discovery, the information conveyed by integrated systems may also be biased by the manner in which integration is preserved. For example, the extent to which prey richness and abundances are preserved surely influences the interpretation of the dynamics of molluscan drilling predation, and differences of generation times bias relative abundances (Leighton 2002; Vermeij and Leighton 2003). It is therefore important that we analyze the ways in which fossilization potentially biases our interpretation of community and ecosystem function and ecology. In this chapter, we explore the ways in which the retention and loss of data affect our reconstructions of attributes important to measuring system dynamics. Using a high-resolution food web of a modern Jamaican coral reef ecosystem (Roopnarine and Hertog 2013), we simulate its fossilization and document subsequent changes of taxon richness and interspecific interactions. We then measure how those changes alter important quantitative measures of food web structure and function, specifically predator dietary breadth, the web's food chain lengths and trophic level distribution, guild richness and evenness, and the modularity of the system. Finally, we assess the reliability with which we could reconstruct those features of a paleocommunity by assuming that our starting point is the fossilized reef system, then constructing a guild-level representation of the fossil community and contrasting the implied trophic structure to that of a similarly resolved modern reef.

Endangered Coral Reefs

Threats to modern coral reef ecosystems bear several similarities to conditions from intervals in the Earth's past, e.g., increasing CO_2, ocean acidification, and oceanic temperatures. Coral reefs may thus be one of the best proxies we have for predicting future ecological changes and biodiversity loss in the oceans. The ocean is a large sink for anthropogenic carbon dioxide today (~30% of total), and a predicted increase of CO_2 concentrations in the coming century is expected to adversely affect marine organisms in a multitude of ways, particularly by decreasing biocalcification (Kleypas et al. 1999; Sabine et al. 2004). Evidence already exists for such a scenario, which is apparent in decreasing calcification rates of individual species and communities, especially in coral reef environments (Albright et al. 2016). In addition, global sea surface temperatures (SST) have risen in the past century (0.4–0.8 °C), with warming predicted to accelerate in the near future (Joachimski et al. 2012). Increasing temperatures have had a large effect on the marine realm, affecting ocean circulation, benthic and planktonic diversity and abundance, productivity, and overall invertebrate physiology (Walther et al. 2002; Parmesan 2006). Coral reefs are extremely sensitive to changes in temperature and pH, frequently expelling their zooxanthellae (photosynthetic algal symbionts) when physiologically stressed, resulting in coral bleaching (Hoegh-Guldberg et al. 2007). During the writing of this chapter (March–April, 2016), Australia's Great Barrier Reef system is experiencing unprecedented bleaching, with approximately 95% of reefs in the ecosystem being affected (A.C. of Excellence Coral Reef Studies 2016). However, corals are not the only organisms under threat in reef ecosystems. Other invertebrates and vertebrates are also in decline due to overfishing, degradation of their coral habitat, and pollution (Nyström et al. 2000; Jackson et al. 2001; Bellwood et al. 2004; Pandolfi et al. 2005; Hardt 2009; Cramer et al. 2015).

Caribbean reefs in particular have suffered significant losses, with reports of considerable reduction (~80%) of coral reef cover since the 1970s in addition to frequent human-induced degradation (Gardner et al. 2003; Mora 2008). Furthermore, in the early 1980s, a massive disease-induced die-off of the urchin, *Diadema antillarum*, resulted in a macroalgal bloom that persists to this day (Lessios et al. 1984; Andres and Witman 1995; Mumby et al. 2006). The less biodiverse algal-dominated state of Jamaica's reefs in particular is exacerbated by historical overexploitation of herbivorous fish (Hardt 2009). It appears, however, that *Diadema* has been functionally replaced to some extent by parrotfishes, highlighting the potential importance of functional redundancy in coral reef and other ecosystems (Andres and Witman 1995; Mumby et al. 2007; Hardt 2009; Nash et al. 2016). Regardless, local increases of parrotfish in various areas of the Caribbean have not reversed the algal-dominated state of the reefs, although studies have indicated that when parrotfish are able to escape overfishing inside marine reserves, they were able to increase grazing intensity and reduce macroalgal cover (Mumby et al. 2006). Recent reports do show a slight recovery of *Diadema* and a subsequent increase in coral recruitment and survivorship, though not nearly as rapidly as expected, with current populations at only 11.62% of their premortem density (Idjadi et al. 2010; Lessios 2016). Jamaican reefs have come to represent an unhealthy reef system, frequently disturbed by anthropogenic (e.g., overfishing and pollution) and non-anthropogenic (i.e., disease, hurricanes) stresses (Hughes 1994).

Today, climate change and other human-induced disturbances are progressing at an unparalleled and alarming rate, increasing the need for studies that use novel analogs and proxies from Earth's past. Data from the fossil record give us a long-term historical perspective from which we can test the influence of extreme environmental conditions on ecological dynamics and community structure. Fossil food web studies are particularly needed due to recent studies indicating that functional extinction of apex predators, large herbivores, or ecosystem engineers in coastal ecosystems may occur several decades to centuries after the onset of ecosystem degradation, resulting in potential collapse of trophic webs (Jackson et al. 2001; Lotze et al. 2006). Thus, deep-time studies have much to contribute to evaluating potential losses of biodiversity, stability, and sustainability in marine ecosystems due to current and future climate change. For example, Aronson et al. (2007) determined that Antarctic benthic food web structure was established 41 million years ago when the climate was much cooler, resulting in predators (i.e., sharks, crabs) being pushed from Antarctic waters. An increase in current temperatures as a result of climate change may result in the the invasion of such durophagous predators, profoundly affecting benthic food web structure. As such, paleoecological perspectives are vital to modern conservation strategies in order to establish how we can buffer anthropogenic and climate-related stress in fragile modern ecosystems currently and in the future.

2 Fossilizing a Coral Reef

The Jamaican coral reef food web describes species interactions from coral reef and adjacent seagrass habitats within Jamaica's marine geopolitical territory, including the offshore Pedro Bank (Fig. 1). The food web model is an amalgamation of data drawn from multiple specific localities. Differences between localities, however, are expected to be ephemeral and changing; this is supported by the significant compositional overlap both among localities within Jamaica, as well as with similar food webs constructed for the neighboring Cayman Islands and Cuba (Roopnarine and Hertog 2013). This northern Caribbean region represents a common regional species pool, and the Jamaican dataset is thus a sample of that pool, integrated over the spatial variation present among the Jamaican localities. Taxon composition was determined from published compilations and reports, including Fishbase (FishBase 2016), and the REEF (Reef Environmental Education Foundation Volunteer Survey Project Database 2011) survey database compiled up to 2011. Details of sources and methods used to determine species interactions are given in Roopnarine and Hertog (2013), and the complete data are archived in the DRYAD database (Roopnarine and Hertog 2012). Together, the resources represent approximately 50 years of data, a

Fig. 1 Map of the Greater Antilles, showing the regions covered by the coral reef data (adapted from Roopnarine and Hertog (2013)). Data treated in this study are from Jamaican geopolitical areas, outlined in red

temporal resolution which is likely much finer than any available in the fossil record. Comparison to sub-fossil and archaeological data from Jamaica, however, suggest that compositionally the data would be congruent with fossil data time-averaged on at least a millennial scale (Hardt 2009). The complete dataset documents the interactions of 749 species in the northern Caribbean, ranging from single-celled protists to multicellular macrophytes and metazoans, of which 728 have records of occurrence in Jamaica. Multiple species were collapsed into trophospecies when they shared exactly the same prey and predators (i.e., had exactly the same interactions in the food web). This process resulted in 265 trophospecies, of which 249 are present in Jamaica, with a total of 4105 inter-trophospecific predator–prey interactions.

Fossilization of the community was simulated with a simple binary filter at the genus level. The occurrence of the genus to which each taxon in the food web is assigned was checked for occurrence in the fossil record using the Paleobiology Database (downloaded September, 2014) and Sepkoski's Compendium of Fossil Marine Genera (Sepkoski et al. 2002) and was considered fossilized if the genus has a documented fossil record. The presumption is based on the premise that characteristics which promote fossilization, such as morphology, life habits, and habitat, can be extended to all members of the genus. If the genus does not have a documented fossil record, then the food web taxon was eliminated from the dataset. This simulated fossilization resulted in a reduced dataset comprising 433 species aggregated into 172 trophospecies, and 1737 inter-trophospecific interactions. We tested whether the likelihood of species fossilization is uniform among the trophospecies, suspecting that biases would exist because of the usual vagaries of fossilization, including biases against soft-bodied taxa, small body size, easily disarticulated skeletons, and depositional environment. The expected number of species fossilized in a trophospecies is estimated simply as the species richness of the trophospecies times the overall fraction of preservation for all the trophospecies,

$$\texttt{E(preservation)} = (\texttt{species richness}) \times 433/728 \qquad (1)$$

Thus, the probability that the number of species lost (i.e., not fossilized) from a single trophospecies is consistent with uniform probabilities of non-preservation among all trophospecies is given by the hypergeometric probability

$$\texttt{P(observed)} = \binom{L}{l_X}\binom{S-L}{s_X - l_X}\binom{S}{s_X}^{-1} \qquad (2)$$

where S is food web species richness, L is the number of species lost during simulated fossilization, s_X is the richness of trophospecies X, and l_X is species loss from X during fossilization. A hypothesis of uniform, unbiased levels of preservation could not be rejected for 130 trophospecies. The remaining trophospecies had either improbably low or high levels of preservation (Fig. 2), and all can be explained by body composition and body size. For example, trophospecies with unexpectedly poor preservation include nanno-zooplankton ($p = 0.0006$), epibenthic sponges ($p = 1.02 \times 10^{-8}$), micro-zooplankton trophospecies such as cyclopoid copepods ($p = 0.0003$), gorgonians ($p = 8.91 \times 10^{-6}$), and sphenopid and zoanthid anthozoans ($p = 8.91 \times 10^{-6}$). In contrast, trophospecies with unexpectedly good preservation include mixotrophic scleractinian corals ($p = 3.95 \times 10^{-15}$) and soft-sediment dwelling, infaunal suspension feeding bivalves ($p = 0.0001$).

A loss of biotic interactions is expected to accompany the loss of taxa with lower probabilities of preservation. We examined the distribution of lost interactions among the trophospecies, expecting that the loss would be a function of the number of interactions of a trophospecies; that is, more connected trophospecies would lose a proportionally greater number of interactions. Biased preservation of species within some trophospecies, however, would also render other trophospecies more poorly connected, or disconnected, than expected based on their numbers of interactions alone. We therefore examined incoming, that is, predatory interactions for each trophospecies using a hypergeometric probability similar to that explained above for taxon preservation. Given that 1737 trophospecies interactions out of 4105 were preserved, the expected number of interactions (prey) retained by a trophospecies is estimated as

$$\texttt{E(preservation)} = (\texttt{interactions}) \times 1737/4105 \qquad (3)$$

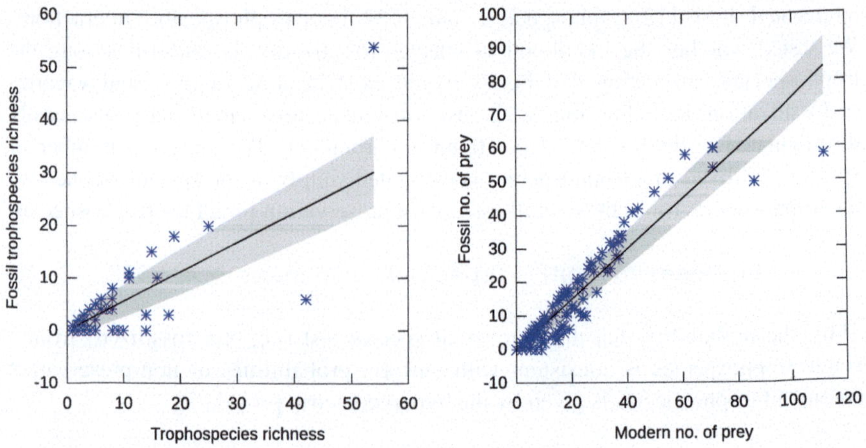

Fig. 2 Expected and observed losses of taxa (left) and interactions (right). Values after simulated fossilization are plotted against the modern values. The expected values, based on an assumption of a uniform probability of preservation among all trophospecies, is plotted as a line. Grey regions represent 1.96 standard deviations around the expected values (see Appendix 1)

The probability of observing the number actually lost is then

$$\texttt{P(observed)} = \binom{L}{l_X}\binom{I-L}{i_X - l_X}\binom{I}{i_X}^{-1} \tag{4}$$

where I is the total number of interactions in the food web, i_X is the number of incoming interactions, or prey, of trophospecies X, and L is the total number of interactions lost. Trophospecies that lost more incoming interactions (i.e., predatory interactions) than expected include scleractinian corals ($p = 0.003$ for both mixotrophic and fully heterotrophic trophospecies), the butterflyfish *Chaetodon capistratus* ($p = 0.004$), and a multi-taxon trophospecies including the fish blue chromis (*Chromis cyanea*), brown chromis (*C. multilineata*), and royal gramma (*Gramma loreto*) ($p = 0.003$). Zooplankton constitute a major or sometimes total portion of the diet of all these taxa, and the significantly low preservation probabilities of zooplankton trophospecies result in their predators being either poorly connected or completely disconnected from the fossil food web. In contrast, a number of fish trophospecies retain more interactions than expected, primarily because benthic invertebrates with hard body parts, and hence high probabilities of preservation dominate their diets. This set of consumers includes the seabream (*Diplodus caudimaculatus*) ($p = 0.00007$), the pufferfish (*Sphoeroides spengleri*) ($p = 3.1 \times 10^{-7}$), and the squirrelfish (*Sargocentron vexillarium*) ($p = 0.00004$).

The reconstruction of taxon level trophic properties is obviously affected by the differential probabilities of preservation of taxa, based primarily on body composition and body size, and consequently the preservation of trophic interactions. The low preservation of a major group such as the copepod zooplankton not only generates a negative bias against the inferred dietary breadths of their predators, but also creates a positive bias toward taxa whose prey have exceptionally high probabilities of preservation. These results are not surprising given what we know about the vagaries and biases of fossil preservation, but the relevant question here is what are the biases created when viewing the community as an integrated system, and not merely as a collection of taxa and interactions. We therefore examined several system-level properties, including the distribution of dietary breadths ("in-degree" distribution), food chain lengths, trophic levels, modularity, and guild structure and diversity.

Dietary Breadth

The distribution of dietary breadths is the distribution of the number of prey species per consumer species, or in the case of aggregated food webs such as this, the number of trophospecies preyed upon by each consumer trophospecies. This distribution is commonly referred to in the food web literature as an in-degree distribution, where degree refers to the number of interactions per species, trophospecies, or trophic guild. Surveys of these distributions (Williams and Martinez 2000; Dunne et al. 2002; Roopnarine 2009) show that the overwhelming majority are "decay" distributions, where the density of the distribution is biased toward low dietary breadths, meaning that there are more taxa with specialized diets in the web than there are species with generalist diets. The precise nature of a web's distribution is unknowable unless all interactions have been recorded, but estimates from a variety of webs and models suggest that the distributions may be exponential, power law, or a mixture of decay distributions (Williams 2010). One interesting feature is that the tails of these distributions are long (hyperbolic) and occupied by generalist species with broad diets. Topological analyses of food web networks suggest that the presence of such highly connected species provides community robustness against the cascading effects of random extinctions because those taxa are less likely to lose all their connections and thus provide some insurance to the community (Dunne et al. 2002). That conclusion must be tempered, however, by a corresponding weakness of many of those links because a general theoretical conclusion is that strong interactions destabilize species interactions and hence communities in general (May 1973; McCann et al. 1998; McCann 2000). Furthermore, empirical measures of interaction strengths between modern species show that the majority of those interactions are relatively weak (Paine 1980). Reconstructions of Lagerstätten food webs, with presumably high probabilities of preservation, have also yielded hyperbolic distributions (Dunne et al. 2008). Models of paleocommunity food web dynamics have used hyperbolic decay distributions, that is, fat-tailed distributions, but the majority of interactions in the community models are typically weak (Roopnarine et al. 2007; Mitchell et al. 2012; Roopnarine and Angielczyk 2015).

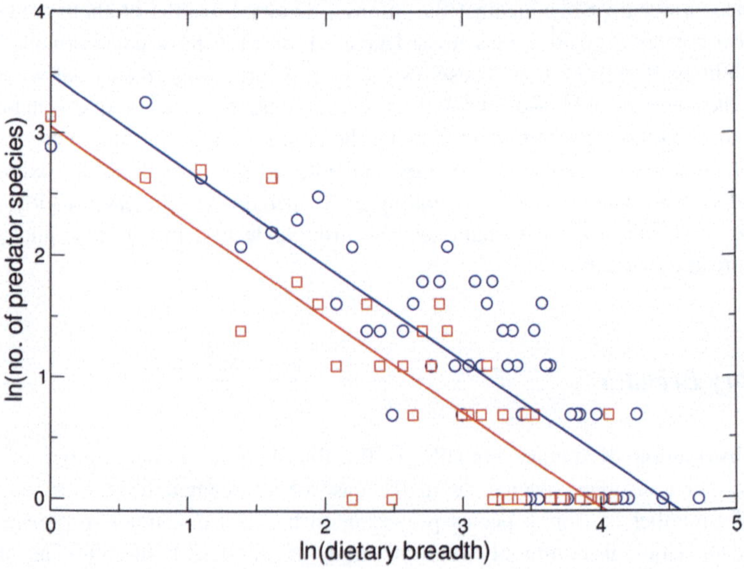

Fig. 3 Trophic in-degree, or dietary breadth, distributions. The number of predatory species with a specific number of prey or resource species (y-axis) is plotted against that specific number of prey (x-axis). Both the modern (open circles) and fossil (open squares) distributions are fit significantly by logarithmic power functions

Given the above observations of taxon and interaction loss during fossilization, however, can paleocommunities be reliable sources for the reconstruction of in-degree distributions? A comparison of the modern and simulated fossil distributions for the coral reef community shows that, at least in this case, a reconstruction based on the paleocommunity would be very accurate (Fig. 3). The modern distribution can be described best with a power function, $\ln(p) = 3.48 - 0.78[\ln(r)]$ where r is the number of prey trophospecies per consumer trophospecies, that is consumer in-degree, and p is the number of consumer trophospecies of that degree ($F(1, 54) = 130.74$, $r^2 = 0.702$, $p < 0.0001$). The fossil distribution is essentially identical, $\ln(p) = 3.05 - 0.78[\ln(r)]$, ($F(1, 37) = 95.41$, $r^2 = 0.713$, $p < 0.0001$), and statistically indistinguishable (Student's $t = 0.01$, $p = 0.990$). Thus, despite the loss of taxa and interactions, including the structurally and functionally important zooplankton trophospecies, one can predict the number of interactions per fossil trophospecies with a high degree of confidence.

Trophic Chains and Levels

Trophic level is an important characteristic in both species and system conservation efforts (Pace et al. 1999; Sergio et al. 2005; Becker and Beissinger

2006). It is generally accepted that high trophic level species, that is, predators with superior capabilities and few predators of their own, are often at greater risk from anthropogenic actions. The reasons for this vary between marine and terrestrial habitats. In the former, high trophic level species are often targeted for harvesting because of high biomass, while terrestrially they are viewed as threats to domestic livestock or competitors for resources, and are often sensitive to habitat alteration or destruction. The impact of overfishing on oceanic predators has been devastating, and populations throughout the Caribbean are in decline (Jackson et al. 2001; Hardt 2009). Scleractinian reefs, however, originated and evolved during a series of changing predatory clades, including Mesozoic marine reptiles, and the radiations of modern chondrichthyans, teleosts, and teleost clades specialized to reef environments (Wood 1998). The potential richness and complexity of coral reef systems might therefore vary with the relative proportions of different life history strategies present, and different suites of trophic levels would represent alternative, yet persistent ecosystem regimes.

Trophic level is broadly understood to describe, in some manner, the position of a species within a food web, if species are arranged hierarchically from basal, primary producers, up to apex predators. Species positions in the food web determine the flow of energy to their populations, and the feasibility of their persistence depends on a productive basal component, sufficient to support all those populations further up the food chains (Williams and Martinez 2004). Energy is lost thermodynamically along each step of a food chain, however, because of respiration and the inefficiency of energy assimilation from consumed material, and this loss is thought to constrain the lengths of food chains (Yodzis 1984). Food chain length may also be constrained by the likelihood of decreasing dynamical stability as the length of a chain grows (Pimm and Kitching 1987), and by numerous factors specific to a particular food web, its composition and environmental context (Post 2002).

All these limitations would restrict the diversity and complexity of predation within a community, but there is variability driven by organismal variation. Yodzis, thinking in terms of webs rather than chains, suggested that increased productivity would make both increased chain length and predator diversity feasible (Yodzis 1989). This in turn means that any predator toward the top of a food chain would have to be a super-generalist, spreading its effort over a greater diversity of prey, in order to garner sufficient energy. This effort could be focused on other predators near the top, or by omnivory, where the predator would feed at multiple levels of the chain thus circumventing some of the thermodynamic loss. Finally Pimm (1982), noting the decreasing trend of efficiency of energy conversion moving from ectothermic invertebrates, to ectothermic vertebrates, to endothermic vertebrates, suggested that invertebrates should support longer chains (Pimm 1982). The generally larger population sizes of smaller species, and greater rates of invertebrate population growth, would also increase the likelihood of dynamical stability (Post 2002).

Variability of complexity is driven in modern Caribbean reefs largely by anthropogenic factors (Jackson et al. 2001; Bascompte et al. 2005), but over longer timescales this variability could be a function of clade diversity dynamics and macroevolutionary trends. Identification of any such trends, and the establishment of reliable historical baselines for modern reefs, depends on whether trophic chains and levels can be reconstructed from fossil and sub-fossil data. As such, here we introduce a method for calculating trophic position from food web network data, network trophic level, and use it to compare the distribution of trophic levels in the modern and fossilized reefs. The network trophic level (ntl) of a species or trophospecies is the average shortest distance of its prey species to a primary producer. Primary producers are assigned an ntl of 1.0, and the ntl of a consumer species i is calculated as

$$\mathtt{ntl}_i = 2 + \frac{1}{r_i}\sum_{j=1}^{S} a_{ij} l_j \tag{5}$$

where r_i is the number of prey species of i, $a_{ij} = 1$ if i preys on species j, and zero otherwise, and l_j is the shortest path length of j to a primary producer. Ntl differs from the prey-averaged trophic level of Williams and Martinez (2004) by a factor of one for consumers. Trophic level is measured in various ways, ranging from simplistic integer values corresponding to discrete categories such as "primary," "secondary," "tertiary" consumer, and so forth, to inferences of the number of steps to a consumer as measured by stable isotopic composition of consumer tissues. A common measure used for fish is fractional trophic level (ftl). Ftl is based on the proportion of specific prey species in a consumer's diet and is a weighted average of the prey ftl values. Romanuk et al. (2011) presented a global database of empirically measured and inferred ftl values for fish. Using that database, Roopnarine (2014) showed that ftl and ntl are correlated significantly (Fig. 4), concluding that ntl is a reliable measure of trophic level, and that a significant proportion of trophic level variance is based only on position in the food web.

Consumer ntl ranges from 2 to 5.5 in the Jamaican reef food web, with a mean value of 2.89 which increases to 3.29 if primary consumers (ntl = 2) are excluded. Almost all trophospecies with ntl values between 4 and 5 are vertebrates, primarily predatory fish with a broad range of body sizes. Carnivorous ophiuroids is the single invertebrate trophospecies in this range, the included species being both benthic deposit feeders and polychaete carnivores (e.g., *Ophiocoma echinata*). All trophic levels of five and above, however, are occupied by invertebrate trophospecies, specifically corallivorous polychaetes and gastropods, e.g., *Hermodice carunculata* and *Coralliophila caribbaea* (ntl = 5.0), and gastropod predators of polychaetes, e.g., *Conus regius* (ntl = 5.5). These very high trophic levels are the result of very long food chains which extend the phytoplankton–zooplankton food chain, but which are also very simple, meaning that the prey breadths of trophospecies along the chain tend to be low. The high ntl invertebrate taxa are also not apex predators, but instead are subject to predation by trophospecies that feed at multiple trophic levels and along multiple food chains. For example, *C. regius* is preyed upon by a trophospecies of carnivorous crustaceans, including *Penaeus duorarum*, with a ntl of 3.18. The assessment of trophic level is therefore complicated by the branching topology of a food web. Ntl, as perhaps with other trophic level measures such as

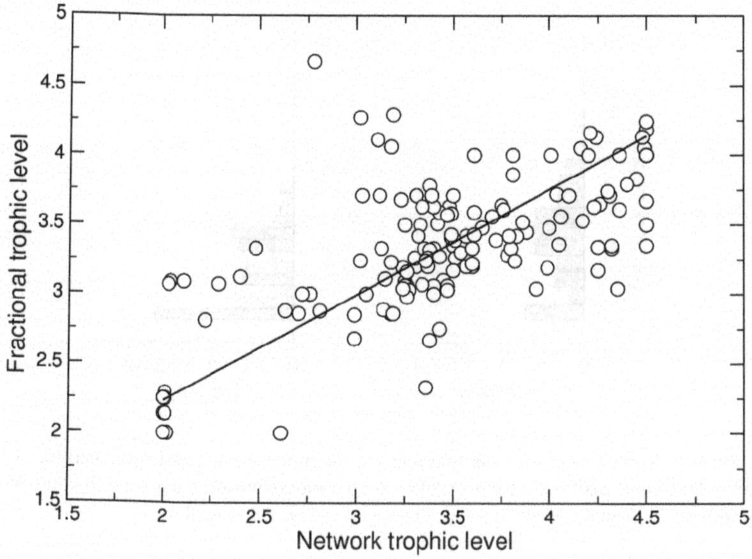

Fig. 4 The relationship between network trophic level and fractional trophic level fit with a reduced major axis function

stable isotopic composition (Post 2002), is thus not a linear ordination of taxa among multiple food chains. They are, however, reliable measures of the average distances of consumers from the productive base of a web. The ntl measures reported here are consistent with other observations that invertebrates tend to occupy longer food chains than vertebrates (May 1983; Yodzis 1984; Pimm et al. 1991). Moreover, although high ntl vertebrate predators frequently prey on invertebrates along long food chains, those vertebrates also feed in much shorter food chains and thus again have lower ntl values. For example, the top predator Caribbean reef shark *Carcharhinus perezi* is of ntl 3.86. The fact that the most powerful predators in a food web network will not be the furthest removed from primary producers is a warning against the common practice of simplifying food web structures into discrete, or nearly discrete trophic levels.

The ntl distribution of the modern reef is significantly higher than the simulated fossil reef (Fig. 5; Kruskal-Wallis, $\chi^2 = 30.70$, $p = 0.0001$). Maximum ntl in the modern reef is 5.5, but only 4 in the fossil reef. The reduction could be the result of a loss of high ntl trophospecies, but there is no significant difference between the ntl distributions of those trophospecies that are preserved and those that have no fossil representation (Student's $t = 1.612$, $p = 0.108$). The difference of ntl distribution is in fact attributable to the reduction of ntl of preserved trophospecies. Many of those trophospecies are specialized predators with poorly preserved prey, such as carnivorous gastropod predators of gorgonians or polychaetes, or zooplanktivorous fish. Those consumers' ntl values have collapsed to two, implying incorrectly that they are primary consumers. Other trophospecies ntl values are reduced because

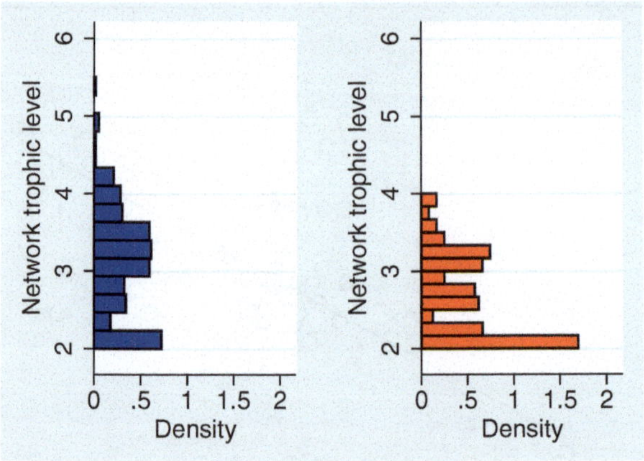

Fig. 5 Network trophic level (ntl) distributions for the modern (left) and simulated fossil (right) food webs. The fossil distribution is truncated, with a maximum ntl value of 4 in contrast to the modern web's 5.5. Primary producers (ntl = 1) are excluded from the plots

they have no preserved prey and their ntl values are 1.0. Thus, the poor preservation of key trophospecies such as zooplankton and soft-bodied taxa has a significant impact on estimates of fossil trophic level, and the impact will always lead to an underestimation of trophic level and food chain lengths. There is essentially no way in which data can be recovered to address this issue, and the evolution and historical baselines of trophospecies trophic level in coral reef ecosystems, and indeed all paleoecosystems, are obscured by biases of preservation. It is conceivable, however, that information is retained, and can be inferred, at higher levels of system organization that might be less subject to bias, and we turn to those in the following sections.

Modularity

A module is a subset of nodes within a network which have more interactions with each other, and fewer with other nodes, than would be expected if interactions occurred at random (Newman 2006). Obvious examples occur in social networks where persons within a family or circle of friends may represent a module within the larger society. A module within a food web would therefore comprise species sharing more interactions with each other than they do with other species in the community, a condition also referred to in the food web literature as compartmentalization. In offering this definition, we distinguish this usage of "module" and "modularity" from an alternative use which refers to pairs or trios of interacting species without regard to their other interspecific interactions (e.g., Kondoh (2008)). We also discount trivial compartmentalizations that result in discrete, non-overlapping sets of interactions, for example, as might occur across strong habitat boundaries. Such compartments are in effect independent food webs. We focus instead on compartments embedded in, and sharing interactions with the rest of the network.

Interest in food web modularity stems from May's theoretical work (May 1973) on the relationship between the local stability of a community (i.e., its ability to return to a static equilibrium after a minor perturbation) and the community's richness, connectance (the density of interspecific interactions), and average interaction strength. May noted that for random networks or food webs, the probability of stability decreases with an increase of any of those community parameters, calling into question the long-held hypothesis of a positive relationship between community complexity and stability. May also pointed toward possible "solutions" to this seeming paradox, including a hypothesis that many food webs might consist of compartments, or modules, which would increase the probability of stability. We suggest here that in addition to stable dynamics, modular structure of a food web would indicate ways in which the energy supplied to a community is partitioned, and the extent to which the community could then be viewed as energetically integrated or compartmentalized.

Results of subsequent searches for food web modularity have been mostly equivocal (Dunne et al. 2005), possibly because of the relatively low resolution of many current food web datasets. Another probable cause is the difficulty of identifying modules in highly resolved, complex food webs. In a community of *S* species, the number of modules could range from 1 (all species within a single module) to *S* (each species represents a separate module). All modular arrangements of size between these two extremes would represent all possible combinations of partitioning schemes of *S*, and that number grows very rapidly with increasing *S*! There are no objective and exact analytical approaches short of an exhaustive search of all those possible combinations of species clusters. Heuristic approaches do exist, however, and here we applied the modularity algorithm due to Newman (2004), which is an optimization of the property Q, where $Q =$ (fraction of interactions within modules) – (expected fraction of such interactions). We used a fast approximation of Newman's algorithm developed by Blondel et al. (2008), as implemented in the network visualization software Gephi (Bastian et al. 2009), to generate repeated modularity measures on the same network. We further tested the null hypothesis that the community network is indistinguishable from a random network of equal species richness and number of interactions, by comparing the food web to such equivalent random networks using the netcarto program (Guimerá and Amaral 2005, 2005a; Guimerà et al. 2007).

Both the netcarto and Blondel et al. implementations support the sub-division of the food web into four modules (Fig. 6), a relative modularity value of 0.282, and with the netcarto randomizations failing to support the null hypothesis of no unique modularity (z-test; mean relative modularity of 1000 random networks = 0.139, sd = 0.002, $p < 0.0001$). The coral reef community thus comprises four modules or sub-communities, and that substructure is highly unlikely to arise by any random assembly of species and interactions. The collapse of 249 trophospecies into

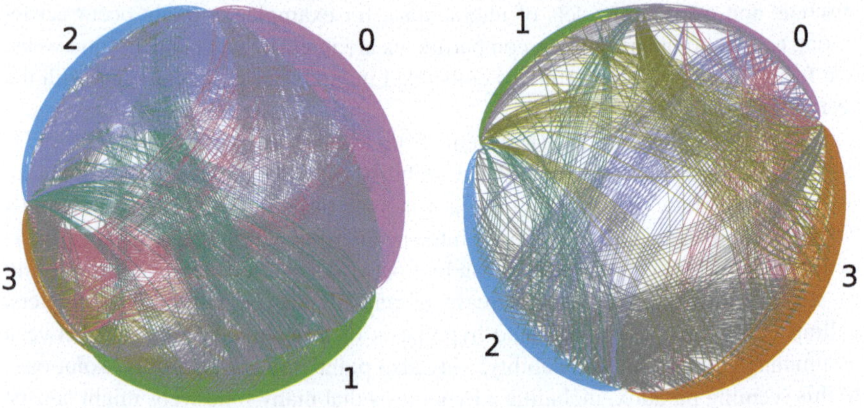

Fig. 6 Modularity of the modern (left) and simulated fossil (right) food webs. Numbers indicating each module are explained in the text, and correspond to the same module between the webs. Visually, modules are identifiable by the density of within-module interactions which are shown on the periphery, in contrast to between-module interactions which cross the interior of the circularly arranged webs

four modules is striking, but do the modules make any sense ecologically? Our expectation that the food web would be partitioned among different conduits of energy flow is partially correct, but factors of habitat and modes of life also play a role. Figure 7 illustrates summaries of the modules, presented as food chains and featuring examples of the trophospecies within each module.

Module 0 (modules are numbered here 0, 1, 2, and 3), composed of 34 trophospecies, is a basal energetic module in the sense that it includes all the macrophytic trophospecies, including macroalgae and seagrasses. It also includes the sponges and herbivorous fish as well as low trophic level omnivores that consume macroalgae and benthic sponges, such as the parrotfish *Scarus iserti*. The highest trophic level trophospecies in the module includes piscivores that specialize on benthic grazing fish, e.g., the tiger grouper *Mycteroperca tigris*, and generalist predatory piscivores, including the apex predator Caribbean reef shark *Carcharhinus perezi*.

Module 1, comprising 66 trophospecies, is dominated by benthic food chains, including benthic foraminifera and metazoan deposit feeders, and grazers such as lytechinid echinoids. Food chains are extended through the module primarily by omnivorous and carnivorous benthic macrocrustaceans, e.g., pagurid crabs, and both benthic invertebrate predators and teleost predators of those taxa, including gonodactylid stomatopods. The food chains are again capped within the module by high trophic level piscivores such as the king mackerel *Scomberomorus cavalla*.

Module 2 does not have a notable primary producer base, containing only encrusting coralline algae, but the module size is nevertheless substantial, comprising 74 trophospecies. Much of this module's diversity is dominated by benthic macroinvertebrates, including herbivorous, omnivorous, and carnivorous grazers,

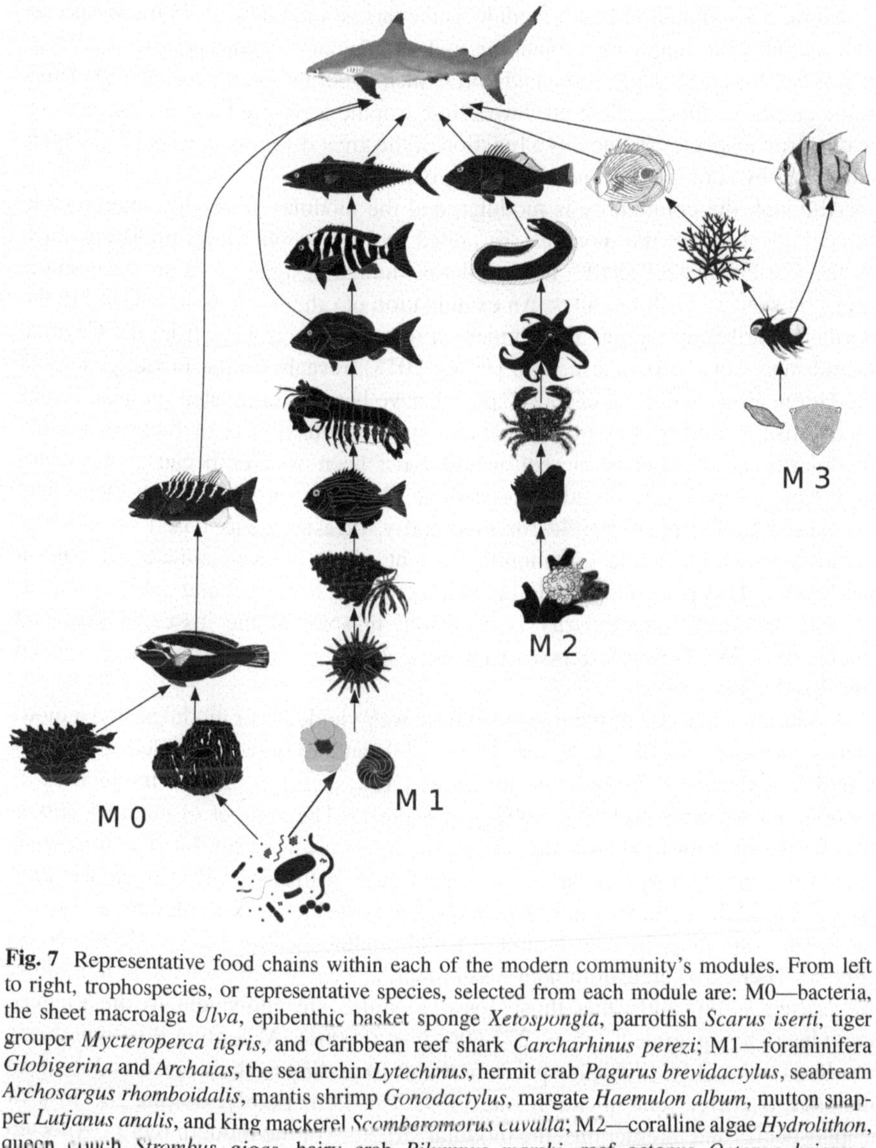

Fig. 7 Representative food chains within each of the modern community's modules. From left to right, trophospecies, or representative species, selected from each module are: M0—bacteria, the sheet macroalga *Ulva*, epibenthic basket sponge *Xetospongia*, parrotfish *Scarus iserti*, tiger grouper *Mycteroperca tigris*, and Caribbean reef shark *Carcharhinus perezi*; M1—foraminifera *Globigerina* and *Archaias*, the sea urchin *Lytechinus*, hermit crab *Pagurus brevidactylus*, seabream *Archosargus rhomboidalis*, mantis shrimp *Gonodactylus*, margate *Haemulon album*, mutton snapper *Lutjanus analis*, and king mackerel *Scomberomorus cavalla*; M2—coralline algae *Hydrolithon*, queen conch *Strombus gigas*, hairy crab *Pilumnus marshi*, reef octopus *Octopus briareus*, purplemouth moray *Gymnothorax vicinus*, and red hind *Epinephelus guttatus*; M3—epiphytic diatoms, copepods (*Podon* sp. shown), staghorn coral *Acropora cervicornis*, reef butterflyfish *Chaetodon capistratus*, and Atlantic spadefish *Chaetodipterus faber*

such as the queen conch *Strombus gigas*, the hairy crab *Pilumnus marshi*, and the reef octopus *Octopus briareus*, respectively. Higher trophic level predators in the module are primarily fish specialized to predation in the coral habitat, such as moray eels and their predators.

Module 3, a plankton-based module, is the largest module with 75 trophospecies and includes the major phytoplankton and zooplankton trophospecies, as well as major benthic and pelagic zooplanktivores, such as corals and the Atlantic spadefish *Chaetodipterus faber*. There are fewer high trophic level predators in the module, and its trophospecies richness is a function of the great diversity of trophic strategies employed by zooplankton and zooplanktivores.

Although the community is modular, and the modules are highly interpretable in ecological terms, the modules are united by high trophic level predators, such as the Caribbean reef shark. This predator, though assigned to a single module, preys on species in all modules. An examination of other large shark species in the northern Caribbean region, as documented in similar food webs from the Cayman Islands and Cuba (Roopnarine and Hertog 2013), reveals similar broad, generalist predation. However, most of those species have been exterminated on local scales by overfishing and are rare in regional reef systems today. Prior to their extirpation, modularity of the Jamaican reef would have been weaker because of greater high trophic level, cross-module predation. Therefore, in spite of the theoretical expectations of a positive role for modularity, measurements based on modern, anthropogenically altered communities might result in overestimates of natural modularity. This possibility could be addressed with historical and paleontological records, but only if modularity is preserved in spite of the losses of taxa and interactions. We therefore repeated the modularity analyses using the simulated fossilized community.

Modularity analysis of the fossilized food web yielded four modules, and a modularity measure of 0.287. Comparison to 1000 random networks failed to support a null hypothesis of insignificant modularity (z-test; mean relative modularity of random networks = 0.169, sd = 0.002, $p < 0.0001$). The number of modules equals that for the modern food web, and the modularities are very similar. The important question though is how comparable are the modules in the two webs? To answer this, we compared the membership of preserved trophospecies in both modern and fossil modules, examining the distribution of fossil modules within each modern module (Fig. 8). There is a single dominant fossil module occurring in each modern module, suggesting significantly that those fossil modules are equivalent to the modern modules (Chi-square test, $\chi^2 = 273.606$, $p < 0.0001$). Accepting this hypothesis results in only 13 fossil trophospecies not corresponding to their presumed modern modules, a correct classification rate of 92.4%. Remarkably, despite the loss of trophospecies and interactions, the modular structure of the modern reef is expected to be preserved in fossil reef communities.

3 Guild Structure and Diversity

Food web reconstructions often aggregate species into groups on the basis of presumed ecological similarities, such as "benthic macroinvertebrates," and "salt marsh plants," etc. Aggregation is perhaps justified when more precise data are

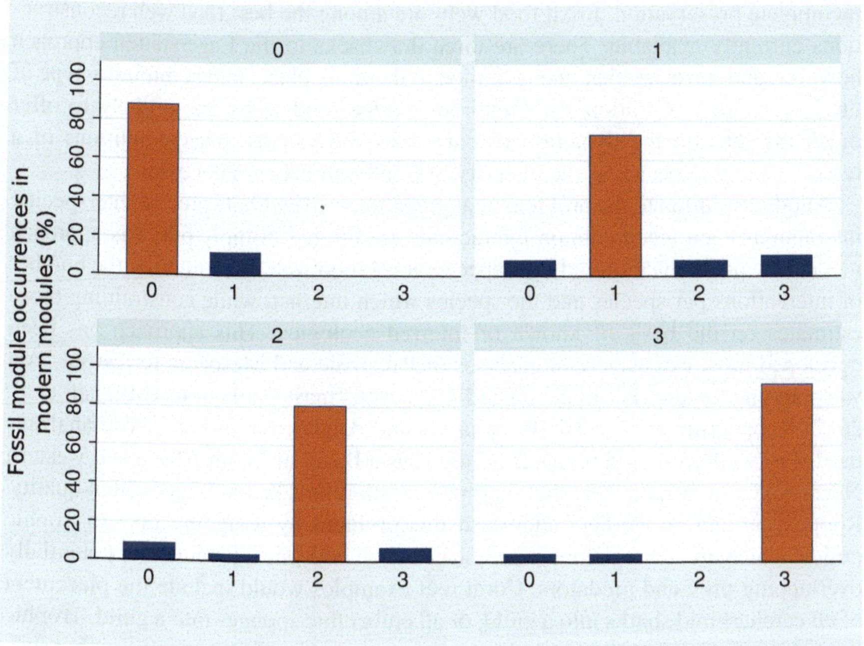

Fig. 8 The distribution of modules identified in the simulated fossil food web, within each module of the modern food web. The statistically dominant occurrence of a single fossil module in each modern module supports the equivalency of the webs' modular structures. Each modern module is plotted separately, as indicated at the top of each plot. Modules in the fossil food web are shown on *x*-axes, numbered 0–3, to indicate significant statistical equivalence to their modern counterparts

not available, or more precise data are unsuitable for the analytical models being employed. The extent to which aggregation obscures or distorts food web structure and dynamics is an unsettled argument, but one solution would be to simply increase the precision with which food web networks are constructed. A version of this solution employs the trophospecies, or "trophic species" approach, in which species aggregates comprise taxa with exactly the same trophic interactions (Martinez 1992). Trophospecies simplify a food web without sacrificing precision and is the approach taken in the construction of the modern Caribbean coral reef community. The trophospecies approach has also been applied to the fossil record in at least three instances, including reconstructions of Cambrian marine food webs of the Burgess Shale and Chengjiang faunas (Dunne et al. 2008), and an Eocene terrestrial food web of the Messel Shale (Dunne et al. 2014). All these reconstructions take advantage of fossil Lagerstätte, where fossil preservation is exquisite, probabilities of preservation are high, and some traces of trophic interactions are available. Furthermore, although taxon preservation is almost certainly incomplete, the Messel Shale food web includes 700 taxa, and with the current coral reef food web represents one of the largest food web reconstructions available. Thus, in spite of incomplete preservation, fossil food webs are among the best food web reconstructions currently available. There are three drawbacks to the Lagerstätten approach however, and those are that one is limited to the time, place, and community type of the Lagerstätten. Additionally, Cambrian marine food webs are unlikely to offer much insight into the dynamics of coral reef ecosystems under conditions of a rapidly changing ocean, as the majority of Cambrian taxa are not extant.

Another solution to the problem of aggregation is to estimate precise interspecific interactions even when data on interactions are limited. Simply put, this approach models the precision required for a species-level food web by estimating the number of interactions per species and the species which interact, while constraining those estimates on the basis of known or inferred ecologies. This approach has been developed most extensively in analyses of Paleozoic and Mesozoic terrestrial food webs around the end-Permian and end-Cretaceous mass extinctions (Mitchell et al. 2012; Roopnarine et al. 2007; Roopnarine and Angielczyk 2015). Those analyses used very well preserved biotas from the Karoo Basin of South Africa and western North America, respectively, but their preservation falls below Lagerstätten quality. Roopnarine and co-workers addressed this problem by assigning taxa to trophic guilds, which are aggregations based on common body size, habitat, and potentially overlapping prey and predators. Coral reef examples would include the placement of all carcharhinid sharks into a guild, or all epibenthic sponges into a guild. Trophic interactions between guilds lack the precision of interspecific interactions, but they are in actuality sets of interspecific interactions, and each interspecific interaction belongs to a single interguild interaction. Given a community of species, there is a finite number of food webs that can be constructed, but many of them would not be consistent with the ecologies and interactions of the community. Guild-level food webs, also termed metanetworks, limit this number to a subset of species-level food webs that are consistent with ecological reality. The dynamic terrestrial models of Roopnarine et al. (2007) and Roopnarine and Angielczyk (2015) sample species-level food webs from the metanetwork subset.

An unresolved question is how well guilds and metanetworks actually capture the functional structure of a community. We address this here by first identifying the guild structure of the reef, and then taking the perspective that one is required to reconstruct the guild structure given only those taxa that would be preserved in the simulated fossil reef. The reliability of the fossil metanetwork is then evaluated by comparing its diversity and evenness to the modern metanetwork, and examining whether the reef's distribution of trophic levels can be reconstructed from the fossil metanetwork.

Identifying Guilds in a Food Web

The purpose of a trophic guild is to aggregate species with overlapping prey and predators. Trophospecies are therefore a type of guild, one in which members share exactly the same prey and predators. Membership can be extended, while maintaining exact overlap, to species with different numbers of prey and predators, but where the interactions of some of the species are subsets of others, i.e., the interactions of one species are nested within those of another. For example, prey of the porcupine fish *Chilomycterus antennatus* is a subset of the trophically more general triggerfish, *Balistes vetula*. A requirement of exact overlap, however, becomes problematic under two conditions. First, increasingly precise data are liable to uncover small differences between species, thereby separating species which were formerly assigned to the same trophospecies. Second, if most interactions are indeed relatively weak, then how much weight should be assigned to interactions that are not shared between species, versus those that are shared? In both cases, species can be ecologically similar, and yet their aggregation in a food web be ambiguous. The questions could potentially be answered for modern communities by ever-increasing empirical work although these are very difficult data to obtain. Furthermore, the questions simply cannot be answered for paleocommunities. Here, we propose a heuristic solution where we use the overlap of interactions among species to recognize guilds, but limit the expected overlap according to the limits of fossil preservation.

The procedure begins with a pairwise measure of interaction overlap between all species in the food web. Overlap is measured separately for both types of interactions, predator–prey and prey–predator, as

$$o = \frac{r_{ij}}{\min(r_i, r_j)} \tag{6}$$

where r_i is the number of prey (predators) of species i, and r_{ij} is the number of prey (predators) that i and j have in common. The value of o is thus zero if the species have no prey in common, and 1 if they have all prey in common or one of the prey sets is nested completely within the other. Predator overlap is measured similarly, counting number or predators for $r_i$and r_{ij} , such that o equals zero when the species have no predators in common. Each set of measures yields a $S \times S$ matrix of overlap measures, where S is the species richness of the community. The two matrices are then combined for each pair of species as the simple product of prey and predator overlap, producing a single matrix of overlap indices, **O**. The resulting elements of **O** then range from 0, where either prey or predator interactions, or both, fail to overlap between the two species, to 1, where overlap or nestedness is complete. Within the **O** matrix will be groups of species which overlap among themselves more strongly than they do with other species, in effect forming modules. Guilds can therefore be identified by examining the modularity of the **O** matrix and would be equivalent to **O**'s modules.

Given the concerns expressed above regarding the precision of overlap, we examined the matrix's modularity, and hence guild composition, at multiple levels of overlap by applying thresholds, where values below a threshold would be excluded from guild recognition. We proceeded from a threshold value of 0.1, where species sharing an overlap greater than 0.1 could potentially be assigned to the same guild, to 1, where species within a guild would have perfectly nested sets of interactions.

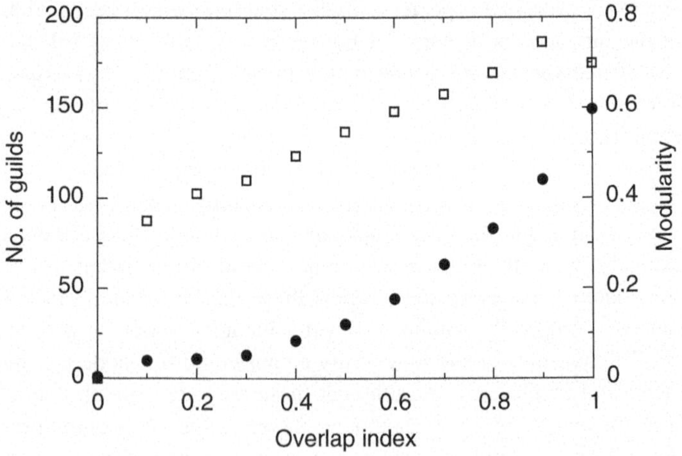

Fig. 9 The number of guilds, or modules, recognized as the overlap or nestedness of interactions between species. The extent of overlap, or threshold, is indicated on the *x* axis, and the corresponding number of guilds (filled circles), and strength of the modularity (open squares), are indicated on the left and *y* axes, respectively

Guild inclusivity thus decreases as the threshold increases, and the number of guilds should therefore increase. These expectations are indeed met, as the number of guilds in the coral reef community, and the strength of overlap as measured by Newman's modularity index, increase as the threshold increases, with modularity being at a maximum when the threshold is 0.9 (Fig. 9). The number of guilds increases from 10 at a threshold of 0.1 to 149 at a threshold of 1. The 728 species in the Jamaican coral reef food web, represented by 249 trophospecies, may therefore be nested within 149 guilds. This number is comparable to work by Bambach (Bambach et al. 2007), in which 118 modes of life were found to exist for recent marine metazoans. Given the greater resolution available in our data, 118 most likely represents a lower limit of the number of guilds.

4 Reconstructing the Community

Imagine that our starting point is not a modern coral reef, but a fossil coral reef instead. Can an understanding of how community information is transformed during fossilization be combined with fossil taxon richness to reconstruct a paleocommunity food web? The easiest reconstruction is at the guild level, where species are aggregated into sets of overlapping interactions. Not all the 149 guilds of the modern reef would be preserved, however, and many interactions are also lost. The losses cannot be accounted for in many cases, for example, the absence of a guild of corallivorous polychaetes, because not only do those species lack a fossil record, but it is doubtful that their existence could otherwise be inferred from preserved taxa. There are two rich and highly connected groups, however, whose existence can be inferred even though they are absent from the fossil record: epibenthic sponges and zooplankton. The sponges do not have a taxonomic record, but their presence is recorded as large numbers of disarticulated spicules. The sponges are a key link between pelagic microorganisms on which they prey and guilds of spongivorous macroinvertebrates and vertebrates. Major groups within the zooplankton guilds are also absent from the fossil record, e.g, copepods, or have no records which include coral reef-dwelling members, e.g., scyphozoans. There are large numbers of preserved zooplanktivorous species and guilds, however, such as corals and many other cnidarians and zooplanktivorous fish. We could therefore insert guilds representing the missing groups into the set of preserved guilds with defensible confidence and connect them to appropriate prey and predator guilds; but two problems must be addressed.

First, whereas the epibenthic sponge trophospecies is assigned to a single guild out of 149 in the modern food web, micro-zooplankton are distributed among six different guilds, and macro-zooplankton are assigned to two guilds, one of which also contains micro-zooplankton. Given the absence of zooplankton from our fossil data, we would have no idea how many zooplankton guilds should be inserted into the reconstructed community. The best and most conservative answer would be to use a threshold of overlap at which a single zooplankton guild first emerges, thereby aggregating all zooplankton into a single guild, yet distinguishing them from other guilds. We therefore examined guild structure at each threshold as described above (Fig. 9), and the zooplankton first emerge as a single guild at the 0.6 level as thresholds increase from 0.1 to 1. There are 42 guilds at this threshold (see Appendix 2, Table 1), and we consider it to be the best resolution that could be inferred from our knowledge of fossil taxa, in the absence of modern data. Thirty three of the 42 guilds include fossilizable taxa.

Second, how many species should be assigned to the sponge and zooplankton guilds? Deriving sponge richness from spicule diversity is notoriously difficult and unreliable (Jones 1984), and the zooplankton essentially have no fossil record in the coral reef system. We addressed this problem by referring to a similar situation in terrestrial paleocommunities with insect faunas. Mitchell et al. (2012), in their reconstructions of Late Cretaceous North American communities, estimated insect richness based on a positive relationship between the insect richness of well-preserved faunas spanning the Phanerozoic, and the richness of vertebrate predation on those faunas. The relationship is consistent over this time interval. The logic of the relationship is that predator and prey richnesses may often be related in communities over evolutionary time, since a greater diversity of prey could support a greater diversity of predators, both energetically and by reducing competition among the predators. Similarly, the presence of predators generally supports a greater diversity of prey species. We therefore examined this relationship for a zooplankton food chain at the 0.6 threshold. Guilds on this food chain include "phytoplankton," "nanno-zooplankton," "foraminifera," and other heterotrophic protists, "zooplankton," "gorgonians," and "pelagic planktivores." The set of observations

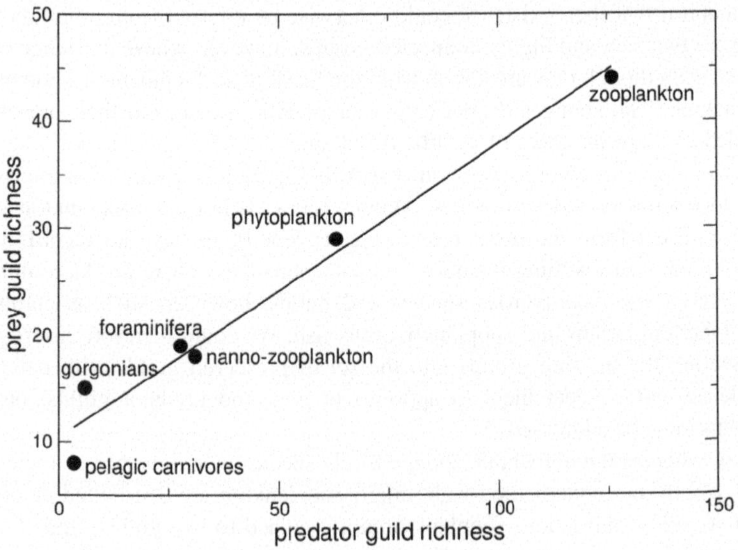

Fig. 10 The relationship between predator guild richness and prey guild richness for guilds along the major "zooplankton" food chain. The line indicates a fitted reduced major axis function. Prey guilds are indicated at each corresponding data point

consists of the richness of a prey guild, and the sum of predators supported by that guild, weighted by number of prey guilds linked to each predatory guild. That is, predator richness

$$|\texttt{predators}| = \sum_{j=1}^{G} \frac{a_{ij}|G_j|}{r_j} \tag{7}$$

where $|G_j|$ is richness of a guild of predators linked to prey guild i, and r_j is the number of guilds preyed upon by predatory guild j. The relationship was fit using reduced major axis (RMA) regression both because neither variable can be considered independent, and neither is measured without error (Fig. 10). Furthermore, a RMA function is symmetrical, meaning that the function is invariant to the choice of which variable is treated as the predictor. The RMA function

$$|\texttt{prey}| = 0.275|\texttt{predators}| + 10.362 \tag{8}$$

is significant ($r^2 = 0.97$, $F = 270.468$, $p = 0.00005$) and predicts a zooplankton guild richness of 45 species. This is in excellent agreement with the actual richness of zooplankton species (44).

A similar function could not be constructed successfully for the sponges because the sponge guild-level food chain at 42 guild resolution includes two highly aggregated guilds, the "hard benthic macroinvertebrates" and "large vertebrate macrophyte and invertebrate grazers" which confound any more specific numerical relationships between prey and predator richnesses. The overall fraction of species preserved in the simulated paleocommunity is 0.59 (433 out of 728 species). Sponge richness was therefore estimated using the fraction of overall preservation (0.59), yielding (0.59 × 53) = 31 species, which is greater than the 16 species (all endolithic) expected to be fossilized, but closer to the original richness of 53 species.

Diversity and Evenness

In order to explore the similarities and/or differences between the modern and fossil coral reef food webs, we adopted methods commonly employed in paleoecology, calculating the Simpson Index of Diversity (1-D) and the Shannon Index (H′) to examine the richness and distribution of species across guilds (Shannon 1948; Simpson 1949). The Simpson Index of Diversity (1-D) is a common metric for quantifying taxonomic diversity and abundance and is used here to estimate the taxonomic richness of guilds in the modern and fossil coral reef communities. Hence, the higher the value of the Simpsons Index of Diversity, the greater the taxonomic diversity of the guilds. The Shannon Index (H′) is used to calculate the evenness ($E = H'/H'_{max}$) of the distribution of species across guilds, with a maximum value of 1.0 indicating a perfectly even distribution and a minimum value of 0 indicating a highly uneven distribution.

In applying these metrics to the modern Jamaican coral reef community, consisting of 728 species within 42 guilds, we found that the community had very high taxonomic diversity, with a Simpson Index of 0.92. Meanwhile, the fossilized coral reef community, consisting of 33 guilds and 433 species, had a very similar taxonomic diversity, with a Simpson Index of 0.88. The modern coral reef community was also fairly even in its distribution of species across guilds, with an evenness (*E*) value of 0.75; the fossil reef community was comparable with $E = 0.72$. In addition, as previously discussed, one can safely assume via fossil evidence (e.g., sponge spicules) that a zooplankton and a sponge guild are likely to be present. By replenishing those two guilds (34 guilds present, 493 species) and recalculating the diversity metrics, we found that the community now had a Simpson Index (1 D) of 0.90 and an evenness (*E*) value of 0.74. As such, it is apparent that by restoring these unfossilized taxa to the fossil community, the taxonomic diversity values become almost identical to that of the modern community.

However, while it is reassuring that these two diversity metrics are comparable between the modern and fossilized communities, we have the foresight to be aware that ten guilds that appear in the modern coral reef community are not preserved in the fossilized community. Thus, it would appear that caution must be taken in relying on these metrics to quantify guild structure; other metrics and analyses, such as in-degree distribution, modularity, and trophic levels are needed in order to fully characterise food web structure. Nonetheless, these metrics show that both the modern and fossil coral reef communities had high taxonomic richness though potentially lacking some redundancy within guilds. These results agree with previous hypotheses indicating that coral reef ecosystems, despite containing a high diversity of species and guilds, are extremely vulnerable to environmental and anthropogenic perturbations given the potential for the loss of key ecosystem processes and decreased resilience due to an uneven distribution of species across guilds (Bellwood et al. 2003; Micheli and Halpern 2005; Guillemot et al. 2011; Brandl and Bellwood 2014; Nash et al. 2016). Low redundancy within guilds, as implied here, would further exacerbate this vulnerability.

Simulated Food Webs

A primary goal of paleo-food web reconstruction has been the modelling of paleocommunity and ecosystem dynamics (Roopnarine et al. 2007; Mitchell et al. 2012; Roopnarine and Angielczyk 2015; Angielczyk et al. 2005; Roopnarine 2006, 2009). Those models require the simulation of food webs that are consistent with paleocommunity structure, as described by resolvable guild-level networks. The reconstruction of the coral reef at the guild level presented here shows that guild structure is preserved during the simulated fossilization, within limits set by the absence of data on at least two significant groups, the zooplankton and epibenthic sponges. Furthermore, the data can be improved with reliable estimates of species for those missing guilds, with analyses of diversity and evenness supporting the improvement. A concluding step in this exercise is then the derivation of a species-level food web from the preserved guild structure. We did this by assigning in-degrees (number of prey) to each consumer species with random draws from the in-degree equation derived above (section “Dietary Breadth”), constrained to the total richness of the guilds upon which the species could potentially prey. The actual prey species are then assigned randomly to the consumer (see Roopnarine (2009) for further details). The resulting simulated food web consists of 398 connected species, with 965 interactions. The latter number is far below the 4105 inter-trophospecific interactions of the modern community, but recall that the simulated food web is based on an aggregation into 44 guilds. A second food web was simulated, but in this case using a mixed exponential-power law in-degree distribution used commonly in previous paleo-food web dynamics studies (Roopnarine et al. 2007; Mitchell et al. 2012; Roopnarine and Angielczyk 2015; Angielczyk et al. 2005; Roopnarine 2006). This distribution takes the form $P(r) = e^{-r/\varepsilon}$, where r is the in-degree of the consumer, $\varepsilon = e^{(\gamma-1)\ln(M)/(\gamma)}$, M is the total number of prey potentially available to a consumer, and γ, the power law exponent, is 2.5. The value of the exponent is the mid-point of a range previously explored for this distribution when applied to Permian-Triassic paleocommunities from the Karoo Basin of South Africa (Roopnarine et al. 2007). Distributions within the range were found to construct species-level food webs with linkage densities and connectances comparable to those reported for modern food webs (Dunne et al. 2002; Roopnarine 2009). Applying the distribution to the system of fossil coral reef guilds here yields a web with a greater density of interactions, with 8924 interspecific interactions.

It is conceivable that, despite the disparities of interactions among the webs, they could nevertheless yield similar dynamical properties. Such similarity would be possible, however, only if the patterns of interaction are similar among the webs (Roopnarine and Angielczyk 2015). This was checked for the species-level fossil and modern webs by comparing their network trophic level (ntl) distributions. Recall that those distributions differed significantly between the modern and simulated fossil webs (Fig. 5). The question addressed here is whether the reconstructions, based upon the reconstruction of guild structure, and restoration of unfossilized zooplankton and epibenthic sponges mitigate any of that loss of structure. The ntl distribution of the simulated web based on the observed in-degree power law distribution differs significantly from the modern web (ANOVA; df = 2, 478; $F = 30.25$, $p < 0.0001$), and a majority of taxa are below ntl 3.0 (Fig. 11). The ntl distribution of the web based on the mixed exponential-power law distribution, however, is statistically indistinguishable from the modern web (Scheffe's multiple comparison test, $p = 1.0$) and reconstructs both the mean (3.29 in both cases)

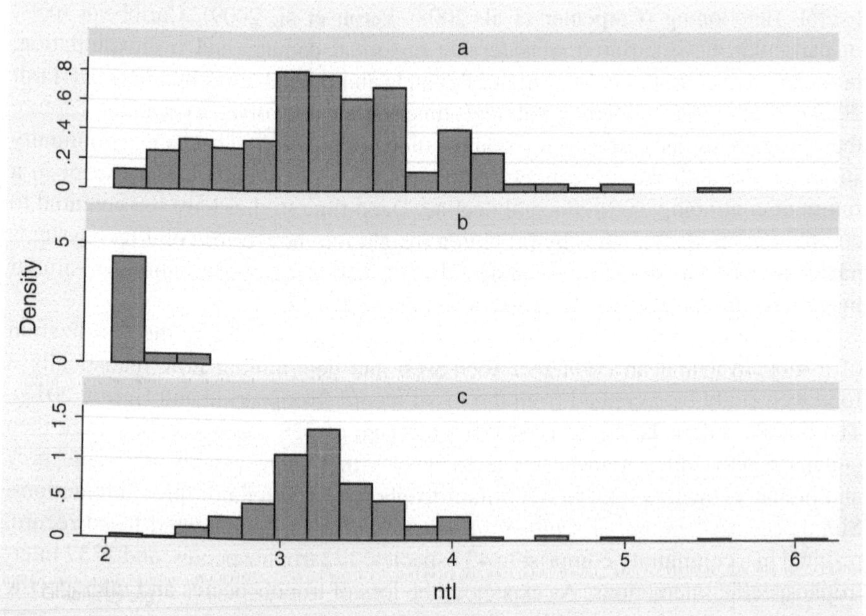

Fig. 11 Observed and reconstructed network trophic level (ntl) distributions for all secondary and greater consumers. (**a**) Observed distribution measured from the modern food web. (**b**) The ntl distribution based on a species-level food web reconstructed from the fossil guild network, and using the power law in-degree (dietary breadth) distribution of the modern food web. (**c**) A similar reconstruction, but made using a mixed exponential-power law in-degree function. The power law-based reconstruction in **b** fails to recreate the trophic level distribution, yielding a highly truncated and short-chained food web. The mixed distribution produces a food web with trophic level distribution statistically indistinguishable from the modern food web

and maximum values (5.5 and 6, respectively). The mixed exponential-power law distribution predicts denser interactions than does the power law, in effect being "fatter-tailed." This feature compensates for the loss of interactions due to the loss of taxa and obviously reconstructs the hierarchical structure of the modern food web with significant fidelity. We predict that dynamic analyses of fossil coral reef communities based on this type of reconstruction would be comparable to analyses of modern communities.

5 Summary

The ongoing, unprecedented, global, anthropogenically driven degradation of natural systems has led to a need for conservation strategies that account for whole ecosystem structures. Coral reefs have received much attention in regard to the effects of anthropogenic and non-anthropogenic stressors on ecosystem functioning, as recent surveys have indicated a global decline in reef diversity, coral cover, and overall functioning (Carpenter et al. 2008; Veron et al. 2009). Caribbean reefs, in particular, have suffered considerable historical damage and overexploitation, potentially since initial colonization of islands and coastal areas by humans (Hardt 2009). Studies of modern coral reef functioning and diversity cannot capture the shifted baselines of species composition, population sizes, and community structure although these are likely required to forecast what we can expect in a future of continuing environmental decline. Deep-time studies have the potential to compliment modern studies by providing insight into how prior coral reef systems have responded to similar environmental perturbations, or perturbations of similar magnitude, in Earth's past (Greenstein and Pandolfi 1997).

Here, we tested the plausibility for such studies by simulating the fossilization of a modern Jamaican coral reef food web and determining how realistically a food web could be recreated from the fossil record (Roopnarine and Hertog 2013). The initial modern Jamaican food web consisted of 728 species, which were then collapsed into 249 trophospecies (i.e., species that share exactly the same prey and predators) with a total of 4105 inter-trophospecific predator–prey interactions. Simulated fossilization, or removal of genera without a documented fossil record, resulted in a community comprising 433 species, 172 trophospecies, and 1737 inter-trophospecific interactions. As expected, the loss of trophospecies and interspecific interactions resulted in a significant bias, both against poorly preserved species, and in favor of those species that had better than average fossil records. Particularly significant was the almost complete loss of zooplankton trophospecies, which resulted in their predators being either poorly connected or completely disconnected from the fossil food web. While the absence of zooplankton in the simulated fossil data places a negative bias against zooplanktivorous species in terms of biotic interactions, the importance of taxa whose prey have exceptionally high probabilities of preservation, e.g. durophagous crustaceans and fish, may be overestimated.

A positive consequence of the latter bias, coupled with the preservation of more than half of the species in the food web, is that several important features are retained by the simulated fossil community, namely the distribution of dietary breadths among consumers, and the modularity or compartmentalization of the community. The distribution of dietary breadths, or in-degree distribution, describes the number of resource species per consumer species. Compilations based on a variety of modern food webs suggest that those distributions are generally of a decay type and hyperbolic. This means that more species are specialists, consuming relatively few species, and fewer species are generalists; yet the hyperbolic nature of the distributions also implies that generalist species occur at frequencies greater than would be expected were dietary breadths distributed normally. The coral reef meets these expectations, being fit significantly with a power function. The distribution for the reduced, simulated fossilized web is statistically indistinguishable from the modern web's distribution. Nevertheless, the loss of interactions does result in a significant alteration of the trophic level structure of the community. Many species have lower trophic levels in the fossil community, particularly those involved in very long food chains that include taxa with low probabilities of preservation. The result is that the distribution of trophic levels in the simulated fossil community is significantly truncated, and there is no possibility of recreating trophic levels based on taxon composition alone.

Modularity has been proposed to be important to the stability or resistance of food webs to perturbations, yet the search for modules in food webs has yielded equivocal results. Here, we demonstrated that the Jamaican coral reef food web is indeed modular. Modularity is most distinct toward the bases of food chains, and the four modules identified in the community are based primarily on the differential utilization of basal and low trophic level resources. High trophic level predators, however, such as the Caribbean reef shark, feed across modules, both uniting modules via top-down effects, and reducing the modularity of the system. Furthermore, many of those predatory species are involved in very strong biotic interactions (Bascompte et al. 2005). The apparent modularity of the modern community might therefore be an anthropogenically amplified effect, because many high trophic level predators have been either extirpated from reef communities throughout the Caribbean, or are now present in very low numbers. The fossilized food web retains the modularity of the modern web, perhaps because it too lacks many higher trophic level species, though for preservational reasons. This emphasizes the caution which should be exercised in the analysis of modern, post-disturbance communities, and it is useful to speculate that assessments of reef community modularity based on historical and sub-fossil records (Hardt 2009) could yield significantly different results. It is also worth considering, in this context, whether the introduction of the invasive lionfish, *Pterois volitans*, into the Caribbean has restored some of the effects of predation now lacking in those communities.

We tested our ability to recreate a coral reef food web from fossil data only, by assuming the simulated fossil reef as a starting point. Following procedures developed for terrestrial paleocommunities (Roopnarine et al. 2007), trophic guild structure was used as the basis to aggregate species into biotically interacting units.

We used a heuristic approach to first identify trophic guilds in the modern food web. If we limit our ability to recognize guilds according to the limits of fossil preservation, we find a total of 42 guilds, whereas the maximum number identified in the modern community is 149. Nevertheless, Simpson (1-D) and Shannon (H′) Indexes for modern and fossil guild diversity were almost identical despite a loss of guilds in the fossil community.

Finally, species-level food webs were reconstructed from the fossil and guild data using stochastic techniques described in (Roopnarine 2009). One reconstruction, based on the in-degree distribution of the modern web, fails to recreate the trophic level structure, and hence hierarchical arrangement of the modern community. In contrast, a second reconstruction, based on a model in-degree distribution developed in previous studies (Roopnarine et al. 2007), creates a trophic level and hierarchical structure statistically indistinguishable from the modern community. The greater success of the latter in-degree distribution is a consequence of its overestimation of species dietary breadths, thereby compensating for losses incurred during fossilization. The important implication of this successful recreation is that coral reef communities can indeed be recreated from fossil data and subjected to the types of dynamic analyses essential for forecasting the behavior of those communities under broad magnitudes of environmental and biotic disturbance. This emphasizes the validity and usefulness of paleoecological data to the field of conservation biology. The creation of deep-time paleocommunity food webs has the ability to enrich and advance our current knowledge of how natural systems behave, especially in response to future environmental changes.

Acknowledgements We wish to thank Courtney Chin, Chrissy Garcia, and Rhiannon Roopnarine for the food chain illustrations and other graphical assistance. We also thank the CAS Paleoecology Discussion Group, including Allen Weik, for many useful discussions which contributed to this chapter. Finally, thanks to Kenneth Angielczyk and Carol Tang for asking difficult questions.

Appendix 1

Hypergeometric Variance

The hypergeometric distribution describes the probability given a population with a featured subset of observations, and a sample from the population, that a random selection of individuals from the population will include a certain fraction of the sample comprising individuals with the featured characteristic. In this case, the population consists of 728 species (N), of which 433 are featured as fossilized (K). A sample is the richness of a trophospecies (n), of which a subset are observed as fossilized (k). The expected fossilization or hypergeometric mean value is

$$\mathrm{E}(\hat{k}) = n\frac{K}{N} \tag{9}$$

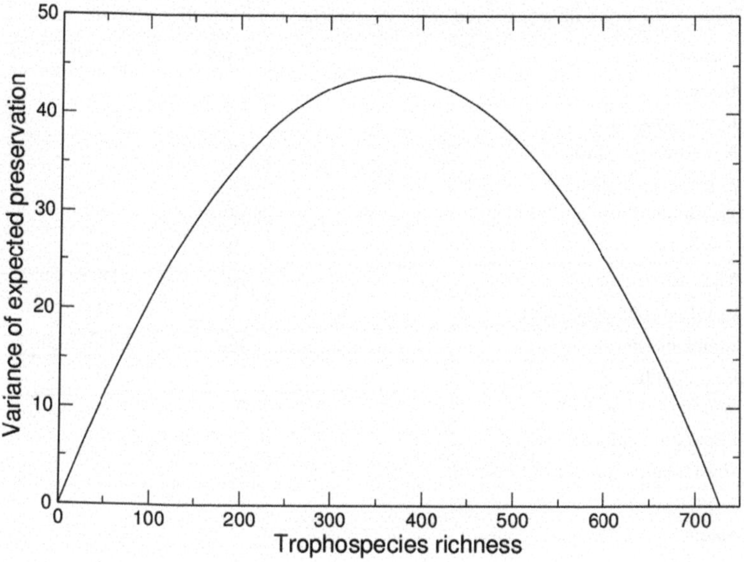

Fig. 12 Hypergeometric variance given a total population of 728 species, and trophospecies of increasing size

The variance of the expectation is given as

$$\sigma^2(k) = n\frac{K}{N}\frac{N-K}{N}\frac{N-n}{N-1} \tag{10}$$

This value grows initially as n because the number of unique ways in which k objects may be selected also grows. The value declines, however, as $n \to N$ (Fig. 12).

Appendix 2

See Table 1.

Table 1 List of guilds recovered at the 0.6 threshold of interaction overlap

Guilds	Modern	Fossil
Bacteria	–	–
Phytoplankton	29	15
Nanno-zooplankton	18	3
Macrophytes, diatoms	28	14
Sponges	53	31*
Corals	70	64
Micro-detritivores	19	18
Corallivorous polychaetes	2	0
Zooplankton	44	45*
Gorgonians	15	0
Benthic carnivores	22	0
Eucidarid echinoids	1	1
Hard benthic macroinvertebrates	123	90
Cypraeids	3	3
Sinum	1	1
Charonia	1	1
Epiphyte-grazing gastropods	3	3
Bryozoans	25	20
Endolithic polychaetes	3	3
Soft benthic macroinvertebrates	80	75
Polychaete-consuming gastropods	6	6
Molluscivorous crustaceans	1	1
Macroinvertebrate predators I	60	25
Stomatopods I	2	2
Large grazers of macroalgae and invertebrates	79	47
Macroinvertebrate predators II	4	4
Octopus	1	0
Macroinvertebrate predators III	11	2
Stomatopods II	3	3
Hemiramphus brasiliensis	1	0
Trachinotus goodei	1	0
Pelagic piscivores I	8	8
Mycteroperca bonaci	1	1
Gymnothorax moringa	1	0
Lutjanus analis	1	0
Scomberomorus cavalla	1	1
Green sea turtle	1	1
Hawksbill turtle	1	1
Loggerhead turtle	1	1
Blacktip shark	1	1
Oceanic whitetip shark	1	1
Caribbean reef shark	1	1

Modern and fossil guild richnesses are shown. Guilds containing single taxa have those taxa listed by taxonomic or unique common name. Fossil guild richnesses with an asterix * were estimated

References

A.C. of Excellence Coral Reef Studies (2016) Coral Bleaching Taskforce documents most severe bleaching on record. http://www.coralcoe.org.au/media-releases/coral-bleaching-taskforce-documents-most-severe-bleaching-on-record

Albright R, Caldeira L, Hosfelt J, Kwiatkowski L, Maclaren JK, Mason BM, Nebuchina Y, Ninokawa A, Pongratz J, Ricke KL et al (2016) Reversal of ocean acidification enhances net coral reef calcification. Nature 531(7594):362

Andres N, Witman J (1995) Trends in community structure on a Jamaican reef. Oceanogr Lit Rev 10(42):889

Angielczyk KD, Roopnarine PD, Wang SC (2005) Modeling the role of primary productivity disruption in end-permian extinctions, Karoo Basin, South Africa. The nonmarine Permian. N M Mus Nat Hist Sci Bull 30:16–23

Aronson RB, Thatje S, Clarke A, Peck LS, Blake DB, Wilga CD, Seibel BA (2007) Climate change and invasibility of the Antarctic benthos. Annu Rev Ecol Evol Syst 38:129–154

Bambach RK, Bush AM, Erwin DH (2007) Autecology and the filling of ecospace: key metazoan radiations. Palaeontology 50(1):1

Bascompte J, Melián CJ, Sala E (2005) Interaction strength combinations and the overfishing of a marine food web. Proc Natl Acad Sci U S A 102(15):5443

Bastian M, Heymann S, Jacomy M et al (2009) Gephi: an open source software for exploring and manipulating networks. ICWSM 8:361

Barnosky AD, Hadly EA, Bascompte J, Berlow EL, Brown JH, Fortelius M, Getz WM, Harte J, Hastings A, Marquet PA et al (2012) Approaching a state shift in Earth's biosphere. Nature 486(7401):52

Becker BH, Beissinger SR (2006) Centennial decline in the trophic level of an endangered seabird after fisheries decline. Conserv Biol 20(2):470

Bellwood DR, Hoey AS, Choat JH (2003) Limited functional redundancy in high diversity systems: resilience and ecosystem function on coral reefs. Ecol Lett 6(4):281

Bellwood DR, Hughes TP, Folke C, Nyström M (2004) Confronting the coral reef crisis. Nature 429(6994):827

Blondel V, Guillaume J, Lambiotte R, Lefebvre E (2008) Fast unfolding of communities in large networks. J Stat Mech Theor Exp 2008:10008

Brandl SJ, Bellwood DR (2014) Individual-based analyses reveal limited functional overlap in a coral reef fish community. J Anim Ecol 83(3):661

Cardinale BJ, Duffy JE, Gonzalez A, Hooper DU, Perrings C, Venail P, Narwani A, Mace GM, Tilman D, Wardle DA et al (2012) Biodiversity loss and its impact on humanity. Nature 486(7401):59

Carpenter KE, Abrar M, Aeby G, Aronson RB, Banks S, Bruckner A, Chiriboga A, Cortés J, Delbeek JC, DeVantier L et al (2008) One-third of reef-building corals face elevated extinction risk from climate change and local impacts. Science 321(5888):560

Cramer KL, Leonard-Pingel JS, Rodríguez F, Jackson JB (2015) Molluscan subfossil assemblages reveal the long-term deterioration of coral reef environments in Caribbean Panama. Mar Pollut Bull 96(1):176

De'ath G, Fabricius KE, Sweatman H, Puotinen M (2012) The 27–year decline of coral cover on the Great Barrier Reef and its causes. Proc Natl Acad Sci U S A 109(44):17995

Dunne JA, Williams RJ, Martinez ND (2002) Food-web structure and network theory: the role of connectance and size. Proc Natl Acad Sci U S A 99(20):12917

Dunne JA, Williams RJ, Martinez ND (2002) Network structure and biodiversity loss in food webs: robustness increases with connectance. Ecol Lett 5(4):558

Dunne JA, Brose U, Williams RJ, Martinez ND (2005) Modeling food web dynamics: complexity-stability implications. In: Aquatic food webs: an ecosystem approach. Oxford University Press, Oxford, pp 117–129

Dunne JA, Williams RJ, Martinez ND, Wood RA, Erwin DH (2008) Compilation and network analyses of Cambrian food webs. PLoS Biol 6(4):e102
Dunne JA, Labandeira CC, Williams RJ (2014) Highly resolved early Eocene food webs show development of modern trophic structure after the end-Cretaceous extinction. Proc R Soc Lond B: Biol Sci 281(1782):20133280
FishBase (2016) (World Wide Web electronic publication). www.fishbase.org
Franklin JF (1993) Preserving biodiversity: species, ecosystems, or landscapes? Ecol Appl 3(2):202
Gardner TA, Côté IM, Gill JA, Grant A, Watkinson AR (2003) Long-term region-wide declines in Caribbean corals. Science 301(5635):958
Greenstein BJ, Pandolfi JM (1997) Preservation of community structure in modern reef coral life and death assemblages of the Florida Keys: implications for the Quaternary fossil record of coral reefs. Bull Mar Sci 61(2):431
Guillemot N, Kulbicki M, Chabanet P, Vigliola L (2011) Functional redundancy patterns reveal non-random assembly rules in a species-rich marine assemblage. PloS One 6(10):e26735
Guimerá R, Amaral LAN (2005) Functional cartography of complex metabolic networks. Nature 433(7028):895
Guimerá R, Amaral LAN (2005) Cartography of complex networks: modules and universal roles. J Stat Mech Theor Exp 2005(02):P02001
Guimerà R, Sales-Pardo M, Amaral LAN (2007) Module identification in bipartite and directed networks. Phys Rev E 76(3):036102
Hardt MJ (2009) Lessons from the past: the collapse of Jamaican coral reefs. Fish Fish 10(2):143
Harnik PG, Lotze HK, Anderson SC, Finkel ZV, Finnegan S, Lindberg DR, Liow LH, Lockwood R, McClain CR, McGuire JL et al (2012) Extinctions in ancient and modern seas. Trends Ecol Evol 27(11):608
Hewitt JE, Ellis JI, Thrush SF (2016) Multiple stressors, nonlinear effects and the implications of climate change impacts on marine coastal ecosystems. Glob Chang Biol 22(8):2665–2675
Hoegh-Guldberg O, Mumby PJ, Hooten AJ, Steneck RS, Greenfield P, Gomez E, Harvell CD, Sale PF, Edwards AJ, Caldeira K et al (2007) Coral reefs under rapid climate change and ocean acidification. Science 318(5857):1737
Holling CS (1973) Resilience and stability of ecological systems. Annu Rev Ecol Syst 4:1–23
Hughes TP (1994) Catastrophes, phase shifts, and large-scale degradation of a Caribbean coral reef. Science 265(5178):1547
Idjadi J, Haring R, Precht W (2010) Recovery of the sea urchin *Diadema antillarum* promotes scleractinian coral growth and survivorship on shallow Jamaican reefs. Mar Ecol Prog Ser 403:91
Jackson JB, Kirby MX, Berger WH, Bjorndal KA, Botsford LW, Bourque BJ, Bradbury RH, Cooke R, Erlandson J, Estes JA et al (2001) Historical overfishing and the recent collapse of coastal ecosystems. Science 293(5530):629
Joachimski MM, Lai X, Shen S, Jiang H, Luo G, Chen B, Chen J, Sun Y (2012) Climate warming in the latest Permian and the Permian–Triassic mass extinction. Geology 40(3):195
Jones WC (1984) Spicule dimensions as taxonomic criteria in the identification of haplosclerid sponges from the shores of Anglesey. Zool J Linnean Soc 80(2–3):239
Kidwell SM (2001) Preservation of species abundance in marine death assemblages. Science 294(5544):1091
Kidwell SM, Bosence DWJ (1991) Taphonomy: releasing the data locked in the fossil record. In: Taphonomy and time-averaging of marine shelly faunas. Plenum, New York, pp 115–209
Kleypas JA, Buddemeier RW, Archer D, Gattuso JP, Langdon C, Opdyke BN (1999) Geochemical consequences of increased atmospheric carbon dioxide on coral reefs. Science 284(5411):118
Knowlton N (2004) Multiple "stable" states and the conservation of marine ecosystems. Prog. Oceanogr. 60(2):387
Knowlton N, Jackson JB (2008) Shifting baselines, local impacts, and global change on coral reefs. PLoS Biol 6(2):e54

Kondoh M (2008) Building trophic modules into a persistent food web. Proc Natl Acad Sci U S A 105(43):16631
Leighton LR (2002) Inferring predation intensity in the marine fossil record. Paleobiology 28(03):328
Lessios H (2016) The Great *Diadema antillarum* Die-Off: 30 Years Later. Annu Rev Mar Sci 8:267
Lessios HA, Robertson D, Cubit J (1984) Spread of *Diadema* mass mortality through the Caribbean. Science 226(4672):335
Lotze HK, Lenihan HS, Bourque BJ, Bradbury RH, Cooke RG, Kay MC, Kidwell SM, Kirby MX, Peterson CH, Jackson JB (2006) Depletion, degradation, and recovery potential of estuaries and coastal seas. Science 312(5781):1806
Mace GM, Norris K, Fitter AH (2012) Biodiversity and ecosystem services: a multilayered relationship. Trends Ecol Evol 27(1):19
Martinez ND (1992) Constant connectance in community food webs. Am Nat 139:1208–1218
May RM (1973) Stability and complexity in model ecosystems, vol 6. Princeton University Press, Princeton
May RM (1983) Ecology: the structure of food webs. Nature 301:566
McCann KS (2000) The diversity–stability debate. Nature 405(6783):228
McCann K, Hastings A, Huxel GR (1998) Weak trophic interactions and the balance of nature. Nature 395(6704):794
McClanahan TR (2002) The near future of coral reefs. Environ Conserv 29(04):460
Meffe G, Nielsen L, Knight RL, Schenborn D (2012) Ecosystem management: adaptive, community-based conservation. Island Press, Washington
Micheli F, Halpern BS (2005) Low functional redundancy in coastal marine assemblages. Ecol Lett 8(4):391
Mitchell JS, Roopnarine PD, Angielczyk KD (2012) Late Cretaceous restructuring of terrestrial communities facilitated the end-Cretaceous mass extinction in North America. Proc Natl Acad Sci U S A 109(46):18857
Mora C (2008) A clear human footprint in the coral reefs of the Caribbean. Proc R Soc Lond B: Biol Sci 275(1636):767
Mumby PJ, Hedley JD, Zychaluk K, Harborne AR, Blackwell PG (2006) Revisiting the catastrophic die-off of the urchin *Diadema antillarum* on Caribbean coral reefs: fresh insights on resilience from a simulation model. Ecol Model 196(1):131
Mumby PJ, Dahlgren CP, Harborne AR, Kappel CV, Micheli F, Brumbaugh DR, Holmes KE, Mendes JM, Broad K, Sanchirico JN et al (2006) Fishing, trophic cascades, and the process of grazing on coral reefs. Science 311(5757):98
Mumby PJ, Hastings A, Edwards HJ (2007) Thresholds and the resilience of Caribbean coral reefs. Nature 450(7166):98
Nash KL, Graham NA, Jennings S, Wilson SK, Bellwood DR (2016) Herbivore cross-scale redundancy supports response diversity and promotes coral reef resilience. J Appl Ecol 53:646–655
Newman ME (2004) Fast algorithm for detecting community structure in networks. Phys Rev E 69(6):066133
Newman ME (2006) Modularity and community structure in networks. Proc Natl Acad Sci U S A 103(23):8577
Nyström M, Folke C, Moberg F (2000) Coral reef disturbance and resilience in a human-dominated environment. Trends Ecol Evol 15(10):413
Pace ML, Cole JJ, Carpenter SR, Kitchell JF (1999) Trophic cascades revealed in diverse ecosystems. Trends Ecol Evol 14(12):483
Paine RT (1980) Food webs: linkage, interaction strength and community infrastructure. J Anim Ecol 49(3):667
Pandolfi JM, Jackson JB (2006) Ecological persistence interrupted in Caribbean coral reefs. Ecol Lett 9(7):818
Pandolfi JM, Jackson JBC, Baron N, Bradbury R et al (2005) Are US coral reefs on the slippery slope to slime? Science 307(5716):1725

Parmesan C (2006) Ecological and evolutionary responses to recent climate change. Annu Rev Ecol Evol Syst 37:637–669

Pimm SL (1982) Food webs. Springer, New York

Pimm SL, Kitching R (1987) The determinants of food chain lengths. Oikos 50:302–307

Pimm SL, Lawton JH, Cohen JE (1991) Food web patterns and their consequences. Nature 350(6320):669

Post DM (2002) The long and short of food-chain length. Trends Ecol Evol 17(6):269

Reef Environmental Education Foundation Volunteer Survey Project Database (2011) (World Wide Web electronic publication). www.REEF.org

Romanuk TN, Hayward A, Hutchings JA (2011) Trophic level scales positively with body size in fishes. Glob Ecol Biogeogr 20(2):231

Roopnarine PD (2006) Extinction cascades and catastrophe in ancient food webs. Paleobiology 32(01):1

Roopnarine PD (2009) Conservation paleobiology. Paleontological society papers, vol 15, GP. Dietl, KW. Flessa (eds). The Paleontological Society, Tokyo pp 195–220

Roopnarine PD (2009) Ecological modeling of paleocommunity food webs. Paleontol Soc Pap 15:196

Roopnarine PD (2014) Humans are apex predators. Proc Natl Acad Sci U S A 111(9):E796

Roopnarine PD, Angielczyk KD (2015) Community stability and selective extinction during the Permian-Triassic mass extinction. Science 350(6256):90

Roopnarine P, Hertog R (2012) Data from: detailed food web networks of three greater antillean coral reef systems: the Cayman Islands, Cuba, and Jamaica. doi:10.5061/dryad.c213h. http://dx.doi.org/10.5061/dryad.c213h

Roopnarine PD, Hertog R (2012) Detailed food web networks of three greater Antillean coral reef systems: the Cayman Islands, Cuba, and Jamaica. Dataset Papers in Science 2013

Roopnarine PD, Angielczyk KD, Wang SC, Hertog R (2007) Trophic network models explain instability of early triassic terrestrial communities. Proc R Soc Lond B: Biol Sci 274(1622):2077

Sabine CL, Feely RA, Gruber N, Key RM, Lee K, Bullister JL, Wanninkhof R, Wong C, Wallace DW, Tilbrook B et al (2004) The oceanic sink for anthropogenic CO2. Science 305(5682):367

Schwartz MW, Brigham C, Hoeksema J, Lyons K, Mills M, Van Mantgem P (2000) Linking biodiversity to ecosystem function: implications for conservation ecology. Oecologia 122(3):297

Scheffer M, Carpenter S, Foley JA, Folke C, Walker B (2001) Catastrophic shifts in ecosystems. Nature 413(6856):591

Seddon AW, Mackay AW, Baker AG, Birks HJB, Breman E, Buck CE, Ellis EC, Froyd CA, Gill JL, Gillson L et al (2014) Looking forward through the past: identification of 50 priority research questions in palaeoecology. J Ecol 102(1):256

Sepkoski JJ Jr (2002) A compendium of fossil marine animal genera. Bull Am Paleontol 363(16):1–560

Sergio F, Newton I, Marchesi L (2005) Conservation: top predators and biodiversity. Nature 436(7048):192

Shannon CE (1948) A mathematical theory of communication. Bell Syst Tech J 27:379

Simpson EH (1949) Measurement of diversity. Nature 183:688

Vermeij GJ, Leighton LR (2003) Does global diversity mean anything? Paleobiology 29(01):3

Villeger S, Novack-Gottshall PM, Mouillot D (2011) The multidimensionality of the niche reveals functional diversity changes in benthic marine biotas across geological time. Ecol Lett 14(6), 561

Veron J, Hoegh-Guldberg O, Lenton T, Lough J, Obura D, Pearce-Kelly P, Sheppard C, Spalding M, Stafford-Smith M, Rogers A (2009) The coral reef crisis: The critical importance of <350 ppm CO2. Mar Pollut Bull 58(10):1428

Walther GR, Post E, Convey P, Menzel A, Parmesan C, Beebee TJ, Fromentin JM, Hoegh-Guldberg O, Bairlein F (2002) Ecological responses to recent climate change. Nature 416(6879):389

Williams RJ (2010) Simple MaxEnt models explain food web degree distributions. Theor Ecol 3(1):45

Williams RJ, Martinez ND (2000) Simple rules yield complex food webs. Nature 404(6774):180
Williams RJ, Martinez ND (2004) Limits to trophic levels and omnivory in complex food webs: theory and data. Am Nat 163(3):458
Wood R (1998) The ecological evolution of reefs. Annu Rev Ecol Syst 29:179–206
Yodzis P (1984) Energy flow and the vertical structure of real ecosystems. Oecologia 65(1):86
Yodzis P (1989) Introduction to theoretical ecology. Harper & Row, New York

Exploring the Species–Area Relationship Within a Paleontological Context, and the Implications for Modern Conservation Biology

Matthew J. Pruden and Lindsey R. Leighton

Abstract Human-driven habitat loss and fragmentation is one of the largest threats to modern biodiversity. The relationship between species and area is one of the oldest patterns observed in ecology; models based on the species–area relationship (SAR) have been used to estimate both local and global extinction rates. A critical difficulty for modern conservation studies is that they are forced to predict the future; empirical testing of these predictions puts at risk the very species the predictions are designed to protect. To help mitigate this problem, we explore the use of the species–area relationship using the paleontological record, which provides empirical data from both before and after species loss. Using species presence/absence data, and a contiguous minimal area grid system, species–area curves (SACs) were constructed for one biofacies in each phase of a transgressive–regressive package. A regression observed for the Pliocene Etchegoin Formation of the San Joaquin Basin (California) inland sea, and the effect of sea-level fall on the perched molluscan fauna, was used as an analogy for modern habitat loss. Using the equation of the species–area curve of the transgressive interval, we predicted the number of species to go extinct in the subsequent regression. To examine extinction severity, we compared these predictions to the data observed for the regressive curve. The results showed that the regressive curve had fewer species than the prediction based on the transgressive curve, suggesting that species loss was more severe than predicted. The results demonstrate an example of how the species–area relationship can be used in the fossil record, and that the paleontological record could be used to provide relevant information for modern conservation problems.

Keywords California · Kettleman Hills · Pliocene · Sea level · Habitat loss · Extinction

---

M. J. Pruden (✉) · L. R. Leighton
Department of Earth and Atmospheric Sciences, University of Alberta, Edmonton, Alberta, Canada
e-mail: mpruden1@ualberta.ca

C. L. Tyler, C. L. Schneider (eds.), *Marine Conservation Paleobiology*, Topics in Geobiology 47, https://doi.org/10.1007/978-3-319-73795-9_7

1 Introduction

The Earth is on the brink of a sixth mass extinction, whereby species extinction rates are estimated to be 1000–10,000 times that of the background rate (Pimm et al. 1995; Millennium Ecosystem Assessment 2005), and the leading cause is human-driven habitat loss (Leakey and Lewin 1996; Millennium Ecosystem Assessment 2005; Barnosky et al. 2011). Preventing the loss of species due to human habitat destruction is one of the top priorities for conservation biologists. Ecologists have been building models to quantify and predict the extent of species loss for decades (MacArthur and Wilson 1967; Levins 1969; Richter-Dyn and Goel 1972; Strebel 1985). Ecologists and conservation biologists have frequently used species–area relationships (SARs) to predict the number of species at risk due to current habitat destruction (National Research Council 1980; Ehrlich and Ehrlich 1981; WCMC 1992). The SAR is a positive logarithmic relationship between species and area (Rosenzweig 1995). For example, Wilson (1989) used deforestation rates, in terms of area lost, of the world's rain forests to predict that 0.2–0.3% of the global species would go extinct per year, or approximately 125,000 species over a 10-year period. Thomas et al. (2004) used SARs to predict that 15–35% of species will be extinct by 2050. A more recent study predicted that 35% of tropical African species, 20% of tropical Asian species, and 15% of tropical American species are set to be extinct by 2050 (Millennium Ecosystem Assessment 2005). However, most modern ecological studies attempting to predict extinction are necessarily limited to models, which can only be tested empirically by the passage of time—at which point, vulnerable species may already have been lost. The goal of this study is to review the use of SARs by studying them in a paleontological context and to explore the use of the paleontological record in studying modern conservation models. This study constructs species–area curves (SACs) for marine molluscs from the Pliocene Kettleman Hills in California; and then uses the curves to compare predicted and observed extinctions due to habitat loss, in order to assess the relationship between species–area curves and extinction severity.

The relationship between species and area is one of the oldest ecological patterns, first described more than 100 years ago by Watson in 1859, or possibly even earlier in 1835 (Dony 1963; Williams 1964). The basic idea of the species–area relationship is that as area increases, so does the number of species (Rosenzweig 1995). Among proposed mechanisms for the SAR, Williams (1943) introduced the habitat diversity hypothesis; larger areas contain a greater array of habitats and therefore can support more species. The disturbance hypothesis claims that populations on smaller "islands" are more susceptible to disturbances because smaller areas are likely to have fewer refuges from the disturbances, suggesting that as area decreases, so should the number of species due to the increase in local extinction (Williams 1943). These explanations for the species–area relationship are similar to the well-known island biogeography model (MacArthur and Wilson 1967), island being a loose term used to describe geographically isolated areas that still allow for immigration and emigration, such as actual islands, lakes, mountain tops, forest fragments, and inland seas.

There are two fundamental (and mathematically equivalent) equations of SAR. The power-function (1) $S = cA^z$ is the equation of the logarithmic curve within arithmetic space, whereby S is the number of species, A is the area, c describes the slope of the graph, and z is a constant. More commonly, researchers compare the species–area relationship in log space; the equivalent equation is (2) $\log S = z \log A + \log c$, which is the equation of the straight line within a log-log space. In eq. (2), z describes the slope of the log-log relationship, and log c is the y-intercept (Rosenzweig 1995). Researchers have been trying to find the canonical *z* values, particular values with consistent biological meaning, for decades. Preston (1962) considered values between 0.17 and 0.33, while MacArthur and Wilson (1967) accepted values between 0.20 and 0.35, to fall within the canonical range. Connor and McCoy (1979) examined 100 SACs and found 45% of the *z* values fell between 0.20 and 0.40. There may be a relationship between island (or region) size and *z* values. Smaller islands have *z* values near 0.35, whereas the mainland tends to have *z* values near 0.15 (Connor and McCoy 2001). In addition, Sugihara (1980) suggested that the *z* value provides information on the processes behind the generation of the community structure.

Species-area data can be sampled using a contiguous gridded plot system. Contiguous grids are made up of fixed area cells, which share a border with an adjacent cell. As cells are added, species richness is recorded for progressively larger regions, while always including the previously recorded smaller subsets of area (Scheiner 2003). The species-area data are then used to construct the log-log equation and the *c* and *z* values are calculated. Once *c* and *z* are determined for a large area, it is possible to predict the number of species for any smaller area within the large area; this approach can be used to assess species loss owing to habitat loss.

Two concerns regarding the application of SARs are specific to modern ecological studies. Firstly, as suggested above, empirical data can be collected only after the extinction event has occurred, at which point vulnerable species have been lost. Secondly, in an analysis of the effect of habitat destruction on metapopulations of two species, Nee and May (1992) made two key observations: there was a time-delay on extinction due to habitat loss, and that the first species lost would be the best competitor. To explain these observations, Tilman et al. (1994) coined the term "extinction debt": a time-delayed, deterministic extinction of species due to habitat loss. In Tilman et al.'s model, after a habitat has been reduced, species start to go extinct in order from the best competitor to the poorest. Competitive and dispersal abilities were assumed to be trade-offs; thus, the weakest competitor would have the best dispersal, and would maintain its local population longer through immigration from remaining habitats (Tilman et al. 1994; Ovaskainen and Hanski 2001). However, the basic model also suggested that it may take 40–500 years before the extinction of the top competitor. Because it may take several generations for the species to go extinct, they are in "debt." If extinction debt does exist, then it is a major problem for modern conservation efforts—it would be difficult to convince policy-makers that extinctions based on models are reliable, and often impossible to provide the relevant empirical data within a single human lifetime. This study proposes a solution to this difficult conundrum: we aim to illustrate the use of SAR models in predicting extinction severity by using paleontological data. The paleontological record has extensive species presence/absence data from before, as well as after, an extinction event. This allows for the SAR model to be tested using empirical data without risking the loss of modern species. Also, the fossil record spans timescales of thousands to millions of years, which bypasses any lag period or extinction debt, so that the full extent of the extinction is evident.

While the idea of using the fossil record to provide needed empirical data for conservation is definitely appealing, it is also important that paleontological examples are meaningful. A major difference between past extinctions and the modern biodiversity crisis is that the primary cause of modern habitat loss is due to human expansion (Leakey and Lewin 1996). For example, a growing requirement for land has led to an increase in deforestation worldwide, with an estimated 1% of the world's rain forests being destroyed each year (Wilson 1989). Due to the majority of the world's species residing within the world's rain forests, the destruction of the rain forests has been used to predict the approximate global extinction rate due to habitat loss (Primack 2006). As the rain forest's (or any other habitat's) area is reduced or fragmented, the habitat maintains an edge and an interior environment. The edge is the transitional boundary between different landscapes, such as between a forest interior and a prairie. While the edge may become more extensive with progressive habitat loss or deterioration, the interior of the forest maintains a stable environment (Bannerman 1998). The pattern of extinction from this sort of progressive habitat loss may be very different from that of some of the major extinctions observed in the fossil record. Whereas many of the major mass extinctions may have been caused by widespread habitat deterioration (e.g., a bolide striking the planet), modern habit fragmentation due to human interference results in a decrease in habitat while some habitat still remains viable. In essence, there is a "moving front" of habitat loss and ultimately species loss. Conservation paleobiology thus needs to find appropriate analogs in the fossil record for this sort of habitat loss.

One such analog for modern habitat loss would be regressions observed in an inland sea connected to the open ocean. As the sea-level drops, the total area within the inland sea may decrease, and facies will migrate oceanwards. Even though the edge of the sea and related nearshore habitat may now comprise a greater proportion of the total remaining area, and so more of the region is vulnerable to disturbances and salinity fluctuations from freshwater input, the interior may continue to remain stable; previous facies may persist but in a new location. We propose to build species–area curves from the transgressive and regressive phases of a T-R cycle, as a means of assessing how habitat loss—in terms of loss of area—affects species richness.

An idealized transgressive–regressive package would provide a plausible analog for modern habitat loss, as well as provide an opportunity to construct species–area curves to examine related species loss. Unfortunately, not all transgressive–regressive (T-R) intervals are ideal. Although the idea that regressions in a basin will reduce area seems intuitive, in reality not all regressions are associated with area decrease. Holland (2012) used measurements of modern topography, and simulations of siliciclastic deposition (Holland and Christie 2013), to demonstrate that transgression and regressions may not necessarily correlate with habitat increase and decrease, respectively. Thus, if one wishes to use T-R cycles as an analog for habitat loss, then regardless of the environmental interpretation, it is essential to demonstrate that the habitable area has actually decreased through time.

In addition, a regression would alter the environment/facies at any one locality (to a shallower habitat), probably resulting in an associated change in the community. A perceived loss in diversity from the transgressive to regressive phase might simply be a function of habitat and community change, rather than being due to a loss in area. Comparisons of area and diversity between transgressions and regressions need to control for environment, which would usually entail sampling from the "new" geographic positions of the focal environment, i.e., the regressive samples would need to be not only stratigraphically higher but also lateral to the transgressive samples, in keeping with Walther's Law. By controlling for environment, any changes in diversity can be attributed to changes in area alone.

There has been previous work using SAR models in the paleontological record (e.g., Flessa and Sepkoski 1978; McKinney 1998; Barnosky et al. 2005). Although not strictly speaking an SAR study, Johnson (1974) examined the impact that regressions have had on marine faunal extinctions, in particular, the impact of regressions on the numerous Paleozoic epicontinental seas and their related perched faunas. Perched faunas are faunas living in epicontinental or other similar basins; in other words, they are "perched" on the continent (Johnson 1974). These faunas are thus restricted to a geographic location with very little immediate immigration or emigration in comparison with faunas on open coastlines. Johnson (1974) noted that such faunas were potentially vulnerable to regressions, particularly as shallow, gently sloped, basins would experience major habitat loss during a rapid drop in sea level. However, he focused on general trends at a global scale, spanning longer chronological time frames, usually millions of years, than those of today's rapid habitat loss. Valentine and Jablonksi 1991) studied the effect of sea-level change on Californian Pleistocene molluscs along the Californian coast. They calculated the probable size of the Pleistocene coast, and used predetermined, estimated z values to predict the effect of the fall in sea level during the Pleistocene, and compared their prediction with observed species diversity in the modern. They found that despite the sea-level fall and related loss in area, there was no significant change in species diversity, except they noted that "perched" faunas were more vulnerable.

The present study focuses on the Etchegoin Formation in the Kettleman Hills of the San Joaquin Basin, California, USA. The Kettleman Hills has an extensive molluscan fossil record, which was recorded by Woodring et al. (1940). The fossils represent a perched fauna that lived in a Pliocene inland sea and are phylogenetically and morphologically similar to today's molluscs. Stanton and Dodd (1970) remarked that the faunal lists by zone were representative of the stratigraphic assemblages and could be used for general faunal time trends. Stanley and Campbell (1981) suggested that Pliocene diversity on the open Pacific coast was largely unaffected by Pliocene cooling, but a study by Bowersox (2005), based largely on the Woodring et al. (1940) dataset, demonstrated that the San Joaquin Basin taxa experienced extinction rates nearly double that of the open Pacific. This result is consistent with Johnson's (1974) and Valentine and Jablonksi (1991) predictions that perched faunas would be more vulnerable to extinction, possibly as a consequence of habitat loss. The San Joaquin Basin inland sea experienced several transgressive–regressive cycles (T-R cycles) throughout the Pliocene. We focus on one specific T-R cycle within the upper Etchegoin Formation. The combination of richly fossiliferous units representing a perched fauna in an inland sea with a T-R cycle makes this setting highly suitable for a study of habitat loss, extinction, and the species–area relationship.

2 Geological Setting

The Kettleman Hills are located within the San Joaquin basin in California, USA. During the Neogene, the San Joaquin basin was an inland sea connected to the Pacific Ocean through a narrow strait west of the present day city of Coalinga (Fig. 1) (Stanton and Dodd 1970; Bowersox 2005). The Kettleman Hills include three northwest trending anticlinal domes, referred to as the North, Middle, and South Domes. There are four formations: the Jacalitos and Etchegoin Formations (Miocene-Lower Pliocene), San Joaquin Formation (Upper Pliocene), and Talure Formation (Pleistocene; Woodring et al. 1940). The present study focuses on the uppermost Etchegoin Formation. This part of the formation consists predominantly of marine deposits, primarily brown, muddy to clean sandstone, with local lenses of conglomerate and silt; and some freshwater beds near the top of the formation. This interval is estimated to be late Early Pliocene in age, approximately 4.2–4.0 mya (Bowersox 2005). Within this interval, Woodring et al. (1940) described five faunal zones based on the most abundant taxa: the *Patinopecten*, *Macoma*, *Siphonalia*, upper *Pseudocardium*, and *Littorina* zones. Subsequently, Adegoke (1969) included all of these five zones within his faunal zone 12. It should be noted that the stratigraphic nomenclature and even which beds are a part of the Etchegoin Formation has changed several times since its original description (Anderson 1905). Consequently, our focal interval, identified as uppermost Etchegoin by Loomis (1990) and Bowersox (2005), was considered lower Etchegoin in the earlier work by Woodring et al. (1940).

Each of Woodring et al.'s five zones experienced a local-scale T-R cycle (Stanton and Dodd 1970) within an overall regressive trend (Stanton and Dodd 1970; Bowersox 2005). As the basin was only open to the ocean through the Priest Valley Strait in the northwest during deposition of the Etchegoin Formation (Loomis 1990), facies migrated south (basinwards) during transgressions, and migrated north during regressions (Stanton and Dodd 1970; Perkins 1987; Bowersox 2005). Facies in the southern portion of the basin are typically non-marine; the basin in general became increasingly brackish and eventually non-marine through time (Stanton and Dodd 1970; Perkins 1987; Bowersox 2005).

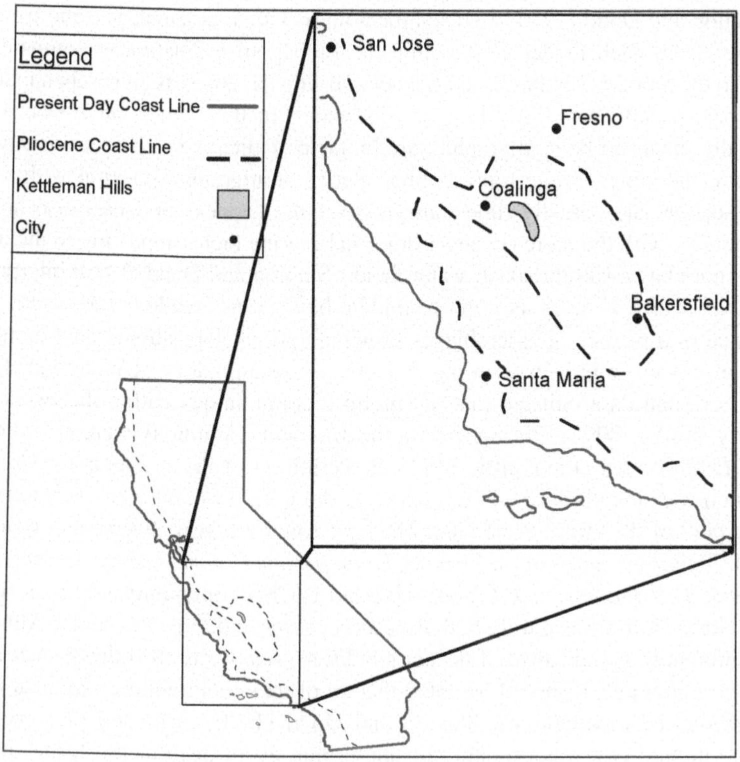

Fig. 1 Pliocene paleogeographic map of the San Joaquin basin California, modified from Stanton and Dodd (1970)

Peak molluscan diversity in the San Joaquin Basin in the Pliocene occurred in the three lower zones (*Patinopecten*, *Macoma*, *Siphonalia*) of the uppermost Etchegoin, when the northern part of the basin was still of normal-marine salinity (Bowersox 2005). These units and associated facies are best exposed in a roughly 200-meter thick interval on and adjacent to the north dome of the Kettleman Hills (Bowersox 2005). The Kettleman Hills do not constitute the entirety of the San Joaquin Basin but they do account for the vast majority of localities in the uppermost Etchegoin, and lateral migration of facies in response to local sea-level change is evident.

Woodring et al. (1940) named the *Patinopecten* zone after the circular pectin found throughout much of the zone. The rocks are predominantly brownish sand silt and silty sand. There are some outcrops of sandstone and conglomerate with scour discontinuity at the base of the overlying *Macoma* zone. This horizon has been used to separate the *Patinopecten* from the *Macoma* zones, but the sandstone conglomerate is not recognized in all areas where the Etchegoin Formation outcrops, and sedimentation appears continuous between the two zones over much of the region. The *Macoma* zone consists of a brown to light-grey tuffaceous sandstone and was originally defined by the abundance of the bent-nose clam, *Macoma*, and the absence of *Patinopecten*.

Stanton and Dodd (1970) used sedimentology, faunal assemblages, and isotopic analyses from each faunal zone within the Etchegoin Formation to compare the zones to the modern San Francisco Bay, classifying the zones as either open marine, outer bay, middle bay, inner bay, or freshwater. In this sense, the zones, while originally meant to be stratigraphic, are biofacies indicative of paleoenvironment. As such, the zones, while representing relative stratigraphic position within any given section, are actually time transgressive: the "zones" or facies also appear horizontally, with the more oceanward/normal-marine facies appearing to the north of the more basinward/brackish-water facies. Stanton and Dodd (1970) interpreted the *Patinopecten* zone as an outer-to-middle bay setting, while the *Macoma* zone consisted of a mosaic of assemblages showing various depositional environments, but primarily in a middle bay setting. Nonetheless, examination of Woodring et al.'s (1940) original data indicate that the mollusc communities within the two zones are very similar; 80% of the species in the *Macoma* community were also present in the *Patinopecten* community. While the definition of the stratigraphic *Macoma* zone indicates the absence of *Patinopecten*, the outer-to-middle bay communities during both of the *Patinopecten* and *Macoma* zones are very diverse and typically include high abundances of the bivalves *Patinopecten*, *Chione*, and to a lesser extent, *Macoma*. This *Patinopecten-Chione-Macoma* (PCM) community was widespread in the North Dome area during both the *Patinopecten* and *Macoma* zones. Although this community is indicative of deeper conditions relative to the other San Joaquin Basin communities, it should be noted that all of these communities are nearshore, shallow-marine communities. Stanton and Dodd (1970) estimated that even the outer-to-middle bay was probably no more than 50 m deep at most, and it may have been less than 15 m deep (Bowersox 2005).

Stanton and Dodd (1970) interpreted the oldest zone, the *Patinopecten*, to be transgressive, with the middle bay community moving southwards (Stanton and Dodd 1970; Perkins 1987); subsequently, the middle bay community moved northward during the more regressive *Macoma* zone. Thus, the two lowest zones of Woodring et al. (1940) in the uppermost Etchegoin Formation contain highly fossiliferous, well-preserved, and diverse samples from a community/biofacies that migrated laterally with local transgressions and regressions. It is plausible that the area occupied by the community also varied with the transgression and regression, and is therefore an appropriate setting to examine the relationship between species, area, and extinction.

3 Methods

We used Pliocene fossil data and outcrop localities within the Etchegoin Formation from Woodring et al.'s (1940) geological survey of the Kettleman Hills (Fig. 1). Using the public land survey system (divides the land into a grid system) and distances from the section lines provided by Woodring et al. (1940), the localities were plotted onto 1:24,000 scale topographic maps of the North Dome. The plots were transferred onto a digital topographic map of the area using the image editing software GIMP 2.0. The physical map was used first in order to keep the scale constant while plotting the localities. Contiguous minimal area cells of 1 km^2 were used to calculate the total area represented by the localities. Contiguous minimal area grids permit the calculation of the number of species as the geographic area increases. Contiguous grids are a series of fixed area cells, whereby each cell shares a common border with the adjacent cell (Scheiner 2003; for examples, see Carey et al. 2006; Rybicki and Hanski 2013). The term "minimal" refers to using as few square cells as possible to measure each area of interest. As the number of cells increases, the contiguous area increases; for each larger area, the preceding area is nested within. The number of species for each progressively larger area was recorded, and the information was used to construct species–area curves by graphing the number of species versus area. Log-log plots were used to determine the equations for the SAR curves and the c and z values. The process was performed for each of the two lower faunal zones in the uppermost Etchegoin Formation.

The strength of the SAR was assessed by calculating the coefficient of determination (denoted R^2) of each graph. R^2 signifies how much of the variation in the y-axis (number of species) is influenced by the x-axis (area). The R^2 value is expressed as a number between 0 and 1; the greater the number, the stronger the relationship. Using the equations defining the SAR of the transgressive *Patinopecten* zone, we predicted the number of species to go locally extinct due to the loss in habitat caused by the subsequent regression in the *Macoma* zone. Then we compared the expected values for the reduced area with those actually observed in the regression. We examined the extinction severity by comparing how well the regression's curve matches the transgression's curve. If the regression's curve generally matches the transgressive curve, and the predicted and observed paleodiversities are similar, that would suggest the SAR model accurately predicted the extinction. If the regression's curve is above the transgression's curve (its observed paleodiversity is greater at equivalent areas), then this suggests that the extinction was less severe than predicted, or that there may be an extinction debt. Finally, if the regression's curve is lower than the transgression's curve (its observed paleodiversity is less than predicted), then the extinction was more severe (Fig. 2). As the paired transgressive and regressive curves do not necessarily have observed data at equivalent areas, we compared the observed species counts at the areas examined on the regressive curve with the predicted species counts from the equivalent areas on the transgressive curve.

4 Results

The total area for the regressive-phase *Macoma* zone was less than that of the transgressive-phase *Patinopecten* zone, thus validating an examination of species loss relative to habitat loss. Both of the species–area curves showed logarithmic relationships, with very large R^2 values and significant slopes (Table 1). The c and z

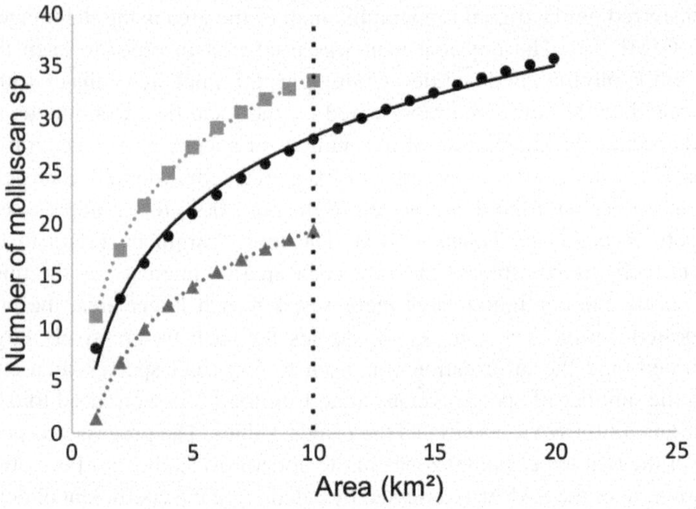

Fig. 2 A hypothetic SAC. The middle line represents a standard logarithmic SAC (transgressive curve). The top line (squares) represents a scenario of an extinction debt, and the bottom line (triangles) represents a scenario in which the extinction is more severe than predicted (regressive curves). The vertical dotted line represents the extent of area lost

Table 1 R^2, c, and z values of the Etchegoin Formations faunal zones

Faunal zone	R^2 value	c value	z value
Patinopecten	0.941	0.175	1.176
Macoma	0.818	0.269	1.632

values for the transgressive and regressive SAR curves are presented in Table 1. The slope of the regressive *Macoma* zone curve was less than that of the transgressive *Patinopecten* zone curve, and the observed species richness at equivalent areas were much less than those predicted using the SAC, not only at the maximum area recorded for the regression, but also for smaller areas as well (Fig. 3).

5 Discussion

The large R^2 values signify that there is a strong logarithmic relationship between the number of species present and area size. This result demonstrates that SAR curves using paleontological data are similar to modern examples of such curves. The observed paleodiversity for the regression's species–area curve was much less than the diversity predicted using the equation defined from the preceding transgressive zone. This suggests that the local extinction in the Pliocene inland sea during this point in Etchegoin time was quite severe. While the long-term regression, closing of the basin, and a related decrease in salinity eventually led to a larger extinction at the end of the Etchegoin (Bowersox 2005), the local loss in species

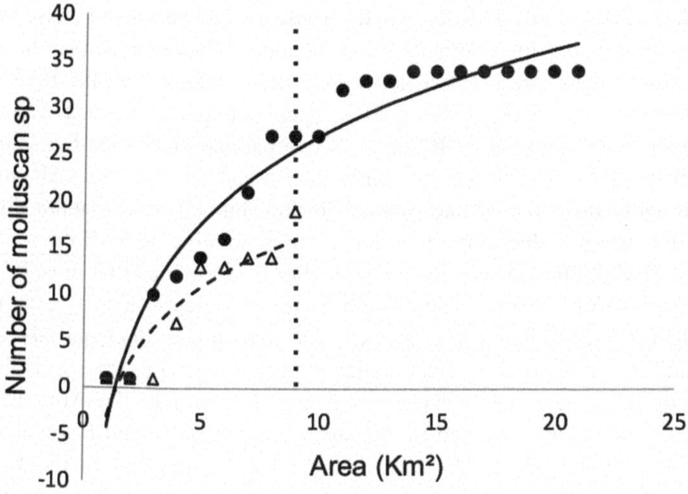

Fig. 3 SACs of the *Patinopecten* zone (transgression, black circles, solid line) and the *Macoma* zone (regression, hollow triangles, dashed line). The vertical dotted line represents the extent of area lost. Note, the graphs were plotted in arithmetic space for clarity, but the c and z values were calculated from log-log space

during *Macoma* zone time detected herein was based on observations from one community, the PCM community, which was interpreted as living in an outer-to-middle bay, normal-marine, environment. This environment migrated northward but changes in salinity or depth probably were not the culprit. Climate cooling might still have affected this system but the decrease in habitat area may also have contributed to species loss. During the subsequent transgressive *Siphonalia* zone, the PCM community and associated outer-to-middle bay environment re-expanded southward, and species richness again increased, only to be lost completely for the remainder of Etchegoin deposition during the severe regression that followed the *Siphonalia* zone. This pattern is consistent with a species-area model. That said, as the transgressive curve serves as a prediction—a null model—the more severe than expected loss of species indicates that habitat loss alone was insufficient to explain the species loss. What is causing the difference? We consider three possibilities.

He and Hubbell (2011) examined the use of SARs in predicting extinctions by comparing them to endemic–area relationships (EAR). Endemic species exist only in a single geographical area (although the size of this area varies among species); in other words, the species' habitat is a single, continuous area (Green and Ostling 2003). The EAR predicts a species will go extinct only when its total single geographic habitat is destroyed, while the SAR assumes all species are randomly distributed, and does not distinguish between endemic and widespread species. If the species are randomly distributed, SAR and EAR curves will be the same. However, most species are non-randomly distributed: many species are gregarious and form "clumped" distributions (Sasaki 1997; Leighton and Schneider 2004). He and Hubbell (2011) built SAR and EAR curves for 279 species of passerine birds from the United States of America's lower 48 states. They noticed that the z values from the SAR curves were much larger than the z values from the EAR curves, i.e., the slopes of the SAR curves were steeper and species richness values were smaller than those from the EAR curves at comparable areas (See Fig. 2 from He and Hubbell 2011). So for non-randomly distributed species, the SAR and EAR curves cannot be used to infer one another. He and Hubbell argue that this difference between the curves is due to the fact that when generating an SAR curve, the area required to find the first species is much smaller than the area that must be lost for the last species to go extinct. Therefore, it requires a greater loss of area to lose a species than it does to add a new species. The SAR model does not take this into account and assumes the area effect on the addition or loss of a species is the same. Based on this argument, SAR curves will always overestimate the extinction debt.

However, the results obtained in this study contradict He and Hubbell's (2011) claim. This may be due to the fact that the local extinction in the Kettleman Hills occurred over an extended period of time, almost certainly longer than any extinction lag effect (i.e., long enough to exceed any extinction debt). However, this process does not explain the sampling error He and Hubbell mentioned, that SARs do not account for clumped distributions of species and instead assumes that distributions are random. This concern may be overstated. First, He and Hubbell (2011) argued that SAR curves will always be steeper than EAR curves, due to the clumped nature of species distributions. The steepness of the SAR slope will project a lower diversity than that of the EAR for any given area when area is small, thus exaggerating the projected loss of species if habitat was reduced to that area. Yet, as the species counts for larger regions (equivalent to gamma diversity) of SAR curves will be the same as those of EAR curves, the slope of an SAR curve, relative to the related EAR curve, is largely determined by the diversity within the first few sample plots, i.e., the smallest plots. If the species distribution is extremely clumped, and the first sample is taken at a point of lowest possible alpha diversity, then the SAR curve may be very steep, and will likely overestimate extinction debt. However, if the first sample is taken at a point of maximum alpha diversity, within a concentration of clumps, the SAR and EAR curves may look very similar, if not identical. The SAR curve will have a gentler slope and would estimate extinction more accurately. He and Hubbell (2011) have thus identified an important but easily fixable concern; workers building SAR curves must use care in choosing those samples representing the smallest areas on the curve.

An additional factor mitigating the He and Hubbell (2011) concern, for paleontological or historical studies in particular, is the process of time-averaging. Just as a function of the normal dispersal of juveniles, the spatial distribution of species from one generation to the next is not identical, varying even (or especially) on smaller scales. Most rock units, beds, or even unlithified sediments are time-averaged, in that even though the depositional environment of the bed did not change significantly, the fauna preserved within the bed represent multiple time-slices, and therefore multiple communities and multiple generations and spatial distributions of species (Olszewski 1999). This effect may be enhanced further by potential post-depositional reworking, and temporal mixing of sediments in some beds. The rock or sediment thus represents an average of time but given the spatial variation of species among generations, the rock or sediments will typically accumulate diversity through multiple time-slices. Time-averaging is sometimes a major concern because over a sufficiently long enough period of time, there is an increasing risk of mixing taxa that never actually lived together. However, such conditions are usually readily identifiable by the beds' stratigraphic context (Kidwell and Holland 2002) and by the faunal assemblages within them (e.g., taxa with equivalent preservation potential are preserved differently in the same bed). In many cases, including the use of SAR curves, time-averaging is a benefit to paleontologist and ecologist alike because it reduces noise (Walker and Bambach 1971; Olszewski 1999), including very localized spatial variation. A time-averaged sample will be less vulnerable to clumping of species because such spatial variation is softened by the averaging of this variation through time within the bed (Leighton and Schneider 2004).

For many fossil beds, the diversity within a time-averaged bed, at least for taxa such as molluscs that have good preservation potential, is potentially the same or greater than if one did a count on a modern plot (or a single time-slice). Studies by Kidwell and Tomašových (2009, 2013) comparing live versus dead molluscan assemblages have demonstrated that the fidelity of species counts for a given assemblage is quite strong, and that dead assemblages have higher alpha diversity than live assemblages. Tyler and Kowalewski (2016), in a similar live versus dead study, also noted that many dead samples had greater diversity than the live counterpart, presumably due to the time-averaging effect described above. So when collecting species-area data, the diversity from the first, smallest, sample is often greater than that of a modern sample, or single time-slice, from the equivalent area. As Kidwell and Tomašových (2009) noted, the beta diversity of a time-slice is effectively being converted to alpha diversity of the paleontological sample. However, eventually as area increases, this effect becomes diminished. When sampling the bed from multiple localities, there will be an area at which maximum diversity is reached, and the curve will flatten out, just as it does in the modern. At these larger scales, species clumping is irrelevant to diversity counts of both modern and fossil samples. If the diversities of the larger areas are the same for both fossil and modern curves, and the diversities of the smaller areas used in the fossil curves are greater (due to time-averaging), then the slope of the fossil curve becomes shallower, essentially negating He and Hubbell's problem. Several other studies have argued against He and Hubbell's claim that SAR's will always overestimate extinctions (Brooks 2011; Fattorini and Borges 2012; Pereira et al. 2012; Thomas and Williamson 2012), but in any event, this problem seems to be less of a concern for a paleontological study, and overestimation of the extinction clearly was not the problem for the present study in the Etchegoin Formation.

Holland (2013) studied the effect of habitat loss on the Pleistocene extinctions on the California coast using numerical simulations to measure relaxation times, the time required for a population to reach equilibrium after an extinction event or change in environment (Diamond 1972). One equilibrium state is when a population equals zero, so there is a strong potential relationship between equilibrium and lag times, or extinction debt. Based on these simulations, Holland suggested several reasons why predicted and observed patterns of species loss may differ. One reason is species longevity: If a species has a low mortality rate, or has a high likelihood of colonizing vacated sites, that species could persist during times of decreased habitat area, which could increase relaxation time. The inverse is true as well; if a species has a high mortality rate or is a poor colonizer, relaxation time could decrease. Variation in speciation rates also may influence relaxation time; after habitat had been reduced, an increase in speciation would lengthen the time required for the biodiversity to fall to equilibrium (Holland 2013), while similarly, a decrease in speciation might shorten the time. These factors could potentially have played a role in the difference between the Etchegoin's predicted and observed species loss.

One additional possible explanation for the apparently more severe local loss of species during the *Macoma* zone is that the SAR model does not directly account for biological interactions, and the effect of the loss of one species on other species. There are reports outlining ecosystem collapses in the modern, sometimes caused by the extinction of keystone species, a species crucial to maintaining diversity and stability in its ecosystem (Paine 1966, 1969; Woodruff 1989; Rohlf 1991; Mills et al. 1993). Typical examples of such keystone species include top predators (Paine 1969) and biofacilitators, such as reef-forming corals (Bruno et al. 2003). An extinction cascade occurs when the extinction of one species leads to the extinction of one or more species, until eventually the ecosystem reaches a tipping point in which too many species have been lost, in turn leading to a trophic level collapse (Ricklefs and Miller 1999). There is still debate as to whether increased food web complexity increases or decreases stability. Initial models (May 1973; Pimm 1979; Plotnick and McKinney 1993) suggested that more complex food webs were more vulnerable to collapse, precisely because of the strong interconnectivity and interdependence among species. More recently, both model (Roopnarine 2006) and empirical (Fox and Olsen 2000; Leighton and Schneider 2008) studies have suggested that high diversity within a given trophic level might stabilize the trophic web and provide a buffer against an extinction cascade. If a single species is lost, its competitors, which are likely to be functionally similar, will compensate for the loss by filling the relevant niche. Regardless of the effects, both good and bad, of increasing connectivity among species in a foodweb, there is some threshold of species loss at which an extinction cascade occurs (Roopnarine 2006). While it is well beyond the scope of the present study to solve this problem, the greater than expected species loss observed in the present study could be an example of an ecosystem that exceeded the threshold of species loss necessary to trigger an extinction cascade. If this is the case, then the loss of habitat in the *Macoma* zone may have been a trigger, rather than the immediate cause. The combination of using SAR curves to predict severity with a detailed examination of which species (including possible keystones) are lost through time is a promising avenue for future research.

The results presented herein are not meant to be the definitive answer to diversity and extinction patterns in this interval of the uppermost Etchegoin Formation of the Kettleman Hills. Rather, the goal was simply to demonstrate that an SAR-based approach could be applied to the fossil record and to show that such research could potentially provide information that is relevant to modern conservation problems. Furthermore, predictions from species–area curves could potentially serve as null hypotheses; if species loss was significantly greater or less than that predicted, then we need to understand why. An important point of the present study is that although the results do not support nor disclaim the existence of extinction debt (nor could they, given the time frame of the study), there is still a relationship between habitat loss and species extinction. Even if there is a lag period, and it may take 40–500 years for extinction to occur, the species affected are still doomed to die, if there is no intervention. This raises an ugly problem for conservationists: convincing policy-makers to consider truly long-term solutions to conservation problems. Hopefully, more paleontological studies showing long-term effects of habitat loss will continue to provide much-needed empirical data. We also need to consider the real possibility that just as modern climate change and habitat loss are accelerating at rates that are exceptionally fast when compared to past rates, the unfortunate result may be that what we have previously considered to be long-term negative consequences on ecosystems may instead occur during a single human generation. For example, the rate of species extinction has already increased substantially in just a few decades (Wilson 2003).

The purpose of this study was to provide a worked example to review the use of the SAR model by studying it in a paleontological context, rather than a definitive case study. As such there are some caveats that need to be addressed for future research. Firstly, this was a meta-analysis using data originally collected for mapping and structural interpretation, rather than for a detailed paleoecological study. Naturally, future research should focus on collecting fossil data specifically for the purpose of ecological studies using SAR curves. Secondly, this study used only one T-R package containing two faunal zones, as these units were the only such zones in the section that were immediately stratigraphically adjacent, i.e., there were no major temporal gaps within the transgressive–regressive cycle, and they were the only such zones that captured the same community/biofacies. While the results are of interest, one result provides only one example. Further research from other times and regions is needed to determine if the extinction severity patterns found in this study are unique only to those zones, or if there is a more general pattern throughout the paleontological record. We hope that this study encourages other workers to explore these possibilities.

6 Conclusion

1. A local extinction within the Californian Pliocene inland sea was significantly more severe than predicted using species–area curves. These results contradict an earlier study which claims that the species–area relationship will always overestimate extinctions.

2. If extinction debt does exist, and it takes 40–500 years for the extinction event to begin, it would be practically impossible to collect empirical data to confirm extinction debt within a single human lifetime. The paleontological record spans a coarser timescale of thousands to millions of years, essentially bypassing any possible lag period, and so provides an opportunity to test extinction severity and related models with empirical data.
3. Modern ecological studies are necessarily limited to using predictive models of species extinction, which can only be empirically tested through the passage of time, at which point species may be irrevocably lost. The paleontological record provides species presence/absence data from before as well as after an extinction event, which allows for the models to be tested with empirical data without the loss of modern species.
4. Many extinctions studied in the paleontological record were caused by catastrophic events, such as a bolide impact or large scale volcanic activity, which lead to widespread habitat deterioration. Modern human-driven habitat loss is different and acts as a "moving front," whereby habitat decreases while the remaining habitat maintains a stable environment. Thus, there needs to be a more appropriate analog in the fossil record. Regressions affecting perched faunas may be an excellent analog for modern habitat loss, and further exploration of such systems in the fossil record has considerable potential to test SAR models and modern extinction predictions.

Acknowledgments The authors would like to thank Steve Holland and Alycia Stigall for their thorough and extremely thoughtful reviews, which greatly improved the manuscript. We thank Kristina Barclay for reading an earlier draft of the manuscript. We also wish to thank Carrie Tyler and Chris Schneider for inviting this contribution. Funding for this research included an NSERC-USRA award to the lead author, and an NSERC Discovery grant to the second author.

References

Adegoke OS (1969) Stratigraphy and paleontology of the marine Neogene formations of the Coalinga Region, California. University of California, Publications in Geological Sciences, Berkeley. 80, p 241

Anderson FM (1905) A stratigraphic study in the Mount Diablo region of California. Proceedings of the California Academy of Sciences, third Series 2:155–248

Bannerman S (1998) Biodiversity and interior habitats: the need to minimize edge effects. Extension note/British Columbia. Ministry of Forests, Research Program, Victoria

Barnosky AD, Carrasco MA, Davis EB (2005) The impact of the species-area relationship on estimates of paleodiversity. PLoS Biol 3:1356–1361

Barnosky AD, Matzke N, Tomiya S et al (2011) Has the Earth's sixth mass extinction already arrived? Nature 471:51–57

Bowersox JR (2005) Reassessment of extinction patterns of Pliocene molluscs from California and environmental forcing of extinction in the San Joaquin Basin. Palaeogeogr Palaeoclimatol Palaeoecol 221:55–82

Brooks TM (2011) Extinctions: consider all species. Nature 474:284

Bruno JF, Stachowitz JJ, Bertness MD (2003) Inclusion of facilitation into ecological theory. Trends Ecol Evol 18:119–125

Carey S, Harte J, del Moral R (2006) Effect of community assembly and primary succession on the species-area relationship and disturbed ecosystems. Ecography 29:866–872

Connor EF, McCoy ED (1979) The statistics and biology of the species-area relationship. Am Nat 113:791–833

Connor EF, McCoy ED (2001) Species-area relationships. In: Levin SA (ed) Encyclopedia of biodiversity, 5 p 397–412

Diamond JM (1972) Biogeographic kinetics: estimation of relaxation times for avifaunas of southwest Pacific islands. Proc Natl Acad Sci USA 69(11):3199–3203

Dony JG (1963) The expectation of plant records from prescribed areas. Watsonia 5:377–385

Ehrlich PR, Ehrlich AH (1981) Extinction: the causes and consequences of the disappearance of species. Random House, New York

Fattorini S, Borges PAV (2012) Species–area relationships underestimate extinction rates. Acta Oecol 40:27–30

Flessa KW, Sepkoski JJ (1978) On the relationship between Phanerozoic diversity and changes in habitable area. Paleobiology 4:359–366

Fox JW, Olsen E (2000) Food web structure and the strength of transient indirect effects. Oikos 90:219–226

Green JL, Ostling A (2003) Endemics-area relationships: the influence of species dominance and spatial aggregation. Ecology 84:3090–3097

He F, Hubbell SP (2011) Species-area relationships always overestimate extinction rates from habitat loss. Nature 473:368–371

Holland SM (2012) Sea level change and the area of shallow-marine habitat: implications for marine biodiversity. Paleobiology 38:205–217

Holland SM (2013) Relaxation time and the problem of the Pleistocene. Diversity 5(2):276–292

Holland SM, Christie M (2013) Changes in area of shallow siliciclastic marine habitat in response to sediment deposition: implications for onshore-offshore paleobiologic patterns. Paleobiology 39:511–524

Johnson JG (1974) Extinction of perched faunas. Geology 2(10):479–482

Kidwell SM, Holland SM (2002) The quality of the fossil record: implications for evolutionary analyses. Annu Rev Ecol Syst 33:561–588

Kidwell SM, Tomašových A (2009) Fidelity of variation in species composition and diversity portioning death assemblages: time-averaging transfers diversity from beta to alpha levels. Paleobiology 34(1):94–118

Kidwell SM, Tomašových A (2013) Implications of time-averaged death assemblages for ecology and conservation biology. Trends Ecol Evol 44:539–563

Leakey R, Lewin R (1996) The sixth mass extinction: patterns of life and the future of humankind. Doubleday, New York

Leighton LR, Schneider CL (2004) Neighbor proximity analysis, a technique for assessing spatial patterns in the fossil record. PALAIOS 19(4):396–407

Leighton LR, Schneider CL (2008) Taxon characteristics that promote survivorship through the Permian-Triassic interval: transition from the Paleozoic to the Mesozoic brachiopod fauna. Paleobiology 34(1):65–79

Levins R (1969) Some demographic and genetic consequences of environmental heterogeneity for biological control. Bull Entomol Soc Am 15:237–240

Loomis KB (1990) Depositional environments and sedimentary history of the Etchegoin formation, west-central San Joaquin Valley, California. In: Kuespert JG, Reid SA (eds) Structure, stratigraphy and hydrocarbon occurrences of the San Joaquin Basin, California. Society of economic Paleontologists and Mineralogists and American Association of Petroleum Geologists Pacific Sections, book 64 and guidebook 65, p 231–246

MacArthur RH, Wilson EO (1967) The theory of island biogeography. Princeton University Press, Princeton

May RM (1973) Stability and complexity in model ecosystems. Princeton University Press, Princeton
McKinney ML (1998) On predicting biotic homogenization: species-area patterns in marine biota. Glob Ecol Biogeogr 7(3):2897–2301
Millennium Ecosystem Assessment (2005) Ecosystems and human well-being: biodiversity synthesis. World Resources Institute, Washington
Mills SL, Soulé ME, Doak DF (1993) The keystone-species concept in ecology and conservation. Bioscience 43:219–224
National Research Council (1980) Research priorities in tropical biology. National Academy of Sciences, Washington
Nee S, May RM (1992) Dynamics of metapopulations: habitat destruction and competitive coexistence. J Anim Ecol 61:37–40
Olszewski T (1999) Taking advantage of time-averaging. Paleobiology 25:226–238
Ovaskainen O, Hanski I (2001) Spatially structured metapopulation models: global and local assessment of metapopulation capacity. Theor Popul Biol 60(4):281–302
Paine RT (1966) Food web complexity and species diversity. Am Nat 100:65–75
Paine RT (1969) A note on trophic complexity and community stability. Am Nat 103(929):91–93
Pereira HM, Borda-de-Água L, Martins IS (2012) Geometry and scale in species-area relationships. Nature 482:E4–E5
Perkins JA (1987) Provenance of the Upper Miocene and Pliocene Etchegoin Formation: Implications for Paleogeography of the Late Miocene of Central California. United States Geological Survey Open-File Report, p 87–167
Pimm SL (1979) Complexity and stability: another look at MacArthur's original hypothesis. Oikos 33:351–357
Pimm SL, Russell GJ, Gittleman JL et al (1995) The future of biodiversity. Science 269:347–350
Plotnick RE, McKinney ML (1993) Ecosystem organization and extinction dynamics. PALAIOS 8:202–212
Preston FW (1962) The canonical distribution of commonness and rarity. Ecology 43:185–215
Primack RB (2006) Essentials of conservation biology, 4th edn. Sinauer Associates, Sunderland
Richter-Dyn N, Goel NS (1972) On the extinction of a colonizing species. Theor Popul Biol 3(4):406–433
Ricklefs RE, Miller GL (1999) Ecology, 4th edn. W.H. Freeman and Company, New York
Rohlf DJ (1991) Six biological reasons why the endangered species act doesn't work and what to do about it. Conserv Biol 5:273–282
Roopnarine PD (2006) Extinction cascades and catastrophe in ancient food webs. Paleobiology 32:1–19
Rosenzweig ML (1995) Species diversity in space and time. Cambridge University Press, Cambridge
Rybicki J, Hanski I (2013) Species-area relationships and extinctions caused by habitat loss and fragmentation. Ecol Lett 16:27–38
Sasaki A (1997) Clumped distribution by neighborhood competition. J Theor Biol 186(4):415–430
Scheiner SM (2003) Six types of species-area curves. Global Ecol and Biogeogr 12:441–447
Stanley SM, Campbell LD (1981) Neogene mass extinction of Western Atlantic molluscs. Nature 293:457–459
Stanton RJ Jr, Dodd JR (1970) Paleoecological techniques: comparison of faunal and geochemical analyses of Pliocene paleoenvironments, Kettleman Hills, California. J Paleontol 44:1092–1211
Strebel DE (1985) Environmental fluctuations and extinction single species. Theor Popul Biol 27:1–26
Sugihara G (1980) Minimal community structure: an explanation of species abundance patterns. Am Nat 116:770–787
Thomas CD, Cameron A, Green RE et al (2004) Extinction risk from climate change. Nature 427:145–148
Thomas CD, Williamson M (2012) Extinction and climate change. Nature 482:E4–E5

Tilman D, May RM, Lehman CL, Nowak MA (1994) Habitat destruction and the extinction debt. Nature 371:65–66

Tyler CL, Kowalewski M (2016) Quantifying the fidelity of the fossil record across multiple higher taxa in nearshore marine macro-invertebrate assemblages (North Carolina USA). Geol Soc America Abstr Prog 48(7)

Valentine JW, Jablonksi D (1991) Biotic effects of sea level change: the Pleistocene test. J Geophys Res 96:6873–6878

Walker KR, Bambach RK (1971) The significance of fossil assemblages from fine-grained sediments: time-averaged communities. Geol Soc America Abstr Prog 3:783–784

WCMC (1992) Global biodiversity: status of the Earth's living resources. WCMC

Williams CB (1943) Area and number of species. Nature 152:264–267

Williams CB (1964) Patterns in the balance of nature. Academic Press, London

Wilson EO (1989) Threats to biodiversity. Sci Am 261:108–116

Wilson EO (2003) The future of life, 1st edn. Random House, New York

Woodring WP, Stewart R, Richards RW (1940) Geology of the Kettleman Hills oilfield, California; stratigraphy, paleontology, and structure. U. S. Geological Survey, Prof. Paper 195 p 170

Woodruff DS (1989) The problems of conserving genes and species. In: Western D, Pearl MC (eds) Conservation for the twenty-first century. Oxford University Press, New York, pp 76–88

Marine Refugia Past, Present, and Future: Lessons from Ancient Geologic Crises for Modern Marine Ecosystem Conservation

Chris L. Schneider

Abstract Refugia are one means of species survivorship during a global crisis. As the Earth is facing a major crisis in the marine biosphere, the study of refugia through past extinctions and other global crises is relevant to creating and maintaining effective marine reserves (including marine protected areas and other formally established havens for conservation). A synthesis of previous studies identifies the following properties common to most definitions of a refugium: (1) During a global crisis, a species can persist in a refugium, which can include a range shift, habitat shift, or migration or contraction to an isolated geographic area. Subsets of isolated geographic refugia include life history refugia (areas necessary for breeding), cryptic refugia (small areas, must remain connected for populations to remain viable), and harvest refugia (defined from the modern literature to escape overfishing pressure). (2) In the refugium, the habitat may remain stressed but is sufficiently habitable for the species to maintain sufficient albeit small populations (relative to pre-crisis population size) over many generations. (3) After the crisis ends, the species emerges from the refugium and expands during the recovery interval. Otherwise, the refugium will become a refugial trap in which the species remains a relict population or ultimately becomes extinct.

The present understanding of refugia from the geologic past comes from three sources, namely fossil data, phylogeographic reconstructions, and species distribution models, the latter two being more common for studies across the last glacial maximum. The synthesis herein suggests several important factors when considering the future of marine reserves. Because climate change is an ongoing process, the present refugia of marine reserves may not be sufficient for the future survival of marine species. Short-term refugia of some present marine reserves may deteriorate because of further climate change and have to be abandoned for new long-term options as new habitats become available. Cryptic refugia of small reserves must remain connected in terms of species' dispersal and exchange, but

C. L. Schneider (✉)
Department of Earth and Atmospheric Sciences, University of Alberta, Edmonton, Alberta, Canada
e-mail: clschnei@ualberta.ca

C. L. Tyler, C. L. Schneider (eds.), *Marine Conservation Paleobiology*,
Topics in Geobiology 47, https://doi.org/10.1007/978-3-319-73795-9_8

must also be flexible, in that cryptic refugia naturally are sometimes ephemeral because of habitat heterogeneity through time. Finally, habitats in marine reserves must be of sufficiently low stress to maintain viable populations, but should frequently be re-evaluated to avoid becoming refugial traps in the future.

Keywords Extinction survivorship · Climate change · Range shift · Habitat shift · Refuge · Species distribution · Conservation paleobiology

1 Introduction

Through all of the global biotic crises in Earth's past, life has found a way to survive, recover, and ultimately, flourish once again. One means of survival through past global crises has been the retreat of taxa into refugia, which are areas where species, lineages, and ecosystems persisted, sheltered from the fatal environmental devastation that culminated in a global mass extinction. Because extinction-causing stresses are present in modern oceans, refugia may again become an important factor in the survival of marine taxa. And because human intervention may be necessary for establishing refugia—in the form of marine reserves, herein including marine protected areas and other places formally established for conservation purposes—modern conservation efforts may be strengthened through a better understanding of why and how ancient refugia were successful through global crises.

Accepting that the Earth's biosphere is facing a mass extinction rivaling those of the geologic past, it is critical that if ecosystems are to survive—and be able to provide the support and services needed by the global human population—then actions to protect ecosystems need to be taken immediately and especially with forethought to the future. In the face of impending climate change that impacts ocean water in terms of warming, oxygenation, acidification, circulation, and other factors, marine reserves as they currently exist may be insufficient to protect systems from a multitude of stressors. The design and establishment of future marine reserves, along with the management of current systems, must take into account future habitat changes while providing refugia in a deteriorating ocean environment.

Models of future changes are hypotheses, not certainties; basing future marine reserves on any one model could result in failure (Makino et al. 2015). The answer to which areas might remain habitable becomes an uncertain prediction, especially without a baseline reference of what constitutes a successful refugium. Meanwhile, the application of the refugial concept to management and design of marine reserves is not a novel idea; increasingly, research is focusing on the possibility that marine reserves can be viable refugia. Although this body of work is producing critical information for conservation efforts, it often lacks perspectives that can only be provided by the fossilized past, specifically how a refugium facilitated survival through a biosphere crisis, from its inception, and very importantly, through its persistence to its successful end.

Although the extensiveness, and even value, of ancient refugia is debated (e.g., refugia create relict populations; Rickards and Wright 2002), refugia through ancient mass extinctions have been shown to exist and contribute to global biosphere recovery (e.g., pre-recovery graptolite diversification in Late Ordovician refugia; Xu et al. 2005). Since the defining work of Vermeij (1986), the understanding of refugia has greatly increased with the recovery of additional fossil and geological evidence, along with a rapidly expanding body of phylogeographic studies across the last glacial maximum and the incipience of species distribution models in paleoecological niche reconstructions. Furthermore, investigations into ancient refugia increasingly emphasize the need to understand why those refugia were successful, sometimes also identifying factors relevant to modern marine biosphere survival. Therefore, a synthesis of investigations into refugia is timely, emphasizing the contributions from the studies of ancient refugia to modern conservation and the prospect for future refugial investigation.

The terms "refuge" and "refugium" are ubiquitous across ecology and paleoecology. Aside from a means of survivorship through environmental crises and mass extinctions, the terms have also been used for ecological phenomena, such as predation refuges or diurnal refuges. Although these are appropriate uses of the term "refuge" and also ensure survivorship, I focus below on the mechanism that allows populations to survive through global environmental perturbations, through multiple generations, and usually, over geologic time scales. Here, I use the Latinized "refugium" (pl., "refugia") to indicate the strategy of extinction survivorship.

2 Defining Refugium

Refugia, as identified across scientific study, are fairly diverse in definitions and types. In general, during an environmental crisis, a species contracts its range to a smaller geographic area, shifts its range to a new region, or takes advantage of a new, but unperturbed, habitat. Then, a species spends time in the refugium, where it can evolve genomes that define an endemic population or, given time, speciate into one or more new taxa, sometimes adapting irrevocably to the specific conditions of the refugium. The refugial interval of survival can last from several generations to millions of years, depending on the severity and duration of the global crisis. Finally, after environmental recovery in extra-refugial habitats, the species expands out of its refugium. Thus, the refugial survival strategy has three stages: entry into the refugium with the onset of environmental perturbation; time spent in the refugium while the global environmental crisis continues elsewhere; and re-expansion out of the refugium with the bettering of adverse environmental conditions.

Other types of refugia, such as physiological or temporal refugia, have been defined in the modern marine literature and depend neither on habitat nor on location. These are outside the scope of this review as they pertain to individuals or single generations. Below, I use the perspective of a single hypothetical species in order to expound on what ideally happens during its time in a refugium. In the survival of a species, the consensus of most definitions of refugia considers the following aspects:

A Species Must Have a Range Contraction, Range Shift, or Migration in Order to Escape the Onset of Global Environmental Degradation That Would Otherwise Cause Extinction of That Species

One general consensus for defining a refugium is that a species must have altered its range in order to escape adverse environmental conditions. A refugium is a place—either isolated and restricted in size (such as the Iberian Peninsula refugium at the last glacial maximum for species that are presently more northerly) or a shift in habitat (such as a deep water refugium to avoid lethal conditions in shallow water) or geographic location (such as a southward shift of North Atlantic temperate species during glacial maxima). The refugial area is generally understood to be smaller in size than that of the original species' range (Maggs et al. 2008; Ashcroft 2010; Zonneveld et al. 2010a). Furthermore, these refugia are areas within species' ranges that remain habitable ("in situ" refugia; Ashcroft 2010; Gavin et al. 2014; "internal" refugia; Shoo et al. 2013) or are acceptable habitats outside of the original range to which species migrate ("ex situ" refugia; Ashcroft 2010; Gavin et al. 2014; "external" refugia; Shoo et al. 2013). Although it may seem obvious, a refugium must be accessible to the species that needs it—the location cannot be a refugium, no matter how pristine the environment, if organisms cannot gain entrance.

Ideally, to identify a refugium, multiple species should inhabit the area in question, each of which has gone through a range contraction into their refugial populations (Vermeij 1986). As a result of the intense study of demographics of species across the last glacial maximum, many of the same refugia have been identified for species of vastly different life modes and trophic levels. For instance, North Atlantic refugia in the English Channel and Iberian Peninsula coast have been identified for macroalgae (Hoarau et al. 2007; Neiva et al. 2014; Assis et al. 2016), invertebrates (Jolly et al. 2006; Campo et al. 2010; Albaina et al. 2012), and fishes (Chevolot et al. 2006; Almada et al. 2012; Mateus et al. 2016). In the ancient fossil record, faunas within refugia through mass extinctions were often more diverse than extra-refugial faunas (e.g., Ordovician brachiopods; Rassmussen and Harper 2011).

One factor in determining which species has a greater chance of refugial survival may be original range size. A species with a broad distribution before an extinction-causing event may be more likely to survive (Kiessling and Aberhan 2007), at least in the short term in the geologic sense (Clapham et al. 2013). In terms of refugial survival, a large range could increase the chances of refugia forming within the species' distribution. However, other studies found no relationship between range size and survivorship, indicating that other factors may be more important in determining survivorship (e.g., Permo-Triassic brachiopods; Leighton and Schneider 2008; Cretaceous molluscs; Myers et al. 2013).

Refugia, as identified in the literature, comprise several different strategies: range and habitat shifts, isolated geographic refugia, cryptic refugia, and life history refugia. In addition, modern anthropogenic pressures on marine species have also created a type of refugium unique to this interval of geologic history, the harvest refugium. Certainly, species may rely on more than one type of refugium for survival through a biotic crisis, and hybrid refugial types may occur. Likewise, marine reserves may be targeted as a specific type of refugium, or may serve as multiple types of refugia, depending on the species escaping the environmental crisis. The studies covered in this synthesis are summarized in tables in the appendix, and the types of refugia are briefly reviewed below.

Range Shifts

The range shift of species is one of the earliest hypothesized methods of survival during a global environmental crisis. For example, Packard (1886) recognized the southward shift of North American species away from the continental ice sheet and used the term "refugia" to denote places where species survived. Darwin (1859) suggested an ecosystem-wide survival mechanism during the ice age, in that species from the same habitat shifted their ranges together with changing climate, thus keeping their communities intact.

In a range shift, changes in the range of a species parallel the changes in the environment. Throughout the range shift, the species maintains its preferred habitat (Fig. 1a, b). A shift in range can be necessary with warming or cooling ocean temperatures or other climate-related changes. For instance, if sea level drops, the species moves with sea level, following its perfect depth range on the reduced shelf area (e.g., contraction to the shelf edge during sea level lowstands of glacial maxima; Holterhoff 1996). Or, if climate warms, a species tracks that depth range and habitat to higher latitudes as the optimal temperature moves away from the equators (e.g., thermal refugia in higher latitudes for shallow-water tropical corals during the last interglacial of Australia; Greenstein and Pandolfi 2008).

As with any refugial strategy, a species escapes lethal conditions; otherwise, it would remain in its original range. A range shift by a species is a series of responses by individuals, such that there is no single, coordinated mass exodus into a refugium during a crisis (Bennett and Provan 2008). Thus, the range shift is a gradual process, with organisms propagating along the habitable leading edge of their ranges and increasing die-off along the trailing edge. In other words, while environmental deterioration occurs in part of a species range, newly opened habitat along another margin of its range allows for local population expansion. Thus, the geographic centroid of the population moves through space and time, tracking its optimal habitat during environmental change. However, this simplified model does not consider potential novel stresses in the newly opened habitats, such as the lack of suitable rocky substrate and increased thermal stress for rockweed with warming climate (Hoarau et al. 2007) or a decrease in the aragonite saturation state as tropical corals shift away from equatorial warming (Descombes et al. 2015), which could alter the process of the range shift.

Range shifts during times of global stress generally coincide with demographic contractions, compared to expansion of ranges in more amenable intervals. For example, as climate cooled during glaciation, temperate and tropical taxa shifted

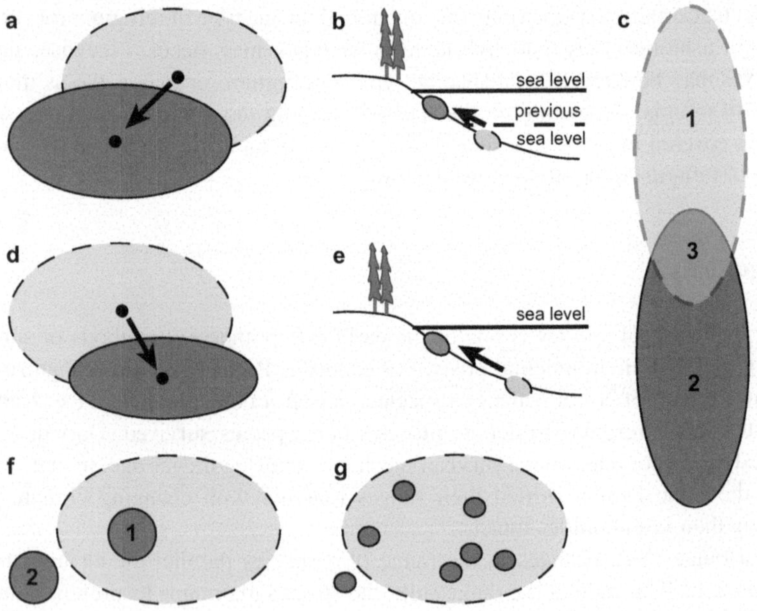

Fig. 1 Types of refugia. (**a**) Range shift: As the population moves, the centroid of the population distribution shifts. (**b**) Range shift: The preferred habitat is maintained; in this example, with sea level rise, the population moves upslope, following the migration of its preferred habitat. (**c**) Range shift in a fluctuating climate, as in a temperate species during glacial–interglacial cycles: (1) range during one climate regime, such as the interglacial range of a northern temperate species; (2) range during the other climate regime, such as the glacial range of a northern temperate species that has shifted south; (3) the "climatic refugium" of Neiva et al. (2014), a stable area of demographic overlap between the two range extremes, where the environment is always habitable and is always occupied by the species. (**d**) Habitat shift: As the population moves, the centroid of the population distribution shifts. (**e**) Habitat shift: The preferred habitat is abandoned for a new habitat; in this example, sea level is static and the population moves upslope, abandoning the deeper water habitat. (**f**) Isolated geographic refugia: the population (1) contracts to one or more smaller areas within its original range or (2) migrates to one or more areas outside of its original range. (**g**) Cryptic refugia: The population either reduces to rare individuals, or reduces to small populated pockets; both types can be within or outside of the original range, and both types constitute a connected metapopulation

toward the equator, often into smaller populations (Ashcroft 2010). Meanwhile, the colder global climate was optimal for northern polar species, which expanded their ranges southward (Almada et al. 2012). Conversely, interglacials are times of thermal stress for polar species. Their present ranges are expected to contract even further poleward with warming climate (Stewart et al. 2010; Ashcroft 2010).

Range shifts between interglacial and glacial periods can result in overlapping areas of potentially continuous habitation (Fig. 1c). Neiva et al. (2014) referred to such areas as "climate refugia" and recognized this type of refugium for the intertidal seaweed *Pelvetia canaliculata* between the Iberian Peninsula and Ireland. Similar types of long-term stable populations across glacial–interglacial changes may occur in long-lived reefs of the Indo-Pacific region (Vermeij 1986). In these types of refugia, although the geographic area is maintained, many species will still need to migrate upslope or downslope with changing sea level.

In general, marine reserves have not yet been established specifically as refugia for species that will shift their ranges because of climate change effects on the ocean environment. Range shift refugia in the form of marine reserves would have to anticipate future demographics of species. For example, populations of tropical species would expand away from the tropics in a warming ocean. Marine reserves currently located in temperate zones and within potential expansion areas may serve as future refugia for tropical species. Likewise, areas similar to the climate refugia of Neiva et al. (2014) may also be adequate to accommodate range shifts, in a location where the marine reserve would overlap the present-day poleward edge of a tropical species' range plus its potential future expansion into currently temperate habitats. Furthermore, a warming global climate will be accompanied by sea level rise, so species that currently inhabit shallow water (e.g., zooxanthellate corals, rocky intertidal algae) will follow their preferred ranges as they move upslope, in addition to any latitudinal shifts in distribution.

Habitat Shifts

Similar to range shifts, during a habitat shift, a taxon disappears from part or all of its range, but instead of moving to a new geographic location *and* maintaining its preferred habitat, a species invades a new habitat (with or without a range shift) (Fig. 1d, e). Although not ideal, the new habitat is survivable compared to the catastrophe in the species' old habitat. Most commonly, habitat shifts in the marine realm involve a change in water depth to find new, tolerable conditions.

Shifts to deeper water habitats are a common means of escaping adverse conditions in shallow water and have been documented across geologic time. Several Late Cambrian trilobite clades survived in a deeper water environment when shallow-water habitats became lethal (Westrop 1989). During hypoxic events, the goniatite *Kokenia* sp. likely survived in deeper marine habitats during the Early Devonian (House 1996). Algeo et al. (2010) reported a deep water refugium for radiolarians across the Late Permian mass extinction when the oxygen minimum zone extended over continental shelves. A move to deep water and proximity to upwelling has been proposed as a survival mechanism for Early Jurassic reefs to escape global warming (Kiessling et al. 2009). Slope through abyssal habitats were once thought to be the main havens for Antarctic shelf species during glacial maxima (Barnes et al. 2006) although this idea is now in contention because of the discovery of cryptic species complexes replacing what were previously thought to be relict populations from prior refugia (Allcock and Strugnell 2012).

Throughout the Phanerozoic, shallow-water refugia were frequently exploited during times of increased oxygen stress in deep water. Particularly relevant during times of global ocean stratification, habitat shifts to shallow water became a critical survival strategy. Some examples of the move to shallow-water refugia to escape adverse conditions in deeper environments include Ordovician brachiopods (Huang et al. 2013), Devonian ostracods (Lethiers and Cassier 1999), and Cretaceous hydrothermal vent faunas (Jacobs and Lindenberg 1998).

A special type of shallow-water refugium during the Early Triassic—the "habitable zone"—was proposed by Beatty et al. (2008, see also Zonneveld et al. 2010a, b, Clapham et al. 2013). In this case, the most habitable area was capped near the base of breaking fair-weather waves, where water was sufficiently oxygenated, and where wave energy and turbidity were tolerable (Beatty et al. 2008). The base of the habitable zone extended to the upper limits of anoxia, where storm wave influence intermittently oxygenated the water. Wide shelves (Zonneveld et al. 2010a) and embayments therein (Beatty et al. 2008) protected the habitable zone refugium from the upwelling of deep, anoxic water.

Other types of habitat shifts include movement into radically new environments. As environmental conditions deteriorate, those species that are more flexible in their environmental tolerance, or are more physiologically plastic, can take advantage of unexploited habitat. For instance, amphipods moved into brackish water glacial lakes of the last glacial maximum along the European coast and from there expanded into freshwater rivers (Krebes et al. 2010). Open marine benthos moved into a frequently perturbed delta system in a humid region to escape fatal warming and oxygen stress of Late Permian and Early Triassic oceans, as recorded in rocks from this interval in France (Gall and Grauvogel-Stamm 2005).

Like range shifts, marine reserves do not currently target future habitat shifts of species. Some species may have experienced habitat shifts in response to adverse conditions (e.g., cold-water kelp survival near the bottom of their photic range in the Galapagos, Graham et al. 2007) or already overlap a marginally acceptable habitat within their present range, such as the potential for preferred survival of mesophotic coral populations in a warming ocean that could disperse to shallow-water populations (i.e., "deep reef refugia," Bongaerts et al. 2010). These examples may suggest the types of habitat shift refugia for future consideration.

Isolated Geographic Refugia

Contraction of ranges to smaller, isolated geographic areas is the most commonly cited refugial strategy in both empirical and theoretical studies. In this type of refugium, a taxon is restricted to one or more locations, each of which is much smaller than the original range (Fig. 1f). During the last glacial maximum, many spatially restricted refugia have been identified in all parts of the shallow ocean, from peri-glacial refugia in the North Atlantic (e.g., Maggs et al. 2008), tropical refugia in the Indo-Pacific (e.g., Vermeij 1986), and refugia in unglaciated areas along the Antarctic shelf (e.g., Graham and Smith 2012). Geographic refugia through ancient mass extinctions tend to be larger regions than those of the last glacial maximum, possibly in part because of the coarse geographic and taxonomic scale of many studies of the ancient past (e.g., regional survivorship of brachiopods across the Late Ordovician mass extinction, Rassmussen and Harper 2011).

Isolated geographic refugia exist primarily because of the inherent heterogeneity of nature. Global deterioration that causes a major crisis in the biosphere tends to be variable across space and time. For instance, spatio-temporal variability in oxygen stress of Permo-Triassic oceans resulted in some areas remaining habitable, while others were fully anoxic (Algeo et al. 2010). Glacial drawdowns of sea level caused major shifts in habitat, such as the reduction of rocky shorelines to a few areas (e.g., North Atlantic refugia of areas along the coast of Ireland, the Hurd Deep, and the Iberian coastline, among others; Maggs et al. 2008) and population isolation resulting from barriers created by a drop in sea level (e.g., Sunda Shelf; Ludt and Rocha 2015).

Isolated geographic refugia sometimes seem to arise from a "luck of the draw." When environmental conditions deteriorate within a species' range, and if one or more areas remain amenable within the range, then those subsets of the original population have the chance of survival "when the world closes in" (Rickards and Wright 2002, p. 3). For instance, Antarctic glaciers at the last glacial maximum reached the shelf edge, save for small, sometimes ephemeral refugia on the outer shelf (Barnes and Kulinski 2010). Organisms that happened to occupy these isolated, unglaciated refugia survived.

Lucky survivors in refugia do not necessarily preclude other taxa from migrating into these open spaces, but certainly competition for resources (particularly food and space) in a geographically restricted refugium could be very tight. For example, incumbent survivors have the advantage of being already established in, and adapted to, the refugium, as interpreted for Patagonian molluscs in nearshore deltas following the Cretaceous-Paleogene mass extinction (Aberhan and Kiessling 2014).

One contentious location of isolated geographic refugia is that of sea mounts. Sea mounts, submerged volcanoes, and other conical islands may increase habitable area in a deteriorating ocean system, particularly in protection from acidification (Jacobs and Lindenberg 1998; Rowden et al. 2010; Tittensor et al. 2010), deep water anoxia (Galil and Zibrowius 1998; Rowden et al. 2010), and global eustatic regressions (Jablonski and Flessa 1986). Sea mounts may have provided refugia to Jurassic corals (Stanley and Beauvais 1994) although corals may have survived better elsewhere (Kiessling et al. 2009). Based on ostracods across the Paleocene-Eocene Thermal Maximum, Yamaguchi and Norris (2015) found that extinction could be greater and recovery longer on these geographically isolated areas than on continents, mainly because of the lengthy "waiting time" for dispersal from distal refugia. Jablonski and Flessa (1986) pointed out that sea mounts and newly exposed islands could act as refugia when sea level fell below continental shelf edges. Conversely, because of the rising sea level predicted for the future, drowned islands, sea mounts, and volcanoes may decrease the availability of these types of isolated geographic refugia.

Most marine reserves are isolated geographic refugia. They are spatially, and sometimes depth, defined, and the species therein acquire partial to full protection from anthropogenic activities. Two recently expanded or established, large marine protected areas include the US Papahānaumokuākea Marine National Monument, which is over 935,000 km^2 of habitat protected from commercial fishing (Obama

2014; The White House 2016) and the 1.55 million km^2 Ross Sea Region Marine Protected Area, which includes no-take and restricted use areas (U.S. Department of State 2016). The smallest reserves are under 1 km^2 and have been shown to have positive effects on fishery stocks (Roberts and Hawkins 1997).

Life History Refugia

Life history refugia refer to those that are necessary only during one portion of a species' life cycle, usually during breeding or for nursery habitats. These types of refugia are mainly used by mobile organisms that can move into the refugium during the necessary interval, and then leave when the habitat is no longer needed. In the fossil record, particularly in demographic reconstructions of last glacial maximum refugia, these are a subset of isolated geographic refugia, in that they are locales that are briefly used, either once (as nurseries) or intermittently (as rookeries).

Breeding refugia are critical for marine animals that need to rear their young on land. Seals, sea lions, marine birds, and sea turtles used breeding refugia through the last glacial maximum (Harlin-Cognato et al. 2006; Pinsky et al. 2010; Naro-Maciel et al. 2014; Younger et al. 2016). Red king crabs, which are a deep water taxon little affected by last glacial maximum climate and sea level changes during their adult stage, needed shallow-water nursery refugia for juveniles, some which were located near a few unglaciated coastal islands of North America and along northeastern Asia (Grant and Cheng 2012).

A shift to a refugium may also result from intolerance of extra-refugial conditions during a life stage. For coccolithophores across the Paleocene-Eocene Thermal Maximum, fatal conditions in the tropics caused them to shift to temperate waters, possibly to avoid stress during one life stage (Gibbs et al. 2016).

Other species may survive in an environmentally poor refugium during a life stage, particularly as spores, propagules, or seeds that could remain in stasis for some period of time (Ashcroft 2010). A dinoflagellate taxon is currently surviving as cysts in the refugium of the warm Indo-Pacific, awaiting an increase in sea surface temperatures greater than that of a normal interglacial period (Mertens et al. 2014).

Many marine protected areas dually serve as isolated geographic and life history refugia. For example, several reserves in New Zealand are key breeding habitats for marine mammals and birds and thus, serve as breeding refugia, such as the Pohatu and Auckland Islands, Motu Maha Marine Reserves (New Zealand Department of Conservation 2016). Protection of breeding habitat is critical for migratory marine animals, such as whales. Along some of the Hawaiian coastlines, the Hawaiian Islands Humpback Whale National Marine Sanctuary was established to protect the overwintering habitat and breeding of humpback whales (National Ocean Service 2016). Likewise, many of the reserves along the west coast of Mexico and southern California also protect overwintering and breeding whales, but also protect other migratory and far-ranging species (Commission for Environmental Cooperation 2011).

Life history refugia in marine reserves often result from protecting the entire ecosystem. For example, marine reserves along the California coast host high populations of the red sea urchin, under whose spines juvenile red abalones find protection from predators (Rogers-Bennett and Pearse 2001). Similarly, a marine reserve for king scallops became more effective with the increase in algal cover, which increased habitat complexity and served as a nursery for juvenile scallops (Howarth et al. 2015).

Cryptic Refugia

Cryptic refugia are populations that comprise rare and sparsely scattered individuals that are still able to breed (Stewart and Lister 2001) or are a subset of isolated geographic refugia in which micro-refugial populations that are not reproductively viable unto themselves are connected as meta-populations (Bennett and Provan 2008) (Fig. 1g). These types of refugia are most commonly defined for terrestrial species (Stewart and Lister 2001; Bennett and Provan 2008), but have applicability to marine refugia. Cryptic refugia are likely difficult to locate in the fossil record (Gavin et al. 2014).

A possible special case of cryptic refugia in the fossil record is that of disaster taxa. Disaster taxa are those that proliferate during and immediately following mass extinctions; where other taxa are in great distress, these taxa thrive (Schubert and Bottjer 1995; Kauffman and Harries 1996). With post-extinction environmental recovery, these taxa are again suppressed into rarity within ecosystems or find a refugium in high-stress habitats (Schubert and Bottjer 1995; Kauffman and Harries 1996). Thus, "classic" disaster taxa—such as the brachiopod *Lingularia* (Schubert and Bottjer 1995; Rodland and Bottjer 2001; although see a re-assessment of Early Triassic *Lingularia*, Zonneveld et al. 2010a, b)—existed in cryptic refugia and other refugia, awaiting times of global environmental catastrophe to once again flourish.

A marine reserve in which one or more species cannot successfully maintain its population is a cryptic refugium for that species. The importance of networks that connect multiple reserves cannot be understated, as most marine reserves are too small to maintain viable population levels and genetic diversity of many species therein. Connectivity inherently occurs outside of reserves, because propagules, larvae, juveniles, and migrants must pass from one reserve to another, which can include passing through openly exploited areas (Steneck et al. 2009). While within-reserve benefits increase with the size of the protected area, connectivity between reserves can be improved by maximizing the number of habitats captured by multiple reserve areas and by spacing reserves at appropriate distances for population exchange of target species, which is often over tens to hundreds of kilometers (Almany et al. 2009). Additionally, a network of marine reserves is insurance against the collapse and loss of one or two protected habitats, which Almany et al. (2009) referred to as "risk spreading."

Another type of potential cryptic refugia is the "artificial marine micro-reserve," defined by Garcia-Gomez et al. (2011, 2014). These are small, commercial, and urban infrastructures that provide areas of colonization. An example they provided is that of the ferruginous limpet, *Patella ferruginea* that is the most endangered invertebrate of the Mediterranean. Although it is rare in most locations, it is highly abundant on man-made structures in the Strait of Gibraltar. Similarly, Inger et al. (2009) suggested that, with careful planning to negate harmful environmental impacts, renewable energy infrastructures built in the ocean may provide "de-facto" marine reserves by providing localized harvest refugia.

Harvest Refugia

Essentially havens from anthropogenic fishing pressures, harvest refugia protect some populations of exploited taxa from potential extinction caused by overfishing. In particular, these refugia commonly shelter source populations that disperse to fished populations as well as maintain the refugial population (e.g., an increase in *Anadara* sp. clam populations outside of the marine reserve; Gell and Roberts 2003) or serve as insurance against fatal depletion of harvested taxa (e.g., increase in reef fish sizes and populations in a no-take marine reserve near South Africa; Maggs et al. 2013). Although harvest refugia have been studied only on the order of years to decades so far, these refugia have considerable importance for the long-term future of fisheries and harvested species.

Before marine reserves, areas that were inaccessible to fishermen remained unexploited and thus, were natural harvest refugia that supported fished populations (Johannes 1978; Gell and Roberts 2003). However, with constant and exponential improvement in resource-gathering technology, naturally occurring harvest refugia are becoming rarer, as areas previously difficult to access become accessible (Gell and Roberts 2003). So, artificially created harvest reserves in the form of marine reserves take on the huge role of supporting populations both within and outside of reserves (Baskett and Barnett 2016).

Most marine reserves are active harvest refugia in that they protect a population of one or more exploited species. Inherent to the harvest refugium of the marine reserve is the requirement not only to maintain intra-reserve population levels, but also to support extra-reserve fisheries and other needs of the local habitat. For example, the San Juan and Upright Channels Sea Cucumber and Sea Urchin Commercial Harvest Exclusion Zones ensure that a dense population of sea urchins, including large individuals, is present to enhance recruitment and breeding, both within the reserve boundaries and in populations outside of the reserves (Washington Department of Fish and Wildlife 2016). Because sea urchins have an impact on kelp forests and sea otter populations, as well as being a commercial resource, population maintenance became a necessary focus in this region. Likewise, many marine reserves in New Zealand are critical harvest refugia, in which species disperse to extra-refugial areas and sometimes into baited traps and fishing lines outside the boundaries of the reserves, thereby supporting local fisheries and recreational fishing (Ballantine 2014).

Studies have repeatedly shown that with age, the marine reserve becomes a more efficient harvest refugium. In the marine reserves of the Philippines, those that had been in place for 10 years are more maintain high yields for harvested fish species which had positive spillover effects (i.e., recruitment of fish into the extra-reserve habitat). A similar study in marine reserves of Europe also found a correlation between the increase of biomass and abundance in marine reserves that were older than 10 years, among other factors (Edgar et al. 2014). Benefits of restocking harvested species outside of refugia can occur in as little as 6 years, but timing varies with trophic level of the target fish (da Silva et al. 2015). Coleman et al. (2015) found that changes in targeted species' abundance and diversity took 5–20 years to manifest, and indirect effects on ecosystems took even longer. Thus, time and continuous stability of protection for target species are important for creating harvest refugia.

The Environmental Conditions of a Refugium Are Sufficiently Habitable Such That the Species' Population Remains Viable During Its Time in the Refugium

Note that this portion of the definition of refugium contains the qualification "sufficiently habitable"; conditions in the refugium need not be pristine, but merely sufficient so that the population can reproduce and recruit. This means that an ideal refugium must have a reduced, yet bearable, level of stress within its boundaries and be isolated from the fatal conditions of the extra-refugial environment. Environmental stress can be countered by beneficial factors typical of local habitats (e.g., large amplitude internal waves providing intermittent cooling of reefs in otherwise fatally warm temperatures; Buerger et al. 2015; Wall et al. 2015; Schmidt et al. 2016), or by the addition of one or more ameliorating factors that boost resistance to stress (e.g., increased nutrients; Vermeij 1986; Graham et al. 2007).

Many ancient refugia were not stress free. The Early Triassic habitable zone and similar environments were stressed by low oxygen and high temperatures in addition to turbidity and shifting sediment (Beatty et al. 2008; Zonneveld et al. 2010a; Clapham et al. 2013; Chen et al. 2015; Song et al. 2015; Zhao et al. 2015). Also across the Late Permian mass extinction, ostracods and other taxa had sufficient oxygen and food within microbial mats, which helped them to survive heat stress in the refugial habitat (Forel et al. 2013; Posenato 2009). The Triassic deltaic cryptic refugia preserved in France lacked the lethal anoxia of the open ocean, but were frequently perturbed by shifting sediment and other stresses inherent to deltaic environments (Gall and Grauvogel-Stamm 2005).

The populations of a species existing on the margin of its range distribution, which usually inhabit more stressful conditions than the main demographic range, may be essentially pre-adapted to life within a refugium (Kauffman and Harries 1996; Rickards and Wright 2002). Refugial habitats are rarely, if ever, matched to species' optimal environmental preferences. Thus, individuals on the margins of a geographic range already tolerate less than optimal conditions, and thus might survive better within refugia that are somewhat stressed.

Likewise, marine reserves are not necessarily stress free. In a changing ocean, conditions that are not particularly stressful for some species can preclude others that would otherwise be present (Ashcroft, 2010). Thus, if marine reserves, or reserve networks, capture a range of habitats, more species are likely be protected (Ballantine and Langlois 2008; Almany et al. 2009; Ballantine 2014). Alternatively, marine reserves established within the climate refugia of Neiva et al. (2014) may protect species that have already adapted to, and persisted through, glacial–interglacial climatic and eustatic fluctuations, and thus may be tolerant to major climate change (Gavin et al. 2014).

In fact, the presence of non-lethal levels of stress in the habitat may be beneficial for some species. Organisms that are adapted to climatically stable habitats are likely more susceptible to perturbations (Ashcroft 2010), but those that are exposed to survivable levels of stress (e.g., zooxanthellate corals and sub-lethally but potentially bleaching warm water masses) and are given sufficient time to recover may adapt to stressful conditions, and thus be positive contributors to species conservation (Chollett et al. 2014). Similarly, species or ecosystems that display resilience and a rapid recovery between perturbations would be appropriate locations for the refugia of marine reserves (Berger et al. 2011).

A Species' Population Is Smaller in the Refugium Than Its Pre-environmental Perturbation Size

A reduced population is rarely an explicit component of any definition of refugium, but usually is assumed or inferred. While in the reduced geographic area of refugia, or in the less optimal habitat such as a shift to deeper or shallower environments, resources and space may be limited or more difficult to acquire. Refugial conditions therefore cause populations to adjust for the limiting conditions of the refugium.

Explicit data on population size in fossil communities is nearly impossible, but studies of demographics through the last glacial maximum show that many populations of intertidal species went through declines. Available habitats were limited by a series of factors: few peri-glacial ice-free zones (e.g., peri-glacial refugia in the North Atlantic; Jolly et al. 2006; Maggs et al. 2008; Almada et al. 2012; Krakau et al. 2012), isolated rocky shorelines at the glacio-eustatic lowstand (e.g., Hurd Deep in the English Channel; Provan et al. 2005; Hoarau et al. 2007; Campo et al. 2010; Doellman et al. 2011; Li et al. 2015), and sufficient nurseries for juveniles (e.g., islands along western North America; Harlin-Cognato et al. 2006; Pinsky et al. 2010; Grant and Cheng 2012) are three common constraints for last glacial maximum intertidal refugia. Within the limited areas of North

Atlantic refugia, populations were constrained to be smaller than their pre-refugial counterparts.

Likewise, with a drop in sea level, shelf area is reduced and species are squeezed into much smaller areas. Although no data exist for population sizes, the constrained areas of refugia around the Antarctic continent would suggest that taxa sequestered in glacial refugia had smaller populations. Similarly, with the glaciation of coastlines, rookery space became constrained; Younger et al. (2016) suggested that the populations of some sea mammals and birds were reduced to bottlenecks of a few thousand breeding females.

A possible alternative to reducing population size is to reduce the size of individuals, or the Lilliput effect (Urbanek 1993; also see Harries and Knorr 2009 for alternate explanations). When resources are decreased and other environmental factors are less than favorable, a reduction in body size compensates for the limitations of the habitat. Thus, a species can maintain a viable population via an overall reduction of body size, if conditions in the refugium cannot otherwise support the species. A body size reduction requires sufficient time in the refugium for the species to adapt in this manner to limited resources. For example, Early Triassic trace fossils were smaller than their pre-extinction and post-recovery counterparts (Beatty et al. 2008; Knaust 2010; Zonneveld et al. 2010a). Although these traces are not body fossils, the smaller size of the traces suggests the Lilliput effect on the animal creating the trace.

As mentioned above for harvest refugia, target species rebound in abundance and biomass with time after the establishment of a marine reserve, especially in areas given high levels of protection or that are designated as no-take. However, recovered species in marine reserves must not only maintain the population inside the reserve, they must also recruit and contribute to the exploited or disturbed population outside of the reserve boundaries.

The Species Remains in the Refugium for Many Generations

Taxa inhabited refugia through the durations of global crises, which are highly variable in duration, depending on magnitude and type of crisis. Time spans range from a few thousand years across the last glacial maximum to several million years during prolonged anoxia and warm climate of the Early Triassic. Thus, for any given species, a multitude of generations passed inside the haven of the refugium. The longer a species inhabited in a refugium, the greater the likelihood of adaptation to the unique habitat of the refugium. With sufficient time, a species can evolve and even diversify within the refugium (Harries et al. 1996). Likewise, time spent in a refugium can lead to evolutionary consequences such as genetic drift, founder effects, and population bottlenecks.

Population bottlenecks in refugia are best understood from the last glacial maximum because of the ability to reconstruct demographic properties (Bennett and Provan 2008; Stewart et al. 2010; Allcock and Strugnell 2012). Bottlenecks caused founder effects in refugial populations, the results of which are still seen in molecular studies of phylogeography across the last glacial maximum (Ludt and Rocha 2015). For instance, extreme reductions in population size and genetic diversity have been reconstructed from molecular data for the intertidal rockweed *Palmaria palmata* along Atlantic Canada (Li et al. 2015). In another example, Antarctic shelf populations of many taxa were reduced to small populations that caused a decline in genetic diversity because of founder effects (Allcock and Strugnell 2012; Gonzalez-Wevar et al. 2013). Likewise, populations of various penguin species were diminished at the last glacial maximum to only 2000 breeding females, based on fossil and phylogeographic data (Younger et al. 2016).

Whether or not genetic bottlenecks occurred, time spent in isolated refugia across the last glacial maximum often resulted in genetic drift. Even now, endemic mitochondrial and nuclear haplotypes remain unique to refugia-proximal populations, regardless of mixing of refugial populations from the last glacial maximum through secondary contact during interglacial expansion (e.g., review and re-analysis of eight North Atlantic intertidal taxa; Maggs et al. 2008). Potential examples are numerous, and refugial-proximal endemic populations have been discovered worldwide. One area where the discovery of endemic haplotypes helped to reveal a specific refugium during the last glacial maximum is that of the Hurd Deep with its rock walls and deep marine lake, where populations of everything from intertidal seaweeds (Provan et al. 2005; Hoarau et al. 2007; Maggs et al. 2008; Li et al. 2015; Assis et al. 2016) to rays (Chevolot et al. 2006) survived in isolation.

Longer time spans spent in the unique conditions of a refugium can result in allopatric speciation (Vermeij 1986; Ludt and Rocha 2015). Theoretically, adaptive radiations can begin in a refugium, where species have time to originate and specialize in the tight, competitive confines (Schubert and Bottjer 1995). However, species can adapt to the specific environmental constraints of a refugium that is not necessarily free of stress, just more survivable than the outside world. Harries et al. (1996) called these evolved populations "endemic refugia species" and pointed out that these species might take longer to expand into their previous ranges, if at all. Species that essentially become evolutionarily "stuck" in refugia can also become relict species, which are those that do not re-expand or otherwise contribute significantly to post-crisis recovery.

Graptolites survival across Paleozoic mass extinctions provides two examples of in-refugial diversification, one contributing to post-crisis recovery, and the other becoming a short-lived relict fauna. In the recovery example, graptolite survivors of the Late Ordovician extinction in South China diversified within a refugium that was recognized as being important to post-extinction expansion (Xu et al. 2005). In the Late Silurian, a series of trough-like refugia in Australia and led to relict graptolite taxa that never contributed to global recovery and ultimately went extinct (Rickards and Wright 2002). In both cases, the reason for survival in refugia was identified—isolation from poor extra-refugial conditions and favorable conditions within the refugium—but the reasons for post-refugial proliferation or demise were not identified.

Evolutionary stasis can also occur in a refugium, at least over a relatively short interval. This was seen in one species of intertidal gastropod, *Acanthinucella spirata*, along the west coast of North America, in which substantial morphological change occurred during expansion out of the refugium rather than in the refugium itself. Although genetic diversity decreased as the species spread northward along the coast, the gastropod evolved a shell shape distinct from southern populations and fossil conspecifics (Hellberg et al. 2001).

As seen in the examples provided for harvest refugia, research in marine reserves emphasize the increasing positive impacts on protected species, diversity, and ecosystems with age of the reserve. Given the decadal scales for species and ecosystems to respond to protection (Ballantine 2014; Coleman et al. 2015), the effects of newly established marine reserves can only be predicted and anticipated.

After the Environmental Crisis Ends, the Species Recovers by Inhabiting Newly Re-opened Habitats, Either Through Population Expansion or Through Adaptive Radiation; Otherwise, the Refugium Became a Trap

The end product of a refugium is not only survival through a global crisis, but ultimately a contribution to global ecological recovery. After survival through a global crisis, refugial populations should ideally expand to repopulate newly available habitat (Holterhoff 1996; Bennett and Provan 2008; Ashcroft 2010). Thus, a post-refugium taxon has a wider geographic distribution and larger population. Vermeij (1986) stressed the importance of understanding the full range of a species before the onset of a global crisis; equally, it is critical to understand the pattern of re-expansion—or not—after the environmental crisis has ended.

A taxon can re-expand into its former range, expand to greater than its former range, or can take over an entirely new habitat. How the recovery occurs after time spent in a refugium depends in part on the degree of adaptation that occurred during its residence and in part on extra-refugial conditions—environmental and biotic—after the global crisis has ended. In ancient refugia through mass extinction events, time spent in refugia can be on the order of millions of years, allowing plenty of time for evolution and diversification (Harries et al. 1996); thus, the taxon or taxa exiting the refugium can be quite different from the ones that were seeking a safe haven.

Thus, conditions within the refugium must be sufficient that a taxon can be ready to expand after a crisis has ended (Vermeij 1986), plus conditions in the recovered environment must be amenable to species expansion. A species, or a population, that cannot recover from its time in a refugium cannot contribute to global rediversification; thus, it becomes a relict species or relict population (or "stranded population" of Kauffman and Harries 1996) and the habitat was not a refugium, but a refugial trap. And, if a taxon becomes extinct while in a purported refugium, even though that taxon might thrive briefly during a global crisis, then that place was also a refugial trap.

Refugial traps are known from geologic time. Over million-year time spans, refugial traps initially appear to be adequate, but eventually, refugial conditions decline and the taxa within either become relicts or cease to exist. For example, the Carboniferous Tindouf Basin appeared to be a prime refugium for taxa across the Serpukhovian extinction event: isolation from oceanic currents and tectonic adjustments to the basin maintained a habitable water depth (Cozar et al. 2014). However, eventually tectonism led to uplift and exposure, and the basin refugium (and populations therein) no longer existed (Cozar et al. 2014).

In another refugial trap, brachiopod taxa that survived the Late Permian mass extinction persisted into the Early Triassic within habitable zone refugia, but shifting anoxic and fatally thermal water masses eventually terminated the refugium and its inhabitants (Clapham et al. 2013). In this case, sudden and lethal environmental perturbations eliminated refugial habitats, causing the extinction of brachiopod taxa, a problem that Clapham et al. (2013) warn is a caution for modern conservation efforts.

Relict populations exist as remnants of more recent refugia. A population of rockweed is trapped on the northwest Iberian Peninsula by oceanographic barriers to the north and south and is threatened by frequent fatal perturbations of heat stress (Hoarau et al. 2007). Other relict intertidal seaweed populations are trapped by winter sea ice and temperatures along the Canadian Atlantic coast (Li et al. 2015). A thriving, but relict, pre-Pleistocene gastropod fauna remains in its refugium in the Gulf of Venezuela, isolated from the rest of the Caribbean by upwelling (Petuch 1981).

Sometimes the reason for the development of a trap from a seemingly good refugium is unknown. The best example is that of the Late Silurian (Wenlock-Ludlow) graptolites that survived in a trough-and-rise system currently in rocks of New South Wales through the mass extinction at the end of the Wenlock (Rickards and Wright 2002). The taxa became a relict fauna, never contributing to the post-extinction recovery (Rickards and Wright 2002). The refugium seemed an adequate habitat, isolated from less habitable conditions in the rest of the ocean by local tectonism, but the reason for the refugium ultimately becoming a trap remains to be known.

However, there is one caveat; relict species may not indeed be "dead-end" taxa, but may remain in refugia, awaiting an environmental change to provide an appropriate habitat. For instance, disaster taxa presumably inhabit refugia in suppressed populations during "normal" environmental conditions (Kauffman and Harries 1996). Polar taxa withdraw to refugia during interglacial periods, in sync with the expansion of temperate and tropical taxa from glacial refugia (Ashcroft 2010; Stewart et al. 2010; Almada et al. 2012). The dinoflagellate *Dapsilidinium pastielsii* is extinct globally except in a warm area of the Indo-Pacific, presumably its current refugium, awaiting warmer sea surface temperatures than Pleistocene interglacials allow (Mertens et al. 2014). Thus, before defining a refugium, it is important to be certain of how a species responded, or will respond, when the need for the refugium has passed.

Currently, all marine reserves are in use as necessary refugia. Nowhere have conditions ameliorated sufficiently that a reserve is no longer necessary, and so the crises that caused the need for marine reserves have not yet ended.

3 Identifying Ancient Refugia

Refugia in deep time are, at best, difficult to identify. More material is not preserved, eroded, or undiscovered than exists in fossil collections. With a spatially, temporally, and preservationally heterogeneous fossil record—combined with highly variable sampling worldwide—identifying undisputable refugia is a challenge. Locating a refugium also depends in part on its geographic extent which influences the chances of encountering the location; in part on the geographic scale of the study; and in part on the quality of preservation of the stratigraphic interval and the fossils within (Westermann 2001).

Many reports of refugia include paleoenvironmental information to support the interpretation of the existence of a refugium. Although ultimately a refugial trap, Cozar et al. (2014) recognized the survival of a reef community through the Serpukhovian-Bashkirian extinction in the Tindouf Basin of Morocco and Algeria, not only through the persistence of fauna that went extinct elsewhere, but also through isolation caused by ocean currents and regional tectonism. Paleoenvironmental and paleogeographic analysis of the habitable zone of Western Canada in the early Triassic explained not only the survival, but also the diversity of shallow-water ichnofauna (Beatty et al. 2008; Zonneveld et al. 2010a, b). In more recent time, interpretations of refugia based on molecular reconstruction of species' demographics through the last glacial maximum are often upheld by independent data that indicate deglaciated areas (e.g., potential Antarctic refugia revealed by geophysical analysis of the sea floor; Graham and Smith 2012).

Tracing the space–time relationships of a lineage through a global crisis would be ideal in identifying a refugium. Unfortunately, such data are very rare; because of sampling or stratigraphic absences, lineages usually disappear from the known rock record through a mass extinction, either to their ultimate demise or to reappear later. The terms "Lazarus effect" and "Lazarus taxa" were coined by Flessa and Jablonski (1983) and further defined by Jablonski (1986) for those taxa that were present before a mass extinction event, then disappeared through the mass extinction interval, to reappear intact later in the fossil record. Stanley and Beauvais (1994) appropriately quip that Lazarus taxa "mysteriously leapfrog over large intervals of geologic time." Lazarus taxa have been described as an extreme form of extirpation, in which the refugia are unknown (Westermann 2001).

Thus, Lazarus taxa and their lengthy disappearance from the rock record indicate the likely presence, somewhere in the world, of undiscovered or unpreserved refugia (Vermeij 1986; Donovan 1989; Erwin and Hua-Zhang 1996). The duration of the

"Lazarus" phase of the taxon (while it is absent from the rock record) corresponds to the length of time of the global environmental crisis (Wignall and Benton 1999). Locating a refugium for a Lazarus taxon is problematic because of decreased area covered by the taxon during demographic contractions compounded by the inherently incomplete nature of the fossil record. Rarely are refugia for Lazarus taxa discovered, such as the Late Ordovician refugium for graptolites located in present-day south China (Xu et al. 2005).

However, the refugial survivorship of Lazarus taxa is in contention (see discussion in Fara 2001); they also could artificially result from their overly theorized existence in evolutionary, phylogenetic, or taxonomic studies (Westermann 2001) or simply be overlooked during sampling because of rarity or size (Erwin and Hua-Zhang 1996; Hautmann and Nützel 2005; Hautmann et al. 2011). Lazarus taxa may also have reduced to population sizes below the preserved resolution of the fossil record (Erwin and Hua-Zhang 1996), such that they would occur as "cryptic refugia" species, those that survive as sparsely distributed and rare individuals across a landscape (Stewart and Lister 2001; Bennett and Provan 2008). The possibility that Lazarus taxa are simply unsampled led Payne (2005) to suggest that they are in fact "Waldo taxa"; that they are present all along, but easily overlooked and require more effort in searching, similar to finding the character Waldo among a cartoon menagerie of people and items.

Aside from the debate of whether Lazarus taxa are actual and represent unfound refugia, there are currently three methods for identifying ancient refugia: analysis of fossil data, phylogeographic reconstructions, and species distribution models. A more thorough review, from the perspective of refugia during the last glacial maximum, was provided by Gavin et al. (2014), but the methods are briefly revisited here:

Fossil Data

The presence of abundant fossils and diverse taxa in a particular area during a global crisis interval might indicate the presence of a refugium. As mentioned earlier, independent paleoenvironmental analysis would provide further support for the existence of a refugium. Furthermore, if an appropriate refugial paleoenvironment was present, it should harbor several taxa, not just one (Vermeij 1986).

The benefit of interpreting refugia from fossil and rock data is the direct examination of the actual location and properties of the refuge, relative to conditions in the rest of the world. Why taxa survived in this location, and under what constraints—ecological, demographic, and environmental—are directly identifiable. Also, fossil data can provide information about ecosystem response, not just surviving taxa, to survivorship in refugia. For instance, Holterhoff (1996) identified one crinoid biofacies that migrated as an intact community through Pennsylvanian-Permian glacio-eustatic fluctuations in the Anadarko Basin of west Texas, but other refugial faunas during sea level lowstands were mixes of taxa from different highstand communities.

Fossil data across the last glacial maximum can also be integrated with molecular analyses and radiocarbon dating for deeper understanding of refugial existence. Only fossils can confirm that a specific region is a refugium; molecular studies can only estimate the location of a nearby refugium (Ludt and Rocha 2015). Fossils confirmed the refugial location of *Nucella lima* gastropods along the northeast Pacific coast during glacial maxima (Cox et al. 2014). Subfossils themselves not only indicate refugial locations, but also can provide the molecular data for phylogeographic studies (e.g., Pinsky et al. 2010).

The obvious caveat of fossil data is the incomplete nature of the fossil record. Preservation of organisms is dependent on the habitat preferences, life mode, taphonomy, and discovery of individual specimens. Transport and reworking of skeletal material can also affect the interpretation of a refugium by taking specimens out of a refugium or concentrating specimens into a high diversity, high abundance assemblage. Of course, the usual taphonomic analyses—comparative taphonomy, abrasion, etc.—can identify potential preservational issues.

One problem with the study of fossils is the resolution of paleontological research. Spatial, stratigraphic, and taxonomic resolution can be coarse (e.g., provincial, stage, and generic levels), which typically averages out high-resolution data like exact refugial boundaries and detailed information about population history. However, the coarse resolution of ancient refugial studies has one advantage: to observe the "big picture" reasons behind the formation and maintenance of refugia, like tectonism and oceanographic changes (e.g., refugia of South China carbonate platforms across the Late Ordovician mass extinction; Huang et al. 2013; Xu et al. 2005). Likewise, the study of long-term refugia can trace the responses of taxa during their isolation, whether refugial evolution and diversification or in becoming relict faunas.

Phylogeographic Studies

Constrained mostly to interpretations based on molecular data of refugia of the recent geological past, phylogeographic studies identify ancient demographic patterns, particularly population constrictions that can help identify refugial remnants. Most studies are based on mitochondrial or chloroplast DNA, but recent advances in molecular analysis has allowed the inclusion of nuclear DNA in demographic reconstructions.

Whichever method is used, the patterns of diversity and endemism interpreted from genetic data enable the identification of demographic contractions that are followed by later expansion and re-mixing. Populations that are remnants of previous refugia usually have high diversity, but also frequently have endemic or "private" haplotypes (Bennett and Provan 2008). Populations that have re-expanded out of refugia that existed during the last glacial maximum are lower in haplotype diversity and lack genetic endemism (Bennett and Provan 2008). Secondary contact, the mixing of populations from two or more refugia, results in high genetic diversity, but like unmixed expansion populations, lack the private haplotypes of refugial populations (Bennett and Provan 2008).

The benefits of phylogeographic studies are the identification of historical migration patterns, population sizes, and the effects of natural selection (Bennett and Provan 2008; Gavin et al. 2014). Comparative studies across taxa can identify shared refugia during the last glacial maximum (Maggs et al. 2008; Gavin et al. 2014). Furthermore, integration of molecular clock interpretations can help determine timing of contraction into refugia and expansion into secondary contact zones (Gavin et al. 2014).

Phylogeographic analyses are limited in that they cannot identify the actual location of refugia (Bennett and Provan 2008; Ludt and Rocha 2015); these areas have long since been drowned by rising sea level, and the remnant population bearing the refugial molecular signal has accordingly adjusted its habitat. However, with the comparison to independent environmental data—such as which coastal areas that remained unglaciated (e.g., unglaciated coastal areas and islands of Antarctica; Allcock and Strugnell 2012; Gonzalez-Wevar et al. 2013; Chown et al. 2015) or the paleo-presence of adequate rocky shorelines for encrusting biota (e.g., Hurd Deep in the English Channel; Provan et al. 2005; Hoarau et al. 2007; Maggs et al. 2008; Doellman et al. 2011)—it is possible to interpret the proximity of modern populations to their refugia during the last glacial maximum.

Species Distribution Models

This theoretical method involves the comparative analysis of the factors that determine the range of a species with environmental data or models, in order to predict its occurrences through space and time. Strengths of using species distribution models are numerous: they can be used to reconstruct past (e.g., intertidal species and environments through the last glacial maximum; Waltari and Hickerson 2013) and identify potential future (e.g., the three possible future refugia of Japanese reefs; Makino et al. 2015) refugia with respect to climate change. Thus, they create testable hypotheses of the existence of past refugia and of possible key conservation areas (Maggs et al. 2008; Gavin et al. 2014). Additionally, species distribution models are relatively inexpensive compared to molecular analyses and field-intensive fossil studies.

A major disadvantage of species distribution models is that results are only as good as the information input into the model. Climate change projections often are too coarse to have meaningful applications and produce results that are too broad for accurately identifying possible past or future refugia (Chollett and Mumby 2013; Gavin et al. 2014; although see van Hooidonk et al. 2015). Furthermore, models assume that species–environment or species–climate relationships are static and predicable through time; in other words, species distribution models rely on the assumption of niche stability of the species and do not accommodate population-scale processes (Gavin et al. 2014).

Another problem with species distribution models, particularly in identifying potential future refugia, is the choice of climate model(s) used in the analysis. Makino et al. (2015) pointed out that, out of multiple climate models available, each provides a different result, and caution that the results based on the input of various climate models do not produce a "no regrets" overlapping area for an ideal refugium that covers all projections.

4 Lessons from the Past for Identifying Future Refugia

Some of the environmental crises facing the marine biosphere today are similar to those of the ancient past: warming sea surface temperatures, ocean acidification, sea level rise, and related effects are some of the stresses previously experienced by the marine biosphere. Presently, these "old hat" stressors are compounded by anthropogenic-related problems; pressures like overharvesting effects on trophic systems and ongoing pollution have never previously been experienced on modern-day scales. The outcome of this unique set of old and new stressors acting on modern marine ecosystems is difficult to quantify, let alone predict. However, the need for viable refugia is clear if global biodiversity is to be maintained.

Worldwide, marine reserves improve short-term survival during environmental perturbations and have the potential to become long-term refugia through a changing ocean system. In the short term, marine reserves have the potential to buffer organisms from the negative impacts of environmental perturbations (Baskett and Barnett 2016). Over longer time intervals, marine reserves can act as refugia, and thus have an inherent conservation value by giving species and ecosystems the ability to persist through climate change and other global crises (Ashcroft 2010). Thus, conservation biologists have a heavy task set for them in managing marine reserves; in addition to protection from immediate stressors, there exists a need to identify where, how, and why future refugia will exist in a changing world.

Refugia from the cold sea surface temperatures and lower sea level of the last glacial maximum are not directly applicable to modern environmental crises, even though studies of population contractions into refugia are most numerous for this period in geologic history. However, past survival through glacial and interglacial periods has proven the resilience of species, which may mean that taxa are more adaptable to climate change than given credit (Gavin et al. 2014), especially if provided with adequate refugia and sufficient time to adapt.

The rock record before recent glacial–interglacial fluctuations has multiple relevant examples of how Earth has fared during global warming, ocean acidification, sea level fluctuations, and biosphere catastrophes. Although these studies are not as numerous as those across the last glacial maximum, they provide examples of how life survived mass extinctions and other global catastrophes. More importantly, some properties of refugia recur in space and time, regardless of the type of environmental stressors, the taxa that find protection, or the scale of the study.

The benefit of studying refugia in the ancient past is that the entire story of a refugium can be told: the contraction or move of a taxon into a refugium, its time spent in the refugium, and its ultimate extra-refugial population expansion (or its demise to a relict population or ultimate extinction, in the case of a refugial trap). The dilemma of modern conservation is that changes in taxon ranges have only begun and that refugia from future environmental deterioration are only beginning to be investigated. The exact locations of future, successful marine refugia, and the final outcome in terms of timing and nature of the post-refugial recovery are yet unknown. The fossil record contains multiple intervals of global crisis in which organisms retreated to refugia, survived, and ultimately recovered, and from these, there are several lessons:

As the Marine Environment Continues to Change, Refugia May Need to Shift

Sometimes one refugium, or one type of refugium, is not appropriate for survival through a crisis. What was an appropriate refugium for initial conditions during the onset of a global crisis may turn into a trap as conditions deteriorate. Thus, initial refugia may no longer be adequate for survival through the later conditions of the crisis. In these instances, refugial survival calls for a succession of range or habitat shifts or a migration between refugia.

As the global climate warms and the oceans transform, the changing needs of taxa may be considered in terms of short-term and long-term refugia. At first, short-term refugia may be needed to support populations through the initial perturbations caused by climate change (e.g., temporary thermal refugia in areas of delayed thermal stress; van Hooidonk et al. 2015). Short-term refugia are areas in the ranges of species where stress levels are not yet fatal, sub-populations are adapted to slightly higher stress thresholds, or recovery intervals are sufficiently lengthy between extreme perturbations. Eventually, with ongoing climate change, short-term refugia will cease to exist as perturbations become too frequent or stress becomes too great, and taxa will need to migrate to long-term refugia (Ashcroft 2010).

An example of potential refugial shifts in coral reefs already exists. Currently, short-term refugia from heat stress for coral reefs are in places where stresses are decreased, such as those within pathways of large-amplitude internal waves (Buerger et al. 2015; Wall et al. 2015; Schmidt et al. 2016), or where corals have adapted to have a slightly higher thermal tolerance, such as in gulfs within the Red Sea (Fine et al. 2013) or in current shadows (McClanahan et al. 2007). While some corals and associated organisms persist in these short-term refugia, reefs in general are shifting their ranges northward (Beger et al. 2014; Descombes et al. 2015; examples of coral range shifts include the Caribbean; Precht and Aronson, 2004; Japan; Yamano et al. 2011; Australia; Baird et al. 2012). Meanwhile, as sea levels continue to rise, reefs can be expected to climb upslope in sync with rising sea level, facing new consequences from changes in nearshore factors like terrestrial runoff (e.g., coral reef shifts during interglacial periods; Pandolfi 1999; Greenstein and Pandolfi 2008). Reefs will also have to contend with changing pollution and fishing pressures with continuous shore-ward migration. Therefore, future long-term reef refugia will need to be protected from anthropogenic-caused stresses, upslope from present-day reef habitats, and in higher latitudes. Finally, habitats that mitigate some of the effects of ocean acidification may become adequate refugia, such as shallow-water sea mount habitats suggested by Tittensor et al. (2010).

Conservation strategies must foresee changes in the refugial needs of targeted taxa or ecosystems. Changing refugial needs are already in consideration by some researchers: short-term and long-term refugia must be accommodated by current and future location and management of marine reserves. Currently, most of the emphasis is on short-term refugia, with or without the realization that many existing conservation areas and populations adapted to currently higher stress-thresholds may be only temporary survival strategies. Thus, all marine reserves as they currently exist support short-term survival, but as environmental conditions change, many taxa and ecosystems may need to move to long-term refugia as they become available.

However, one caution: when considering climate change and shifting refugia, the concept of "plan for the worst, hope for the best" may not be the best strategy for a successful conservation outcome. One study undertaken by Makino et al. (2015) compared the results of three separate models of greenhouse gas trajectories in predicting the future distribution of coral reefs around Japan. Each of the three results differed with little to no overlap in future coral distributions; thus, there was no area to establish what they called a "no regrets" marine reserve that would assure conservation in all three future scenarios. So, implementing a plan required by one climate change scenario could fail, if the actual, final environment differed from predicted.

Refugial Size and Connectivity Can Enhance Survivorship, But Can Also Have Evolutionary Consequences

Size and connectivity of refugia is already understood by conservation biologists in considering marine reserves and proved repeatedly in the geologic past. Obviously, a taxon confined to a single, small refugium (or reserve) is at greater risk of being eliminated by a fatal perturbation than if that taxon is spread across a large area (Almany et al. 2009). If a marine reserve is sufficiently large, it is comparable to a refugium that is adequate to support the entire species for the length of time needed. Multiple small marine reserves are equivalent to cryptic refugia, because in many cases, they must rely on connectivity between reserves or with individuals in unprotected areas.

For taxa with dispersal limitations or specific habitat needs, connectivity becomes difficult and populations are at risk of become fragmented and isolated. From the geologic past, population isolation of these species has been caused by a multitude of factors, such as larval duration (e.g., *Nucella lima* gastropods of the north Pacific; Cox et al. 2014), dispersal (e.g., female dispersal patterns of thornback rays; Chevolot et al. 2006), oceanographic conditions (e.g., Alboran

Sea isolation during the Messinian salinity crisis; Perrin and Bosellini 2013; equatorial split in Azorean barnacle populations, Quintiero et al., 2015), and habitat constraints (e.g., Antarctic intertidal limpet *Nacella concinna* unable to utilize deeper refugia; Gonzalez-Wevar et al. 2013). Ultimately, what is expressed in phylogeographic studies are only those populations that survived. Populations that were insufficient in abundance or genetic diversity, or experienced other constraints such as unsurvivable habitat conditions ultimately vanished, and thus are unrecoverable by genetic analyses. However, of those isolated populations that did survive, molecular analyses frequently highlight the consequences of geographic and demographic isolation, namely genetic drift, founder effects, and bottlenecks.

Isolated geographic refugia across the last glacial maximum resulted in many endemic refugial populations worldwide. Northern temperate and polar taxa, especially those requiring specific habitats (e.g., rocky intertidal zones; Maggs et al. 2008) and those needing ice-free rookeries (e.g., northern fur seals; Pinsky et al. 2010), commonly display endemism in the genomes that originated from refugial populations. Tropical taxa likewise show the genetic signatures of isolated populations (e.g., north and south refugia for Great Barrier Reef corals across the last glacial maximum; Van Oppen et al. 2011), as do populations isolated in the Southern Ocean (e.g., crinoids; Hemery et al. 2012). At greater extremes, bottlenecks appear in some populations. Among the most severe are bottlenecks across the last glacial maximum, in some marine mammals and birds, in which only a few thousand breeding females survived the decline in habitat availability and quality (Younger et al. 2016). What is important about these and other studies of refugia from the last glacial maximum are that founder effects and bottlenecks within refugia occurred only within a few thousand years and under conditions that were free of harvesting pressure and pollutants. Thus, isolation within refugia (whether from natural causes or from fishing pressures outside of a refugium) can lead to evolutionary consequences within a geologically very short time.

Cryptic refugia may mitigate the evolutionary problems of isolated populations. By nature, cryptic refugia are connected in space and time. Reproduction, recruitment, and dispersal within the metapopulation cross refugial boundaries, while the distribution of individuals across the generations shifts as microhabitats emerge or disappear (Stewart and Lister 2001; Bennett and Provan 2008). Furthermore, cryptic refugia can be ephemeral. As conditions in a heterogeneous environment shift, new refugia are created as others disappear. The example of Triassic deltaic refugia of Gall and Grauvogel-Stamm (2005) for marine taxa is apt: as sediments and currents shifted in the delta system, old habitats were destroyed and others opened. Likewise, shifting glacial ice opened and closed ephemeral refugia on the Antarctic shelf which taxa had to colonize or abandon accordingly (Graham and Smith 2012). It is important to remember that cryptic refugia occur on the order of very short geologic time spans, which means that on the order of human life spans, cryptic refugia may appear fairly stable, such as individual reefs that have survived for decades in a warming habitat (e.g., temperature refugia on the Great Barrier Reef; Ban et al. 2012).

Thus, the geological record of refugia proves two important points. (1) Genetic diversity and connectivity must be maintained within and across refugia, a factor well understood by conservation biologists. (2) Cryptic refugia—those areas supporting small populations that are not viable unto themselves, but contribute to a greater network of connectivity—must be able to disappear and appear elsewhere, if circumstances warrant. As with the realization that refugia may need to shift in response to a changing climate, in heterogeneously perturbed areas, cryptic refugia may have to be allowed to disappear and new refugia identified as they become established in new microhabitats. Thus, planning and management must be sufficiently flexible to accommodate shifting habitats.

However, modern refugia will be fully unlike deep-time refugia in one respect: marine reserves must also be sources for the populations harvested outside of protected areas. Thus, reserves must function beyond the support of sufficient populations within their borders, and act as "super-refugia" serving both the demographic needs of the reserves and the resources required by a growing human population. This factor may require refugia that are larger, in area and connectivity, than a simple survival-through-crisis would require.

Conditions Inside Refugia May Not Necessarily Remain Pristine, But Will Need to Be of Sufficiently Lower Magnitude of Total Stress to Maintain Viable Populations

The persistence of a species is affected by the quality of the refugial habitat (Ashcroft 2010). As stated earlier, Vermeij (1986) pointed out that a good refugium does not need to be free of stress, but should have tolerable stress levels, which can include an offsetting or balancing ameliorating factor (such as increased nutrient supply). As shown in earlier examples, many refugia in the geologic past were not pristine, unaffected sanctuaries from the stresses that caused mass extinctions. More often than not, the refugial paleoenvironment was less than perfect.

Although it is beyond the goal of this manuscript to review the effectiveness of no-take and less protected reserves, studies increasingly show that partial protection, when compared to no-take marine reserves, may not be sufficient to support populations or ecosystems (e.g., Kelaher et al. 2014; Nakin and McQuaid 2014; Toth et al. 2014; Sciberras et al. 2015). In modern ecosystems, there is some evidence that, by eliminating harvest pressure, no-take marine reserves increase the resilience of populations to environmental perturbations (e.g., coral survival in temperature refugia within marine reserves on the Great Barrier Reef; Ban et al. 2012). Thus, decreasing a major stressor or introducing a major benefit may create greater tolerance to lesser stresses. However, there also is evidence that the presence of one stress can increase susceptibility to another (e.g., ocean acidification may increase susceptibility to thermal bleaching in corals; Anthony et al. 2008; Fine et al. 2013). Likewise, if environmental perturbations occur within refugia, even if they are not fatal, organisms need sufficient time between stressful events to recover or may become too vulnerable (Connell et al. 1997; Coumou and Rahmstorf 2012; Wenger et al. 2016).

The geologic record of refugia demonstrates the broad range of tolerance and adaptation by survivor species to stressful refugial conditions. In some refugia, stressors were similar to, but reduced from, conditions of the extra-refugial environment. One previously discussed example of reduced, but tolerable, stress in a refugium was the escape of Early Triassic taxa from lethally hot and anoxic surface water through survival the nearshore habitable zone, which had tolerable, albeit warm, temperatures and perhaps also had a non-lethal level of oxygen stress (Beatty et al. 2008; Zonneveld et al. 2010a, b; Clapham et al. 2013; Song et al. 2015; Zhao et al. 2015). Other refugia were very different in the nature stress to be tolerated, as in the shifting sediments of Triassic deltaic refugia, which experienced frequent disturbance from shifting sediment and water currents, but was protected from anoxia and lethal thermal stress in the open ocean (Gall and Grauvogel-Stamm 2005). Sometimes, ameliorating conditions in refugia allowed the species to tolerate stress. For example, oxygen and food production by Late Permian microbial mats allowed ostracods to exist in hot surface water (Forel et al. 2013) and the availability of rocky shores made up for cold, peri-glacial waters for macroalgae across the last glacial maximum (Olsen et al. 2010). Thus, there are a multitude of potential ameliorating factors that might enhance a species' odds of survival in a refugium where conditions are less than pristine. Which ameliorating factors are beneficial to survivorship may depend on responses at the species, and even population, level, and may not be generalizable across taxa.

There has been some suggestion that species that have persisted in refugia through both interglacial and glacial periods may have a high level of tolerance to future climate change (Gavin et al. 2014). However, if climate change follows extreme trajectories because of anthropogenic enhancement (i.e., warming greater than past interglacial levels and ocean acidification), places that have been long-term hotspots of diversity, such as Indo-Pacific reefs (Vermeij 1986; Pellissier et al. 2014), may have to be abandoned for new refugia (Descombes et al. 2015).

Beware the Refugial Trap

Refugial traps can seem, initially, to be excellent refugia. Only as the refugium deteriorates, or when taxa do not expand from the refugium, is the location realized as a trap. The importance for conservation biology is understanding and predicting what constitutes a trap rather than a successful refugium.

The commonality across all refugial traps is that, initially, the location appears to be a suitable refugium. In refugia that have become traps since the last glacial maximum, Pleistocene populations found the needed sanctuary from environmental deterioration, but since then, it is the modern populations that are in distress. For example, some intertidal seaweed populations of Atlantic Canada and the Iberian Peninsula were unable to expand from their refugia because of geographic barriers that occurred with rising sea levels or warming oceans (Hoarau et al. 2007; Li et al. 2015), resulting in one Iberian Peninsula isolated, relict population of *Fucus serratus* rockweed population currently facing acute thermal distress (Hoarau et al. 2007). Likewise, the Bahaman Caribbean blacknose shark population remains isolated by oceanographic barriers since its last glacial maximum refugium (Portnoy et al. 2014). As these studies show, initially successful refugia can lead to later isolation and potentially, populations in distress. In these cases, sea level rise and changes in ocean currents are two of the factors that contributed to population isolation.

Refugial traps from the ancient geologic past remained viable refugia for longer periods of time than those of the Pleistocene and the circumstances that terminated those refugia also tended to be of larger magnitude. Tectonism and eventual drying of an ocean basin were the demise of ecosystems in the Carboniferous Tindouf Basin of Morocco (Cozar et al. 2014) and Anadarko Basin of the USA (Holterhoff 1996). In these cases, the gradual transformation of refugia into traps is apparent in the tectonic uplift and basin fill. More relevant to modern refugia are the spatially and temporally shifting, lethally hot and anoxic water masses that eventually devastated holdover brachiopod faunas in Early Triassic habitable zones (Clapham et al. 2013).

Thus, refugial traps often occur where isolation of populations is inevitable or where conditions are in danger of fatal levels of deterioration. This may be an obvious statement, but the key to understanding whether the refugium will become a trap is in adequately predicting future conditions, such as oceanographic changes with sea level rises, temperature and salinity fluctuations of water masses, plumes of polluted water, heterogeneity of acidification, etc. One factor is the recognition of dead-end basins, which are those areas that are at risk of becoming fatal traps because there is literally no escape route for the organisms therein from lethal conditions. For example, in the Red Sea, although short-term refugia are present because of the higher tolerance of many taxa to salinity and thermal fluctuations (Fine et al. 2013), the basin may eventually become too warm, and thus become inhabitable for many of its endemic taxa in the future. If taxa cannot escape deteriorating conditions in the enclosed basin of the Red Sea, short-term refugia within the basin will become traps

For most of the fossil examples, the actual changes that turned seemingly adequate refugia into traps were geologically gradual. From a human time–space perspective, the development of a trap out of a hospitable environment could be so gradual that the transformation could be too subtle to be easily noticed during a single career of conservation research. In this case, marine reserves may require continuous observation and comparison to their baselines to ensure that the reserves are not becoming traps. However, refugia can also become traps very rapidly, such as shifting anoxic water masses that terminated brachiopod faunas in the Early Triassic habitable zone (Clapham et al. 2013), which may each have occurred suddenly, from the perspective of geological time.

5 Future Directions for Investigating Ancient Refugia

Since the seminal work of Vermeij (1986), paleobiologists have increasingly found potential refugia that allowed species to persist during crises in the marine biosphere. Molecular studies on modern populations have increased the recognition of marine refugia through the last glacial maximum by an order of magnitude, particularly in species that underwent population bottlenecks and founder effects.

Most importantly, research has focused on the "why" of refugia, rather than simply identifying the locales of survivorship; it is the reasons for refugial survival that are pertinent to modern conservation biology.

The fossil and geological record of refugia is a baseline for understanding one way of how the marine biosphere survives a global crisis. Although lacking the level of detail of modern ecological investigation, paleontological study of refugia nonetheless can impart critical information relevant to conservation paleobiology. The properties of taxa that thrived in refugia, why a particular refugium formed, how a refugium and the species therein changed through time, and what caused the successful recovery or fatal trap at the end of the global crisis, are among some of the questions that paleobiology can answer. These answers, in turn, are relevant to understanding what may constitute a successful marine reserve.

For conservation biologists, the synthesis of ancient refugia herein is meant to provide information, as it stands, to help refine current and future marine reserves for best efficacy. For paleobiologists, the purpose of this synthesis is a baseline from which future directions of refugial studies can proceed. This requires not only further research and confirmation of presently hypothesized refugia discussed above, but also the effort to discover and investigate other refugia. Intervals that are characterized by environmental and ecological degradation similar to modern systems may be most appropriate, such as consequences of ocean warming across the Permo-Triassic mass extinction and the Paleocene-Eocene Thermal Maximum, and times of reef demise, such as the Frasnian-Famennian biotic crisis.

6 Conclusions

Throughout geologic history, life repeatedly has survived global crises of environmental deterioration and mass extinction. Life meanwhile will survive the current global crisis, but the important issue to humanity is that species and ecosystems need to persist in a way that allows humanity to not only survive, but thrive. This means that humanity needs a marine biosphere that provides the same, and likely greater, level of services as has always been expected. Thus, it is beneficial to understand how and why organisms survived ancient global crises, and of importance to marine reserves to know how refugia played a role in that survivorship.

As reviewed above, refugia are areas where, during a global crisis, species contract into isolated geographic areas, undergo range and habitat shifts, or find locations for certain life stages, such as where populations can successfully breed. While in the refugium, conditions are not necessarily stress free, but the total sum of detrimental and beneficial environmental parameters in the refugium is sufficient that populations can remain viable. Furthermore, a single refugium may not persist through the entire crisis of global climate change; distinct short-term and long-term refugia may be necessary for some taxa. The ultimate goal of refugia is the contribution of surviving species to global recovery after a crisis has ended. Species will need to re-expand into previous, greater, or new ranges, thus re-populating extra-refugial environments.

Conservation biologists can learn several lessons from ancient extinction survivorship, applicable to the future of marine reserves as refugia. In particular, with a future of ongoing climate change, refugial strategies need to be flexible. Some marine reserves, as they currently exist, may serve as short-term refugia that are adequate for present environmental conditions, but as perturbations become more frequent and intense, these areas may become too deteriorated for the species currently inhabiting them. Short-term refugia will need to be abandoned for long-term refugia, when they become available and as the new stable state of the oceans is reached. These refugia may be located at higher latitudes or upslope with sea level rise relative to the locations of present-day marine reserves. Furthermore, the locations of short-term refugia, whether as reserves or as habitats recognized as less perturbed than others, may also need to be flexible in space and time and therefore, allowed to move elsewhere, especially in areas where environmental deterioration is heterogeneous.

As in the geologic past, environmental and ecological conditions in current marine reserves may not remain pristine, but if they are stressed, they will need to be adequate to support populations that can easily continue to reproduce and recruit. This not only includes reduced levels of stress and tolerable effects of combined stressors—such as decreased pollution, buffering from acidification, and less frequent negative perturbations—but also the amelioration of stress by other favorable conditions, such as the location of a refugium in areas that have high nutrient levels (e.g., kelp in upwelling zones; Graham et al. 2007).

Types, sizes, and population connectivity of refugia matter. Multiple cryptic refugia, similar to smaller marine reserves, can remain connected through inter-refugial areas to support the continuance of one or more species. Furthermore, multiple refugia are potential insurance against the risk of extinction: if one or a few refugia are lost, so long as there are sufficient populations, the species can continue (Ballantine and Langlois 2008; Ballantine 2014). Otherwise, a single refugium that is sufficiently large can also insure a low extinction risk, provided that it covers an adequate selection of habitats and allows space for future range shifts.

"No regrets" marine reserves—those that cover all possible future scenarios of environmental deterioration—will be difficult to identify, if not impossible (Makino et al. 2015). Areas of "last resort" where conditions have not yet deteriorated completely and populations continue to survive must be protected (Chollett and Mumby 2013) as short-term refugia to allow time for the identifying and planning for future, long-term refugia. However, it is important to realize that many of these short-term refugia will also disappear (e.g., Indo-Pacific reef biodiversity hotspots; Descombes et al. 2015), and that efforts will need to shift to identifying and establishing new, long-term refugia before the demise of short-term protected areas.

One important lesson *from* conservation biology to paleobiologists is where paleontology can provide information relevant to modern marine crises. Paleontology and geology have the unique perspective of following a previous and devastating global crisis from start through finish and finally to recovery. Refugia and other extinction survivorship strategies are important to assisting in modern biosphere survival; as Vermeij (1986) suggested, understanding the how and why of survivorship of taxa through mass extinctions is of equal importance as studying the extinction itself.

Whatever the type, a refugium (or multiple refugia) needs to last through the entire crisis. Understanding the length of time that a refugium may be needed is critical. Refugia across the last glacial maximum existed on the order of thousands of years; those across mass extinctions were utilized by survivors and their descendent taxa for millions of years. Modern marine ecosystems are facing stressors that have been experienced by, and survived by, ancient life, but the modern marine system has the unique factors of human-caused environmental crises and the potential for significant human intervention. As the ultimate ecosystem engineer, humanity has the potential to not only create and maintain adequate refugia, it also can undo human-caused environmental degradation. So, a refugium (marine reserve) must last as long as it needs to, which is dependent on how fast and how well society can "fix" the global environmental crisis. As Kauffman and Erwin (1995) compared the modern crisis to ancient mass extinctions, if unfixed, recovery from this newest mass extinction at its greatest could take up to 2–10 million years, unchecked. Thus, the best long-term solutions need to go beyond identifying and establishing refugia and need to reverse the human-caused crisis.

Acknowledgements I am grateful to my reviewers S. Donovan and J.P. Zonneveld who provided advice which greatly improved the manuscript. In particular, they helped improve the flow and relevance of the manuscript in addition to cuing me to additional studies. I also thank L. Leighton and the Leighton Lab at the University of Alberta (D. Molinaro, K. Barclay, S. Mendonca, E. Garcia, M. Provan, A. Rocca, and C. Berendt) who either participated in a discussion about refugia or commented on an early draft of the manuscript. I also wish to acknowledge, thank, and remember my late friend, S. Needlepaws, whose presence during future manuscripts will be sorely missed.

Appendix

A summary of studies cited in the manuscript.

Types of refugia:
R = Range shift
H = Habitat shift
H (D) = Habitat shift to deep water
H (S) = Habitat shift to shallow water
I = Isolated geographic refugia
I (M) = Sea mount refugia
C = Cryptic refugia
L = Life history refugia
LGM = last glacial maximum
LIG = last interglacial
OMZ = oxygen minimum zone
N = north
E = east
S = south
W = west

Reference	Type	Age, where	Taxa	Location	Interpretation
Westrop (1989)	H (D)	Late Cambrian, North America	Trilobites	Upper continental slope	Slope environments expanded during sea level rise
*Rickards and Wright (2002)	I	Late Ordovician, New South Wales	Graptolites	Trough-and-rise tectonic system	A series of deep troughs were isolated from extinction-causing stress in the ocean
Xu et al. (2005)	I	Late Ordovician, South China	Graptolites	Yangtze Platform	Isolated from open ocean; preserved the preferred habitats of graptolites
Rassmussen and Harper (2011)	I	Late Ordovician, various	Brachiopods	Isolated regions	Biodiversity hotspots between first and second phases, Late Ordovician extinction
Huang et al. (2013)	H (D)	Late Ordovician, South China, Norway	*Dicoelosia* sp. brachiopods	Shallow water	Deep water taxon moved to shallow water to avoid "abnormal" conditions
Lethiers and Cassier (1999)	H (S)	Late Devonian, worldwide	Ostracods	Shallow water	Dysoxic-tolerant ostracods spread into near-littoral environments
House (1996)	H (D)	Late Devonian, worldwide	*Kokenia* sp. goniatite	Deeper water	Hypothetical; escape hypoxic events in shallow water
*Cozar et al. (2014)	I	Mississippian-Pennsylvanian, N Africa	Foraminifera, corals, brachiopods, algae	West Tindouf Basin	Position in paleocurrents, tectonism, and emerging land masses isolated the basin
**Holterhoff (1996)	I, R	Pennsylvanian-Permian, SW USA	Crinoids, including one biofacies	Anadarko Basin	Contraction and expansion during glacio-eustatic changes; basin ultimately dried up

Reference	Type	Age, where	Taxa	Location	Interpretation
Forel et al. (2013)	H	Late Permian, equatorial	Ostracods, others	Microbial mats	Microbial mats provide oxygen, unlimited food supply, but remain in hot environments
*Farabegoli et al. (2007)	I	Late Permian, Italy	Benthic fauna	Isolated region	Few million years survival of holdover fauna after decline elsewhere
Waterhouse and Shi (2010)	I?	Late Permian—Early Triassic, Gondwana	Invertebrates	Region, perhaps with refugia	Southern hemisphere Gondwana region provided refugia from anoxia in the northern hemisphere
Schubert and Bottjer (1995)	?	Early Triassic, W USA	Invertebrates	Unknown refugia	Hypothetical existence of refugia from poor environmental conditions
Gall and Grauvogel-Stamm (2005)	I, C	Early-Middle Triassic, France	Invertebrates and others	Delta	Holdover, Lazarus, and progenitor taxa protected from anoxia; fragmented habitats
Beatty et al. (2008)	H (S)	Early Triassic, W Canada	Trace fossils	Habitable zone	Survival in habitable zone between wave base and anoxic deeper water in embayments along coast
*Posenato (2009)	H	Early Triassic, Dolomites	*Orbecoelia, Strepto-rhynchus* brachiopods	Microbial mats	Survival for a few thousand years after decline elswhere in less stressed, nearshore habitats
Algeo et al. (2010)	H (D)	Early Triassic, Japan	Radiolarians	Deeper water	Warm water, low oxygen-tolerant taxa moved to deep ocean to escape upwelling OMZ on shelf
Zonneveld et al. (2010a, b)	H (S)	Early Triassic, W Canada	Trace fossils	Habitable zone	Survival in narrow habitable zone between wave stressed proximal and anoxic distal environments

Reference	Type	Age, where	Taxa	Location	Why refugium exists
Shen et al. (2013)	H (S)	Early Triassic, S China	Acritarchs	Shallow water	Moved to shallow shelf during times of stress
*Clapham et al. (2013)	H (S)	Early Triassic, W USA, Italy, S China	Brachiopods	Habitable zone in carbonate platform	Moved to littoral zone, but eventually overcome by shifting anoxic water masses
Chen et al. (2015)	H (S)	Early Triassic, S China	Brachiopods	Mid-water refugia zone	Heat, hypoxia-tolerant species escaped lethally hot surface and anoxic deep water
Zhao et al. (2015)	H (S)	Early Triassic, S China	Trace fossils	Shallow water	Refugium between lethally hot surface water and anoxic deep water
Stanley and Beauvais (1994)	I (M)	Early Jurassic, North America, Peru	Reef corals	Volcanic island	Long-lived survivors of Karnian and end Triassic extinction events in accreted terrains
Westermann (2001)	I	Early Jurassic, New Zealand	Ammonites	Contraction to region	Contraction from E Pacific to New Zealand, later re-expanded to E Pacific
Kiessling et al. (2009)	I	Early Jurassic, Europe	Reef corals	Continental shelf	Did not need volcanic islands for refugia; continental shelf environment was sufficient
*Yin and Fursich (2008)	I	Early Jurassic, Tibet	Ammonites, bivalves	Isolated region	Several taxa survived in Tibet longer than elsewhere
Jacobs and Lindenberg (1998)	I	Cretaceous, Pacific	Hydrothermal vent faunas	Submarine volcanoes	With high spreading rates, vent faunas escaped deep water anoxia on volcanoes
Gibbs et al. (2016)	H	PETM, worldwide	Coccolithophores	Higher latitudes	Heat-intolerant during one life cycle, but tolerated pH changes at higher latitudes

Reference	Type	Age, where	Taxa	Location	Why refugium exists
*Perrin and Bosellini (2013)	I	Miocene, Alboran Sea	Corals	Range contraction	Initially escaped onset of Messinian salinity crisis; finally only *Porites* and microbialites remained
Calvo et al. (2015)	I	Miocene, Atlantic, E Mediterranean	Reef-building vermetid gastropods	Range shift	Contraction to areas with normal marine conditions
Petuch (1981)	I	Pliocene-present, Gulf of Venezuela	Molluscs	Isolated region	Area isolated by ocean currents and upwelling; colder temperatures than the rest of Caribbean
Vermeij (1986)	I	Plio-Pleistocene, Indo-Pacific	Tropical shallow marine species	Diversity hotspot in Indo-Pacific	Nutrient-rich and comparatively low magnitude of changes in temperature
Greenstein and Pandolfi (2008)	R	LIG, Australia	Reef corals	Higher latitudes	"Temperature refugia" in higher latitudes away from equatorial warming
Barnes et al. (2006)	R, H	LGM, Antarctic shelf	Benthos	Continental slope	Speciation down slope; modern eurybathy among taxa
Barnes and Kulinski (2010)	I	LGM, Antarctic shelf	Bryozoans	Isolated areas	No deep sea refugia, so must be in situ ice-free areas on the shelf
Allcock and Strugnell (2012)	I, H	LGM, Peri-Antarctic areas	Marine taxa	Isolated shelf areas, deep sea	Populations in ice-free shelf refugia went through bottlenecks; deep sea taxa didn't
Graham and Smith (2012)	I	LGM, Alexander Island, W Antarctic Peninsula	–	Isolated areas	Geophysical study; ephemeral and one permanent ice-free areas on the outer shelf
Hemery et al. (2012)	I	LGM, various shelf and islands, Antarctica	*Promachocrinus kerguelensis* crinoid	Isolated populations	Development of genetic endemism of seven populations

Gonzalez-Wevar et al. (2013)	I	LGM, S Georgia Island	*Nacella concinna* limpet	Isolated area	Development of genetic endemism of S Georgia Island population
Younger et al. (2016)	I (L)	LGM, Antarctic coastal areas	Penguins, petrels, elephant seals	Rookeries	Ice free breeding habitats and nearby foraging grounds limited; many genetic bottlenecks
Provan et al. (2005)	I	LGM, N Atlantic	*Palmaria palmate* seaweed	Isolated areas	Development of genetic endemism in Hurd Deep, NW Atlantic, and Ireland populations
Harlin-Cognato et al. (2006)	I	LGM, N Atlantic, N Pacific	*Eumetopias jubatus* Stellar's sea lion	Isolated rookeries	Contraction and expansion across glacial cycles with decline and increase in rookery availability
Chevolot et al. (2006)	I	LGM, NE Atlantic	*Raja clavata,* Thornback rays	Isolated areas	LGM populations centered in Iberia, the Mediterranean, and Hurd Deep
Jolly et al. (2006)	I	LGM, NE Atlantic	Two spp. polychaete tubeworms	Isolated areas	Shared history of vicariant events between the two taxa, including northern peri-glacial refugia
Hoarau et al. (2007)	I	LGM, NE Atlantic	*Fucus serratus* rockweed	Isolated areas	Expansion from Ireland, less so Hurd Deep populations; relict Iberian Peninsula population
Maggs et al. (2008)	I	LGM, N Atlantic	Eight intertidal species	Isolated areas	Nine rocky shoreline refugial populations, including four peri-glacial northern areas
Campo et al. (2010)	I, R	LGM, NW Atlantic	*Pollicipes pollicipes* stalked barnacle	Isolated areas and southerly shift	Two refugial populations in Iberia and Brittany; large N Africa population; expansion during LIG
Olsen et al. (2010)	I	LGM, N Atlantic	*Ascophyllum nodosum* seaweed	Isolated areas	Refugial populations from Atlantic Canada, Brittany rocky intertidal areas

Reference	Type	Age, where	Taxa	Location	Why refugium exists
Krebes et al. (2010)	H	LGM, NE Atlantic, Europe	*Gammarus duebeni* amphipod	Brackish coastal outwash	Refugia in brackish coastal glacial outwash lakes to avoid competition, later adapted to rivers
Doellman et al. (2011)	I	LGM, N Atlantic	*Littorina saxatilis* gastropod	Isolated areas	Many local European refugia; two northern Atlantic and two Atlantic N American refugia
Almada et al. (2012)	I	LGM, N Atlantic	*Taurulus bubalis* longspined bullhead	Isolated areas	Coldwater species; multiple refugia in Atlantic and north sea, including peri-glacial refugia
Krakau et al. (2012)	I, R	LGM, N Atlantic	*Cerastoderma edule* cockle	Isolated area and range shift	Northern peri-glacial refugium and large LGM population British Isles to northwest Africa
Waltari and Hickerson (2013)	I	LGM, N Atlantic	Six intertidal invertebrates	Isolated areas	Spp distribution models identified several European and North American refugia
Naro-Maciel et al. (2014)	I	LGM, Central Atlantic	*Chelonia mydas* green sea turtle	Isolated rookeries	Distinct N and S populations, each with multiple small refugial populations across the LGM
Neiva et al. (2014)	I, R	LGM, NE Atlantic	*Pelvetia canaliculata* fucoid seaweed	Isolated area and range shift south	Spp distribution models: Morocco to Celtic Sea; phylogeography: refugia in Iberia, Great Britain
Quintiero et al. (2015)	I, R	LGM, Azores to Cabo Verde islands, Atlantic	*Megabalanus azoricus* barnacle	Isolated areas	Stable southern refugia during glacial maxima in the islands
Li et al. (2015)	I	LGM, N Atlantic	*Palmaria palmata* red macroalgae	Isolated areas	Multiple European endemic refugial populations; two relict refugial populations Atlantic Canada

Assis et al. (2016)	I	LGM, NE Atlantic	*Laminaria hyperborean* kelp	Isolated areas	Niche modelling; several LGM refugia; southern Iberian populations will disappear with warming
Mateus et al. (2016)	I	LGM, Iberia	European lampreys	Isolated area	Long term isolation of LGM refugial Iberian populations; later expansion northward
Albaina et al. (2012)	I, R	LGM, Adriatic Sea. Iberia	*Nassarius nitidus* whelk	Isolated areas	Refugia in the Adriatic Sea and around the Iberian Peninsula; current expansion front at North Sea
Portnoy et al. (2014)	I, R	LGM, S Gulf of Mexico	Blacknose shark	Isolated area	Five modern populations from one refugial population in southern Gulf of Mexico
Hellberg et al. (2001)	I, R	LGM, southern California	*Acanthinucella spirata* gastropod	Isolated area, range shift	Contraction of population to an area north of Los Angeles
Pinsky et al. (2010)	L, R	LGM, NE Pacific	*Callorhinus ursinus* northern fur seals	Isolated rookeries	High disperser that utilized many available sub-Arctic rookeries, so no endemic populations
Grant and Cheng (2012)	L	LGM, N Pacific	*Paralithodes camptschaticus* crab	Isolated nurseries	Juvenile nurseries near Japan, Kodiak Island, Haida Gwaii
Cox et al. (2014)	I, R	LGM, N Pacific	*Nucella lima* gastropod	Isolated areas	Split between E, W refugia 317 ka; E Pacific population bottleneck during LGM
Van Oppen et al. (2011)	I, R	LGM, Great Barrier Reef	*Acropora millepora* coral	Isolated areas	Northern and southern refugial populations during glacial maxima
Ludt et al. (2012)	I	LGM, Hawaii	Two spp. lagoonal and two spp. reef slope wrass	Isolated areas	Lagoonal populations fragmented with sea level drop; reef slope population remained connected
Pellissier et al. (2014)	I	LGM, Indro-Australian Archipelago	Reef fish	Reefs	Stable reef habitats during past climate changes influence modern reef fish distribution

References

Aberhan M, Kiessling W (2014) Rebuilding biodiversity of Patagonian marine molluscs after the end-Cretaceous mass extinction. PLoS One 9:e102629

Albaina N, Olsen JL, Couceiro L et al (2012) Recent history of the European *Nassarius nitidus* (Gastropoda): phylogeographic evidence of glacial refugia and colonization pathways. Mar Biol 159:187–1884

Algeo TJ, Hinnov L, Moser J et al (2010) Changes in productivity and redox conditions in the Panthalassic Ocean during the latest Permian. Geology 38:187–190

Allcock AL, Strugnell JM (2012) Southern Ocean diversity: new paradigms from molecular ecology. Trends Ecol Evol 27:520–528

Almada VC, Almada F, Francisco SM et al (2012) Unexpected high genetic diversity at the extreme northern geographic limit of *Taurulus bubalis* (Euphrasen, 1786). PLoS One 7:e44404

Almany GR, Connolly SR, Heath DD et al (2009) Connectivity, biodiversity conservation and the design of marine reserve networks for coral reefs. Coral Reefs 28:339–351

Anthony KRN, Kline DI, Diaz-Pulido G et al (2008) Ocean acidification causes bleaching and productivity loss in coral reef builders. Proc Natl Acad Sci USA 105:17442–17446

Ashcroft MB (2010) Identifying refugia from climate change. J Biogeogr 37:1407–1413

Assis J, Lucas AV, Barbara I et al (2016) Future climate change is predicted to shift long-term persistence zones in the cold-temperate kelp *Laminaria hyperborea*. Mar Environ Res 113:174–182

Baird AH, Sommer B, Madin JS (2012) Pole-ward range expansion of *Acropora spp*. along the east coast of Australia. Coral Reefs 31:1063–1063

Ballantine B (2014) Fifty years on: lessons from marine reserves in New Zealand and principles for a worldwide network. Biol Conserv 176:297–307

Ballantine WJ, Langlois J (2008) Marine reserves: the need for systems. Hydrobiologia 606:35–44

Ban NC, Pressey RL, Weeks S (2012) Conservation objectives and sea-surface temperature anomalies in the Great Barrier Reef. Conserv Biol 26:799–809

Barnes DKA, Kulinski P (2010) Bryozoans of the Weddell Sea continental shelf, slope and abyss: did marine life colonize the Antarctic shelf from deep water, outlying islands or in situ refugia following glaciations? J Biogeogr 37:1648–1656

Barnes DKA, Hodgson DA, Convey P et al (2006) Incursion and excursion of Antarctic biota: past, present and future. Glob Ecol Biogeogr 15:121–142

Baskett ML, Barnett LAK (2016) The ecological and evolutionary consequences of marine reserves. Annu Rev Ecol Evol Syst 46:49–73

Beatty TW, Zonneveld J-P, Henderson CM (2008) Anomalously diverse Early Triassic ichnofossil assemblages in northwest Pangea: a case for a shallow-marine habitable zone. Geology 36:771–774

Beger M, Sommer B, Harrison PL et al (2014) Conserving potential coral reef refuges at high latitudes. Divers Distrib 20:245–257

Bennett KD, Provan J (2008) What do we mean by 'refugia'? Quat Sci Rev 27:2449–2455

Berger M, Babcock R, Booth DJ et al (2011) Research challenges to improve the management and conservation of subtropical reefs to tackle climate change threats (Findings of a workshop conducted in Coffs Harbour, Australia on 13 September 2010). Ecol Manag Restor 12:e7–e10

Bongaerts P, Ridgway T, Sampayo EM et al (2010) Assessing the 'deep reef refugia' hypothesis: focus on Caribbean reefs. Coral Reefs 29:309–327

Buerger P, Schmidt GM, Wall M et al (2015) Temperature tolerance of the coral *Porites lutea* exposed to simulated large amplitude internal waves (LAIW). J Exp Mar Biol Ecol 471:232–239

Calvo M, Alda F, Oliverio M et al (2015) Surviving the Messinian Salinity Crisis? Divergence patterns in the genus *Dendropoma* (Gastropoda: Vermetidae) in the Mediterranean Sea. Mol Phylogenet Evol 91:17–26

Campo D, Molares J, Garcia L et al (2010) Phylogeography of the European stalked barnacle (*Pollicipes pollicipes*): identification of glacial refugia. Mar Biol 157:147–156

Chen J, Tong J, Song H et al (2015) Recovery pattern of brachiopods after the Permian–Triassic crisis in South China. Palaeogeogr Palaeoecol 433:91–105

Chevolot M, Hoarau G, Rijnsdorp AD (2006) Phylogeography and population structure of thornback rays (*Raja clavata* L., Rajidae). Mol Ecol 15:3693–3705

Chollett I, Mumby PJ (2013) Reefs of last resort: locating and assessing thermal refugia in the wider Caribbean. Biol Conserv 167:179–186

Chollett I, Enriquez S, Mumby PJ (2014) Redefining thermal regimes to design reserves for coral reefs in the face of climate change. PLoS One 9:e110634

Chown SL, Clarke A, Fraser CI et al (2015) The changing form of Antarctic biodiversity. Nature 522:431–438

Clapham ME, Fraiser ML, Marenco PJ et al (2013) Taxonomic composition and environmental distribution of post-extinction rhynchonelliform brachiopod faunas: constraints on short-term survival and the role of anoxia in the end-Permian mass extinction. Palaeogeogr Palaeoecol 374:284–292

Commission for Environmental Cooperation (2011) North American marine protected areas network. http://www2.cec.org/nampan/. Accessed 05 Nov 2016

Connell JH, Hughes TP, Wallace CC (1997) A 30-year study of coral abundance, recruitment, and disturbance at several scales in space and time. Ecol Monogr 67:461–488

Coumou D, Rahmstorf S (2012) A decade of weather extremes. Nat Clim Chang 2:491–496

Cox LN, Zaslavskaya NI, Marko PB (2014) Phylogeography and trans-Pacific divergence of the rocky shore gastropod *Nucella lima*. J Biogeogr 41:615–627

Cozar P, Vachard D, Somerville ID et al (2014) The Tindouf Basin, a marine refuge during the Serpukhovian (Carboniferous) mass extinction in the northwestern Gondwana platform. Palaeogeogr Palaeoecol 394:12–28

da Silva IM, Hill N, Shimadzu H et al (2015) Spillover effects of a community-managed marine reserve. PLoS One 10:e0111774

Darwin C (1859) On the origin of species by means of natural selection, or the preservation of favoured races in the struggle for life. John Murray, London

Descombes P, Wisz MS, Leprieur F et al (2015) Forecasted coral reef decline in marine biodiversity hotspots under climate change. Glob Chang Biol 21:2479–2487

Doellman MM, Trussell GC, Grahame JW et al (2011) Phylogeographic analysis reveals a deep lineage split within North Atlantic *Littorina saxatilis*. Proc Biol Sci 278:3175–3183

Donovan SK (1989) Palaeontological criteria for the recognition of mass extinction. In: Donovan SK (ed) Mass extinctions: processes and evidence. Bellhaven Press, London, pp 19–36

Edgar GJ, Stuart-Smith RD, Willis TJ et al (2014) Global conservation outcomes depend on marine protected areas with five key features. Nature 506:216–220

Erwin DH, Hua Zhang PH (1996) Recoveries and radiations: gastropods after the Permo-Triassic mass extinction. In: Hart MB (ed) Biotic recovery from mass extinction events. Geological Society Special Publication, London, pp 223–229

Fara E (2001) What are Lazarus taxa? Geol J 36:291–303

Farabegoli E, Perri MC, Posenato R (2007) Environmental and biotic changes across the Permian–Triassic boundary in western Tethys: The Bulla parastratotype, Italy. Glob Planet Chang 55:109–135

Fine M, Gildor H, Genin A (2013) A coral reef refuge in the Red Sea. Glob Chang Biol 19: 3640–3647

Flessa KW, Jablonski D (1983) Extinction is here to stay. Paleobiology 9:315–321

Forel M-B, Crasquin S, Kershaw S et al (2013) In the aftermath of the end-Permian extinction: the microbialite refuge? Terra Nova 25:137–143

Galil BS, Zibrowius H (1998) First benthos samples from Eratosthenes Seamount, eastern Mediterranean. Senckenberg Marit 28:111–121

Gall J-C, Grauvogel-Stamm L (2005) The early Middle Triassic 'Grès àVoltzia' Formation of eastern France: a model of environmental refugium. Comptes Rendus Palevol 4:637–652

Garcia-Gomez JC, Lopez-Fe CM, Espinosa F et al (2011) Marine artificial micro-reserves: a possibility for the conservation of endangered species living on artificial substrata. Mar Ecol 32:6–14

Garcia-Gomez JC, Guerra-Garcia JM, Espinosa F et al (2014) Artificial Marine Micro-Reserves Networks (AMMRNs): an innovative approach to conserve marine littoral biodiversity and protect endangered species. Mar Ecol 36:259–277

Gavin DG, Fitzpatrick M, Gugger PF et al (2014) Climate refugia: joint inference from fossil records, species distribution models and phylogeography. New Phytol 204:37–54

Gell FR, Roberts CM (2003) Benefits beyond boundaries: the fishery effects of marine reserves. Trends Ecol Evol 18:448–455

Gibbs SJ, Bown PR, Ridgwell A et al (2016) Ocean warming, not acidification, controlled coccolithophore response during past greenhouse climate change. Geology 44:60–62

Gonzalez-Wevar C, Saucede T, Morley SA et al (2013) Extinction and recolonization of maritime Antarctica in the limpet *Nacella concinna* (Strebel, 1908) during the last glacial cycle: toward a model of Quaternary biogeography in shallow Antarctic invertebrates. Mol Ecol 22:5221–5236

Graham AGC, Smith JA (2012) Palaeoglaciology of the Alexander Island ice cap, western Antarctic Peninsula, reconstructed from marine geophysical and core data. Quat Sci Rev 35:63–81

Graham MH, Kinlan BP, Druehl LD et al (2007) Deep-water kelp refugia as potential hotspots of tropical marine diversity and productivity. Proc Natl Acad Sci USA 104:16576–16580

Grant WS, Cheng W (2012) Incorporating deep and shallow components of genetic structure into the management of Alaskan red king crab. Evol Appl 5:820–837

Greenstein BJ, Pandolfi JM (2008) Escaping the heat: range shifts of reef coral taxa in coastal Western Australia. Glob Chang Biol 14:513–528

Harlin-Cognato A, Bickham JW, Loughlin TR et al (2006) Glacial refugia and the phylogeography of Steller's sea lion (*Eumatopias jubatus*) in the North Pacific. J Evol Biol 19:955–969

Harries PJ, Knorr PO (2009) What does the 'Lilliput Effect' mean? Palaeogeogr Palaeoecol 284:4–10

Harries PJ, Kauffman EG, Hansen TA (1996) Models for biotic survival following mass extinction. In: Hart MB (ed) Biotic recovery from mass extinction events. The Geological Society, London, pp 41–60

Hautmann M, Nützel A (2005) First record of a heterodont bivalve (Mollusca) from the Early Triassic: palaeoecological significance and implications for the 'Lazarus problem'. Palaeontology 48:1131–1138

Hautmann M, Bucher H, Bruhwiler T et al (2011) An unusually diverse mollusc fauna from the earliest Triassic of South China and its implications for benthic recovery after the end-Permian biotic crisis. Geobios 44:71–85

Hellberg ME, Balch DP, Roy K (2001) Climate-driven range expansion and morphological evolution in a marine gastropod. Science 292:1707–1710

Hemery LG, Eleaume M, Roussel V et al (2012) Comprehensive sampling reveals circumpolarity and sympatry in seven mitochondrial lineages of the Southern Ocean crinoid species *Promachocrinus kerguelensis* (Echinodermata). Mol Ecol 21:2502–2518

Hoarau G, Coyer JA, Veldsink JH et al (2007) Glacial refugia and recolonization pathways in the brown seaweed *Fucus serratus*. Mol Ecol 16:3606–3616

Holterhoff PF (1996) Crinoid biofacies in Upper Carboniferous cyclothems, midcontinent North America: faunal tracking and the role of regional processes in biofacies recurrence. Palaeogeogr Palaeoecol 127:47–81

House MR (1996) Juvenile goniatite survival strategies following Devonian extinction events. In: Hart MB (ed) Biotic recovery from mass extinction events. Geological Society Special Publication, London, pp 163–185

Howarth LM, Roberts CM, Hawkins JP et al (2015) Effects of ecosystem protection on scallop populations within a community-led temperate marine reserve. Mar Biol 16:823–840

Huang B, Rong J-Y, Harper DAT (2013) A new survivor species of *Dicoelosia* (Brachiopoda) from Rhuddanian (Silurian) shallower-water biofacies in south China. J Paleontol 87:232–242

Inger R, Attrill MJ, Bearhop S et al (2009) Marine renewable energy: potential benefits to biodiversity? An urgent call for research. J Appl Ecol 46:1145–1153
Jablonski D (1986) Causes and consequences of mass extinctions: a comparative approach. In: Elliot DK (ed) Dynamics of extinction. Wiley and Sons, New York, pp 183–229
Jablonski D, Flessa KW (1986) The taxonomic structure of shallow-water faunas: implications for Phanerozoic extinctions. Malacologia 27:43–66
Jacobs DK, Lindenberg DR (1998) Oxygen and evolutionary patterns in the sea: onshore/offshore trends and recent recruitment of deep-sea faunas. Proc Natl Acad Sci U S A 95:9396–9401
Johannes RE (1978) Traditional marine conservation methods in Oceania and their demise. Annu Rev Ecol Syst 9:349–364
Jolly MT, Viard F, Gentil F et al (2006) Comparative phylogeography of two coastal polychaete tubeworms in the Northeast Atlantic supports shared history and vicariant events. Mol Ecol 15:1841–1855
Kauffman EG, Erwin DH (1995) Surviving mass extinctions. Geotimes 14:14–17
Kauffman EG, Harries PJ (1996) The importance of crisis progenitors in recovery from mass extinction. In: Hart MB (ed) Biotic recovery from mass extinction events. Geological Society Special Publication, London, pp 15–39
Kelaher BP, Coleman MA, Broad A et al (2014) Changes in fish assemblages following the establishment of a network of no-take marine reserves and partially-protected areas. PLoS One 9:e85825
Kiessling W, Aberhan M (2007) Geographical distribution and extinction risk: lessons from Triassic–Jurassic marine benthic organisms. J Biogeogr 34:1473–1489
Kiessling W, Roniewicz E, Villier L et al (2009) An early Hettangian coral reef in southern France: implications for the end-Triassic reef crisis. PALAIOS 24:657–671
Knaust D (2010) The end-Permian mass extinction and its aftermath on an equatorial carbonate platform: insights from ichnology. Terra Nova 22:195–202
Krakau M, Jacobsen S, Jensen KT et al (2012) The cockle *Cerastoderma edule* at Northeast Atlantic shores: genetic signatures of glacial refugia. Mar Biol 159:221–230
Krebes L, Blank M, Jurss K et al (2010) Glacial-driven vicariance in the amphipod *Gammarus duebeni*. Mol Phylogenet Evol 54:372–385
Leighton LR, Schneider CL (2008) Taxon characteristics that promote survivorship through the Permian-Basal Triassic interval: implications for the change in evolutionary faunas. Paleobiology 34:78–92
Lethiers F, Cassier J-G (1999) Autopsie d'une extinction biologique. Un exemple: la crise de al limite Frasnien-Famennien (364 Ma). Earth Planet Sci 329:303–315
Li J, Hu Z, Duan D (2015) Genetic data from the red alga *Palmaria palmata* reveal a mid-Pleistocene deep genetic split in the North Atlantic. J Biogeogr 42:902–913
Ludt WB, Rocha LA (2015) Shifting seas: the impacts of Pleistocene sea-level fluctuations on the evolution of tropical marine taxa. J Biogeogr 42:25–38
Ludt WB, Bernal MA, Bowen BW et al (2012) Living in the past: phylogeography and population histories of Indo-Pacific wrasses (genus *Halichoeres*) in shallow lagoons versus outer reef slopes. PLoS One 7:e38042
Maggs CA, Castilho R, Foltz D et al (2008) Evaluating signatures of glacial refugia for North Atlantic benthic marine taxa. Ecology 89:S108–S122
Maggs CA, Mann BQ, Cowley PD (2013) Contribution of a large no-take zone to the management of vulnerable reef fishes in the South-West Indian Ocean. Fish Res 144:38–47
Makino A, Klein CJ, Possingham HP et al (2015) The effect of applying alternate IPCC climate scenarios to marine reserve design for range changing species. Conserv Lett 8:320–328
Mateus CS, Almeida PR, Mesquita N et al (2016) European lampreys: new insights on postglacial colonization, gene flow and speciation. PLoS One 11:e0148107
McClanahan TR, Ateweberhan M, Muhando CA et al (2007) Effects of climate and seawater temperature variation on coral bleaching and mortality. Ecol Monogr 77:503–525
Mertens KN, Takano Y, Head MJ et al (2014) Living fossils in the Indo-Pacific warm pool: a refuge for thermophilic dinoflagellates during glaciations. Geology 42:531–534

Myers C, MacKenzie RA III, Lieberman BS (2013) Greenhouse biogeography: the relationship of geographic range to invasion and extinction in the Cretaceous Western Interior Seaway. Paleobiology 39:135–148

Nakin MDV, McQuaid CD (2014) Marine reserve effects on population density and size structure of commonly and rarely exploited limpets in South Africa. Afr J Mar Sci 36:303–311

Naro-Maciel E, Reid BN, Alter SE et al (2014) From refugia to rookeries: phylogeography of Atlantic green turtles. J Exp Mar Biol Ecol 461:306–316

National Ocean Service (2016) Hawaiian islands humpback whale national marine sanctuary. http://hawaiihumpbackwhale.noaa.gov. Accessed 05 Nov 2016

Neiva J, Assis J, Fernandes F et al (2014) Species distribution models and mitochondrial DNA phylogeography suggest an extensive biogeographical shift in the high-intertidal seaweed *Pelvetia canaliculata*. J Biogeogr 6:1137–1148

New Zealand Department of Conservation (2016) Marine reserves A-Z. http://www.doc.govt.nz/marinereserves. Accessed 04 Nov 2016

Obama B (2014) Proclamation 9173—Pacific remote islands national marine monument expansion. Fed Regist 79:58645–58653

Olsen JL, Zechman FW, Hoarau G et al (2010) The phylogeographic architecture of the fucoid seaweed *Ascophyllum nodosum*: an intertidal 'marine tree' and survivor of more than one glacial–interglacial cycle. J Biogeogr 37:842–856

Packard AS (1886) Geological extinction and some of its apparent causes. Am Nat 20:29–40

Pandolfi JM (1999) Response of Pleistocene coral reefs to environmental change over long temporal scales. Am Zool 39:113–130

Payne JL (2005) Evolutionary dynamics of gastropod size across the end-Permian extinction and through the Triassic recovery interval. Paleobiology 31:269–290

Pellissier L, Leprieur F, Parravinci V et al (2014) Quaternary coral reef refugia preserved fish diversity. Science 344:1016–1019

Perrin C, Bosellini FR (2013) The Late Miocene coldspot of z-coral diversity in the Mediterranean: patterns and causes. Comptes Rendus Palevol 12:245–255

Petuch EJ (1981) A relict Neogene caenogastropod fauna from northern South America. Malacologia 20:307–347

Pinsky ML, Newsome SD, Dickerson BR et al (2010) Dispersal provided resilience to range collapse in a marine mammal: insights from the past to inform conservation biology. Mol Ecol 19:2418–2429

Portnoy DS, Hollenbeck CM, Belcher CN et al (2014) Contemporary population structure and post-glacial genetic demography in a migratory marine species, the blacknose shark, *Carcharhinus acronotus*. Mol Biol 23:5480–5495

Posenato R (2009) Survival patterns of macrobenthic marine assemblages during the end-Permian mass extinction in the western Tethys (Dolomites, Italy). Palaeogeogr Palaeoecol 280:150–167

Precht WF, Aronson RB (2004) Climate flickers and range shifts of reef corals. Front Ecol Environ 2:307–314

Provan J, Wattier RA, Maggs CA (2005) Phylogeographic analysis of the red seaweed *Palmaria palmata* reveals a Pleistocene marine glacial refugium in the English Channel. Mol Ecol 14:793–803

Quintiero J, Manent P, Perez-Dieguez L et al (2015) Phylogeography of a marine insular endemic in the Atlantic Macaronesia: The Azorean barnacle, *Megabalanus azoricus* (Pilsbry, 1916). PLoS One 10:e0124707

Rassmussen C, Harper DAT (2011) Interrogation of distributional data for the End Ordovician crisis interval: where did disaster strike? Geol J 46:478–500

Rickards RB, Wright AJ (2002) Lazarus taxa, refugia and relict faunas: evidence from graptolites. J Geol Soc Lond 159:1–4

Roberts CM, Hawkins JP (1997) How small can a marine reserve be and still be effective? Coral Reefs 16:150

Rodland DL, Bottjer DJ (2001) Biotic recovery from the end-Permian mass extinction: behavior of the inarticulate brachiopod *Lingula* as a disaster taxon. PALAIOS 16:95–101

Rogers-Bennett L, Pearse JS (2001) Indirect benefits of marine protected areas for juvenile abalone. Conserv Biol 15:642–647

Rowden AA, Dower JF, Schlacher TA (2010) Paradigms in seamount ecology: fact, fiction and future. Mar Ecol 31:226–241

Schmidt GM, Wall M, Taylor M et al (2016) Large-amplitude internal waves sustain coral health during thermal stress. Coral Reefs 35:869–881

Schubert JK, Bottjer DJ (1995) Aftermath of the Permian-Triassic mass extinction event: paleoecology of Lower Triassic carbonates in the western USA. Palaeogeogr Palaeoecol 116:1–39

Sciberras M, Jenkins SR, Mant R et al (2015) Evaluating the relative conservation value of fully and partially protected marine areas. Environ Evid 16:58–77

Shen J, Lei Y, Algeo TJ et al (2013) Volcanic effects on microplankton during the Permian-Triassic transition (Shangsi and Xinmin, south China). PALAIOS 28:552–567

Shoo LP, Hoffman AA, Garnett S et al (2013) Making decisions to conserve species under climate change. Clim Chang 119:239–246

Song H, Yang L, Tong J et al (2015) Recovery dynamics of foraminifers and algae following the Permian-Triassic extinction in Qingyan, South China. Geobios 48:71–83

Stanley GD Jr, Beauvais L (1994) Corals from an early Jurassic coral reef in British Columbia: refuge on an oceanic island reef. Lethaia 27:35–47

Steneck RS, Paris CB, Arnold SN et al (2009) Thinking and managing outside the box: coalescing connectivity networks to build region-wide resilience in coral reef ecosystems. Coral Reefs 28:367–378

Stewart JR, Lister AM (2001) Cryptic northern refugia and the origins of the modern biota. Trends Ecol Evol 16:608–613

Stewart JR, Lister AM, Barnes I et al (2010) Refugia revisited: individualistic responses of species in space and time. Proc Biol Sci 277:661–671

Tittensor DP, Baco AR, Hall-Spencer JM et al (2010) Seamounts as refugia from ocean acidification for cold-water stony corals. Mar Ecol 31:212–225

Toth LT, van Woesik R, Murdoch TJT et al (2014) Do no-take reserves benefit Florida's corals? 14 years of change and stasis in the Florida Keys National Marine Sanctuary. Coral Reefs 33:565–577

U.S. Department of State (2016) On the new marine protected area in Antarctica's Ross Sea. https://www.state.gov/secretary/remarks/2016/10/263763.htm. Accessed 04 Nov 2016

Urbanek A (1993) Biotic crises in the history of the Upper Silurian graptoloids: a palaeobiological model. Hist Biol 7:29–50

van Hooidonk R, Maynard JA, Liu Y et al (2015) Downscaled projections of Caribbean coral bleaching that can inform conservation planning. Glob Chang Biol 21:3389–3401

Van Oppen MJH, Peplow LM, Kininmonth S et al (2011) Historical and contemporary factors shape the population genetic structure of the broadcast spawning coral, *Acropora millepora*, on the Great Barrier Reef. Mol Ecol 20:4899–4914

Vermeij GJ (1986) Survival during biotic rises: the properties and evolutionary significance of refuges. In: Elliott DK (ed) Dynamics of extinction. Wiley, New York, pp 231–246

Wall M, Putchim L, Schmidt GM et al (2015) Large-amplitude internal waves benefit corals during thermal stress. Proc Biol Sci 282:1–9

Waltari E, Hickerson MJ (2013) Late Pleistocene species distribution modelling of North Atlantic intertidal invertebrates. J Biogeogr 40:249–260

Washington Department of Fish and Wildlife (2016) Marine protected areas within Puget Sound. http://wdfw.wa.gov/fishing/mpa/sanjuan_upright_channels.html. Accessed 04 Nov 2016

Waterhouse JB, Shi GR (2010) Evolution in a cold climate. Palaeogeogr Palaeoecol 298:17–30

Wenger AS, Williamson DH, daSilva ET et al (2016) Effects of reduced water quality on coral reefs in and out of no-take marine reserves. Conserv Biol 30:142–153

Westermann GEG (2001) Modes of extinction, pseudo-extinction and distribution in Middle Jurassic ammonites: terminology. Can J Earth Sci 38:187–195

Westrop SR (1989) Trilobite mass extinction near the Cambrian-Ordovician boundary in North America. In: Donovan SK (ed) Mass extinctions: processes and evidence. Columbia University Press, New York, pp 89–103
White House (2016) Fact sheet: President Obama to create the world's largest marine protected area. https://www.whitehouse.gov/the-press-office/2016/08/26/fact-sheet-president-obama-create-worlds-largest-marine-protected-area. Accessed 04 Nov 2016
Wignall PB, Benton MJ (1999) Lazarus taxa and fossil abundance at times of biotic crisis. J Geol Soc Lond 156:453–456
Xu C, Melchin MJ, Sheets HD et al (2005) Patterns and processes of latest Ordovician graptolite extinction and recovery based on data from south China. J Paleontol 79:842–861
Yamaguchi T, Norris RD (2015) No place to retreat: heavy extinction and delayed recovery on a Pacific guyot during the Paleocene–Eocene Thermal Maximum. Geology 43:443–446
Yamano H, Sugihara K, Nomura K (2011) Rapid poleward range expansion of tropical reef corals in response to rising sea surface temperatures. Geophys Res Lett 38:L04601
Yin J, Fursich FT (2008) Dispersal events of Triassic-Jurassic boundary faunas, and paleoenvironment of Tibetan Himalaya. Sci China Ser D Earth Sci 52:1993–2000
Younger JL, Emmerson LM, Miller KJ (2016) The influence of historical climate changes on Southern Ocean marine predator populations: a comparative analysis. Glob Chang Biol 22:474–493
Zhao X, Tong J, Yao H et al (2015) Early Triassic trace fossils from the Three Gorges area of South China: implications for the recovery of benthic ecosystems following the Permian–Triassic extinction. Palaeogeogr Palaeoecol 429:100–116
Zonneveld J-P, Gingras MK, Beatty TW (2010a) Diverse ichnofossil assemblages following the P–T mass extinction, Lower Triassic, Alberta and British Columbia, Canada: evidence for shallow marine refugia on the northwestern coast of Pangaea. PALAIOS 25:368–392
Zonneveld J-P, McNaughton RB, Utting J et al (2010b) Sedimentology and ichnology of the Lower Triassic Montney Formation in the Pedigree-Ring/Border-Kahntah River area, northwestern Alberta and northeastern British Columbia. Bull Can Petrol Geol 58:115–140

Training Tomorrow's Conservation Paleobiologists

Patricia H. Kelley, Gregory P. Dietl, and Christy C. Visaggi

Abstract Conservation paleobiology (CPB) is committed to the mission of applying geohistorical records to the conservation and restoration of biodiversity and ecosystem services. Fulfilling this mission remains challenging because most conservation paleobiologists have not been trained to translate their science into management practice and policy. Ongoing discussion among conservation biologists provides lessons applicable to training tomorrow's conservation paleobiologists. We offer six recommendations for more effective training in conservation paleobiology: (1) integrate CPB into truly cross-disciplinary conservation curricula; (2) promote a problem-solving and policy-oriented approach to CPB education; (3) implement hands-on experience in real-world settings; (4) promote informal opportunities for cross-disciplinary interaction; (5) establish connections among stakeholders; and (6) modify faculty incentive systems to reward CPB activities. Institutional constraints make implementation of these recommendations challenging, and development of CPB programs will involve tradeoffs. Despite these challenges, the approach we propose will better prepare tomorrow's conservation paleobiologists to function effectively in the conservation world.

Keywords Conservation training · Cross-disciplinary curricula · Interpersonal skills · Problem-solving · Policy-oriented education · Real-world experience

---

Patricia H. Kelley and Gregory P. Dietl contributed equally.

P. H. Kelley (✉)
Department of Earth and Ocean Sciences, University of North Carolina Wilmington, Wilmington, NC, USA
e-mail: kelleyp@uncw.edu

G. P. Dietl (✉)
Paleontological Research Institution, Ithaca, NY, USA

Department of Earth and Atmospheric Sciences, Cornell University, Ithaca, NY, USA
e-mail: gpd3@cornell.edu

C. C. Visaggi
Department of Geosciences, Georgia State University, Atlanta, GA, USA
e-mail: cvisaggi@gsu.edu

C. L. Tyler, C. L. Schneider (eds.), *Marine Conservation Paleobiology*,
Topics in Geobiology 47, https://doi.org/10.1007/978-3-319-73795-9_9

1 Business As Usual Is Not Enough

The past decade has witnessed workshops defining and charting the course of the emerging field of conservation paleobiology (CPB), sessions and presentations on CPB-related topics at professional meetings, an expanding technical and popular literature (Dietl 2016), and even university classes on the new field. Some of us now consider ourselves conservation paleobiologists and have begun to train students to follow in our footsteps. But is it enough to follow in our footsteps, or must tomorrow's conservation paleobiologists blaze a new path? What does it really mean to be a conservation paleobiologist?

Obviously, it means more than being a paleobiologist; otherwise, the modifier would be unnecessary. By adding "conservation" to our moniker, we are committing to a mission—a value-laden framework for our work (Dietl 2016). This mission-driven approach is implicit in definitions adopted for CPB. For instance, a 2011 National Science Foundation (NSF)-funded Conservation Paleobiology Workshop (2012, p. 1; see also Dietl et al. 2015) defined CPB as "the application of geohistorical records to the conservation and restoration of biodiversity and ecosystem services." Although conservation paleobiologists have done a good job extracting and interpreting geohistorical records, as exemplified by papers in this volume, the *application* to the conservation and restoration of biodiversity and ecosystem services remains challenging.

Most of us who claim to be conservation paleobiologists typically conduct the research we would have done anyway; at best, we comment vaguely (usually in the introduction and conclusions sections of our papers) that the work has implications for conservation (e.g., "we studied the Pliocene and the Pliocene was warm, so our results have implications for how biota are likely to respond in a warmer Anthropocene world"). Often the conservation implications of our work are poorly articulated and, even if stated, are only read by our fellow conservation paleobiologists. This ineffectiveness at translating our results into management practice and policy is not surprising because the majority of conservation paleobiologists are traditionally trained paleontologists who are recruits to the new field from our own specialized research disciplines within paleontology. As academic scientists, we have not been trained to interpret geohistorical data in a management context. Our abilities in management and policy, if any, are self-taught or acquired by trial and error, or better yet trial and success (see Flessa 2017).

The need for cross-disciplinary (sensu Ciannelli et al. 2014)[1] training of conservation professionals is recognized widely (e.g., Jacobson 1990; Jacobson and Robinson 1990; Kainer et al. 2006; Moslemi et al. 2009). The field of CPB is often touted as being cross-disciplinary (e.g., Conservation Paleobiology Workshop

[1]Cross-disciplinary work may include multidisciplinary, interdisciplinary, or transdisciplinary collaboration (Rosenfield 1992; Ciannelli et al. 2014) in which workers from multiple disciplines address a common problem with no (multidisciplinary), some (interdisciplinary), or substantial (transdisciplinary) integration of disciplines.

2012), but few of us were trained in a cross-disciplinary manner and our collaborations with non-paleobiologists (beyond the occasional biologist) are limited. CPB is dominated by academic scientists "from strongly single-disciplined, specialized backgrounds" (as described by Jacobson 1990, p. 433, for conservation biologists) who have little experience outside the university and are ill-prepared to train students for the current job market in conservation. Consequently, this conventional emphasis on disciplinary research in CPB training leaves many students poorly prepared to contribute to the field of conservation. This problem is particularly acute because, as noted by Noss (1999), the vast majority of jobs for conservation biologists (and presumably future conservation paleobiologists) are not and will not be in universities, but in governmental land-managing and regulatory agencies, conservation groups (e.g., NGOs), consulting firms, research institutes, and industry. Indeed, Lucas et al. (2017) noted that only 10% of conservation jobs in the current market are in academia. We cannot assume that our students will pursue research-based academic careers as we did. This situation is true for the natural sciences in general; the number of PhD scientists entering academic positions declined and of those taking positions in business and industry increased markedly (from 19 to 32%) from 2004 to 2014 (Kurilla 2016, citing data from the NSF). Doctoral students are trained for research careers, but employers seek candidates with more than research experience (Campbell et al. 2005; Perez 2005; Blickley et al. 2012; Kurilla 2016).

Fortunately (or perhaps unfortunately), we are not alone in our concerns about the lack of adequate training for future conservation (paleo)biology practitioners. The issue of inadequate training of conservation *biologists* to meet real-world challenges is long-standing (e.g., Jacobson 1990; Noss 1999; Clark 2001; Perez 2005; Muir and Schwartz 2009). In this essay, we draw lessons from this ongoing discourse in conservation biology that are applicable to training tomorrow's conservation paleobiologists.

2 A Call to Action

Brewer (2001, p. 1203), in what might be characterized as a "call to action," asked "What do we want conservation biologists in the twenty-first century to know and be able to do?" Similarly, a good place to start in determining how we might train tomorrow's conservation paleobiologists is by considering what a conservation paleobiologist should know and be able to do.

Our students need to be trained to function effectively in the real world of conservation. Such training should enable students to acquire both the appropriate scientific knowledge and the ability to apply it. Traditional academic training focuses on acquisition of knowledge and, especially at the PhD level, development of discipline-focused research skills. However, we do an abysmal job of preparing students to function outside narrowly defined disciplinary research topics, even though the job market in conservation "increasingly requires broadly trained, versatile professionals" (Noss 1999, p. 118). A decades-long lament in conservation biology (Jacobson and McDuff 1998, p. 433) is that most academic curricula "remain rigidly departmental with little opportunity for interdisciplinary training" (Noss 1999, p. 118); depth of knowledge is emphasized at the graduate level at the expense of breadth, even though solving complex environmental problems requires working closely with colleagues from many disciplines (Jacobson 1990, Jacobson and Robinson 1990, Ciannelli et al. 2014).

Jacobson and McDuff (1998, p. 263) warned that, in conservation biology "we may, in fact, be training idiot savants—individuals skilled in certain areas... but largely inept in other aspects of the field." In the case of CPB, our "idiot savants" may be well trained in the technical paleontological methods and theories applicable to conservation, but they lack the ability to translate the results of their research into action. This gap between research and implementation is recognized increasingly in conservation biology, in which "conservation assessments are rarely translated into actions that actually conserve nature" (Knight et al. 2008, p. 610; see also Jarvis et al. 2015). In general, the basic sciences do not value development of essential interpersonal skills in policy processes (Clark 2001), communications (e.g., the ability to resolve conflict or build trust in relationships), and leadership, including the ability to influence others (Jacobson and McDuff 1998). None of these abilities (or "soft skills") is viewed traditionally as part of the professional identity of a paleontologist, and yet they are essential if conservation paleobiologists are to function effectively in the conservation world. Translating Jacobson and McDuff's comments into CPB terminology, a graduate advisor would never send a paleontology student into the field without training in measuring sections, describing lithology, and collecting samples, as appropriate to the project. Yet we are currently sending the first generation of conservation paleobiologists into the conservation field "to confront diverse stakeholders without any training on how to assess and address" real-world conservation issues (Jacobson and McDuff 1998, p. 265). This situation must change if CPB is to have any real-world impact. The next generation of conservation paleobiologists must have truly cross-disciplinary training, including training in the human dimensions of conservation issues, if they are to help solve conservation problems. To achieve this goal, CPB training must adapt in order to produce graduates with the skillset required to navigate the conservation landscape.

3 Bridging the Gap

We offer six specific recommendations, none of which is original to us, but all of which would prepare students better to be effective conservation paleobiologists, i.e., who can bridge the widening research-implementation gap in conservation (Knight et al. 2008; Jarvis et al. 2015). Recommendations are curricular, programmatic, and/or institutional: (1) integrate CPB into truly cross-disciplinary conservation curricula; (2) promote a problem-solving and policy-oriented approach to CPB education; (3) implement hands-on experience in real-world settings;

(4) promote informal opportunities for cross-disciplinary interaction; (5) establish connections among stakeholders; and (6) modify faculty incentive systems to reward CPB activities.

Recommendation 1

Most universities have established programs relevant to conservation. The Society for Conservation Biology[2] reports nearly 600 academic programs worldwide specifically in conservation biology. Of those, 60 are graduate programs. Such programs typically are housed within traditional biology departments or wildlife and natural resources programs (Jacobson 1990). Many schools without formal conservation biology programs offer programs in environmental studies that provide some of the same training. Thus, CPB programs need not "reinvent the wheel"; making connections with existing programs on campus is an ideal way to develop cross-curricular CPB programs.

Jacobson (1990) highlighted 16 exemplary conservation biology programs, the majority of which were cross-disciplinary and included non-biology faculty. She provided a model for cross-disciplinary conservation biology education, in which biology represented one of five general areas. Additional areas considered fundamental to conservation biology education included the physical environment (chemistry, geology, physics, geography), social environment (economics, political science, sociology, anthropology, philosophy), applied management sciences (e.g., wildlife, forestry, fisheries), and the implementational environment (e.g., planning, education, law, communication). Typically, the highlighted programs require a few core courses, with electives allowing students to emphasize areas of interest (Jacobson 1990).

We recommend working with existing programs on campus to design a cross-disciplinary CPB option or track, which could be housed within a geology/geoscience degree program, a conservation biology program, or an environmental studies program as appropriate to the culture of the institution. The CPB track would include the necessary geology courses (e.g., paleoecology, stratigraphy, field and/or research methods) but would also benefit from including courses in such areas as conservation biology, environmental policy, economics, environmental law, and environmental philosophy. Another option would be to implement a cross-disciplinary conservation certificate program, rather than a degree program, as has been done successfully at University of Florida (the "Tropical Conservation and Development" program[3]; Kainer et al. 2006). Even if formal implementation of a CPB track within a degree or certificate program

[2]https://conbio.org/professional-development/academic-programs.

[3]http://www.tcd.ufl.edu/.

is not possible, students with CPB interests should be advised to take courses within those areas to obtain the cross-disciplinary training needed to function as conservation professionals (see also Recommendation 2).

Recommendation 2

Education of future conservation paleobiologists should not only be cross disciplinary but also policy oriented. Jacobson and McDuff (1998) and Muir and Schwartz (2009) argued for inclusion of more of the "human dimension" in conservation training. A survey of conservation alumni employed in the academic and non-academic sectors revealed that, whereas academicians emphasized research skills, the practitioners focused on decision-making, policy, and management skills (Muir and Schwartz 2009). Clark (2001) also recognized that conservation biology curricula need a more policy-oriented approach. Likewise, CPB will only fulfill the first part of its name if tomorrow's conservation paleobiologists receive training in policy making. Critical thinking skills beyond those involved in conducting scientific research are needed; students need to be able to apply these skills to policy problems (Clark 2001). For instance, CPB research can inform decision-making on a host of policy issues, including establishing conservation priorities (e.g., regarding populations, species, regions), selecting biological reserves, and restoring ecosystem function by substituting extant for extinct species in "rewilding" projects (Conservation Paleobiology Workshop 2012). However, for the results of CPB research to be used most effectively, future practitioners need to understand not only the technicalities of the decision-making process and the political context (Flessa 2017) but also the psychology of the human interactions involved. The results of CPB research are rarely translatable directly into policy. In addition, the best solution to a conservation issue from a scientific standpoint may not be the best solution from a societal perspective. Goals and values of policy makers and their constituents play a role that may conflict with recommendations based solely on scientific principles (Kroll 2007; Dietl and Flessa 2018). The scientific principles, however, must be conveyed in a manner accessible to policy makers and to the general public (Brewer 2001), including via appropriate communication through the media (Flessa 2017), e.g., news media, social media, and public-accessible videos.

Policy-oriented training might be accomplished in formal course work. Most institutions offer a wide range of courses that could contribute to a policy-oriented approach to conservation education. For instance, at the University of North Carolina Wilmington[4] (UNCW), relevant graduate-level courses include "Coastal and Environmental Science and Policy," "Decision Making and Negotiation," "Strategic Communication: Analysis of Persuasion and Cultural Principles and Techniques," "Advanced Environmental Law and Policy," "Conservation and Culture," "Envi-

[4]http://www.uncw.edu/gradschool/index.html.

ronmental Management," and "Sociology of Organizations," among many others. Judicious choice of electives by future conservation paleobiologists could help fill a major gap in their education.

Recommendation 3

Education needs to go beyond course work to provide real-world experiences for future conservation paleobiologists. Meffe (1998, p. 260) noted that "Without practical experience [students] could provide only theoretical 'knowledge' of an experience-based activity. Medical schools train their students in real hospitals with actual patients. Auto mechanics learn on real transmissions and clutches. Why should conservation biologists not be similarly trained in the actual complexities and difficulties of real-world issues?" We pose the same question for conservation paleobiologists.

Case studies are used by some conservation programs (Jacobson and Robinson 1990; Clark 2001) to provide real-world context. By giving students the opportunity to participate in devising solutions to a particular environmental issue and interacting with a stakeholder (e.g., Ciannelli et al. 2014), this approach can provide practical experience in the real world (i.e., experiential learning). To a degree, service-learning activities embedded within content courses (Hansen and Fortner 2016) also can provide such experiences. Conservation-related service learning, involving hands-on community service, allows students to engage in local issues. For instance, at Indiana University~Purdue University Indianapolis[5] (IUPUI), students undertake environmental stewardship activities at sites in central Indiana, including community work days focused on ecosystem restoration (Tedesco and Salazar 2006). Such place-based education (Gruenewald 2003; Sobel 2004) capitalizes on the personal attachment that learners have for their surroundings to teach concepts across disciplines; it increases student engagement, academic success, and retention (Tedesco and Salazar 2006; Semken and Butler Freeman 2008; Semken et al. 2009) and provides skills valued by employers, including communication, project management, collaboration, and technical skills (Shriberg and Harris 2012). Such activities would provide useful additions to CPB-focused courses at the undergraduate or graduate level.

The Smithsonian Mason School of Conservation,[6] in which students can enroll for a semester (similar to study abroad enrollment), combines course work with a practicum experience at the Smithsonian Conservation Biology Institute. Other approaches incorporate policy and management components into field or research experiences (Ort et al. 2006; Kelley and Dietl 2012). For example, the San Juan River field course operated by the Environmental Sciences Program at Northern

[5]http://csl.iupui.edu/.

[6]http://smconservation.gmu.edu/.

Arizona University[7] combines classroom work with a cross-disciplinary (geology, biology, and chemistry) field experience with in-depth discussion of related policy and land management issues (Ort et al. 2006). In the NSF-funded "Research Experiences for Undergraduates in Biodiversity Conservation" summer program[8] at UNCW (Kelley and Dietl 2012) in which the three of us were involved, geology, biology, archeology, and environmental science students worked in cross-disciplinary teams to compare the modern marine ecosystem in the Carolinas with communities of the past, using the Plio-Pleistocene fossil record, the archeological record of shell middens, and live-dead analysis in modern environments. We discussed current environmental issues, their policy implications, and how conservation paleobiology might contribute to their resolution, for instance, by determining baselines and the natural range of variability for use in restoration ecology.

Although such experiences are valuable, more immersive experiences in real-world situations are needed. The conservation education literature emphasizes that internships are necessary to provide job experience outside the university (Jacobson 1990; Jacobson and Robinson 1990; Jacobson and McDuff 1998; Soulé and Press 1998; White et al. 2000; Clark 2001; Kainer et al. 2006; Kroll 2007; Moslemi et al. 2009; see also Conservation Paleobiology Workshop 2012 and Flessa 2017). Undergraduate and graduate students who participate in summer or semester-long internships with government agencies (e.g., the National Parks Service Geoscientist-in-the-Park internship), industry, and mainstream environmental NGOs (e.g., The Nature Conservancy, World Wildlife Fund, Oceana) are far better prepared to function in the conservation profession (Conservation Paleobiology Workshop 2012). Internships foster development of skills valued by employers, including teamwork, problem solving, communication, fundraising and project management, policy analysis, and negotiation (Blickley et al. 2012). They allow students to observe the decision-making process (Kroll 2007) and interact with those making the decisions as well as with those affected by the decisions (Clark 2001). Therefore, we feel strongly that an internship or similar sustained, immersive, real-world experience is essential to a future conservation paleobiologist's training, even if the internship does not have an explicit paleontology dimension.

Recommendation 4

We recognize that institutional and other barriers exist to implementing our previous three recommendations (see below), but less formalized program elements can also contribute to CPB training. The Tropical Conservation and Development certificate program at University of Florida includes core courses, but much of the training occurs in "alternative learning spaces" (Kainer et al. 2006, p. 9). In addition to

[7] https://nau.edu/cefns/natsci/seses/sce/.

[8] http://www.nsf.gov/awardsearch/showAward?AWD_ID=0755109&HistoricalAwards=false.

internships, these activities include workshops, seminars, retreats, and visits by professionals from outside the university. Many of these program elements are shared by the Biogeochemistry and Environmental Biocomplexity Program[9] at Cornell University (Moslemi et al. 2009, p. 516). Such activities are not difficult to implement and can be done within the framework of existing university structure.

When students go beyond the minimum expectations of course and thesis/dissertation work, they acquire non-academic skills and experiences that are valued in the job market. Thus, involving students in organizing such seminars and workshops allows them to develop and demonstrate skills in project management, networking, and communication that can prepare them for non-academic conservation careers (Blickley et al. 2012). Seminar series that cross departmental boundaries foster sharing of knowledge among disciplines and cross-disciplinary collaboration. Such venues also can provide opportunities for students to interact with local conservation leaders. Workshops, short courses, and training academies (Kainer et al. 2006; Moslemi et al. 2009; Ciannelli et al. 2014) have also been used to foster interaction across disciplines and students' abilities to work in cross-disciplinary teams. Less formal activities, such as social events (e.g., taking an outside speaker for drinks or dinner), retreats, and reading groups, can also promote a feeling of community among students involved in conservation programs (Moslemi et al. 2009). For instance, at Cornell University, CREST[10]—the "Cornell Roundtable on Environmental Studies Topics"—brings together faculty, graduate students, and community members from the humanities, arts, social sciences, natural and physical sciences to discuss environmental issues.

Recommendation 5

Campbell et al. (2005) have argued that improvement of doctoral education will require establishing connections among stakeholders, including those organizations that fund and hire doctoral students. Students also need to be made aware of the range of employment opportunities outside academia and the skill sets needed in such careers; seminar series that include speakers from outside academia, or meetings with alumni, are useful in this regard (Campbell et al. 2005). Panels composed of alumni have been used successfully at our institutions to provide such guidance.

Partnerships with practitioners can be enhanced by including them as outside members on thesis/dissertation committees (Jacobson and McDuff 1998; Blickley et al. 2012), where they can provide input regarding real-world implications of CPB research. When students address research questions of interest to an external organization, partnership with the organization can "provide practical experience

[9] http://www.geo.cornell.edu/biogeo/beb/.

[10] https://blogs.cornell.edu/crest/.

with the institutional structure, culture, and work processes" (Blickley et al. 2012, p. 32) that will better prepare the student for a career in conservation, although from the perspective of the NGO or government agency the product of such collaborations is more likely to be solving a conservation problem rather than publishing a scientific paper (Boyer et al. 2017). Including practitioners as adjunct faculty or on program advisory boards or visiting committees would also strengthen ties between academic and non-academic stakeholders (Jacobson and McDuff 1998). Indeed, the process of establishing (and subsequently reviewing) a formal CPB program would require input from practitioners.

Brewer (2002) stressed the importance of partnerships with the community, for instance, with local schools or in citizen science. Environmental service learning at IUPUI, described above, is made possible by collaboration with 30 community partners, including the Indianapolis Department of Parks and Recreation Land Stewardship Office (Tedesco and Salazar 2006). Such partnerships are facilitated by the campus-wide IUPUI Center for Service and Learning, with a mission to promote civic engagement. Conservation paleobiology students could benefit from community partnerships as exemplified by recent efforts in community-based planning and research initiatives. In such projects, a researcher works with the community (e.g., NGOs, neighborhood councils, local schools) and students (undergraduate and graduate students, and often K-12) to set goals, collect data, and discuss implications; perspectives of all participants are included and valued. Students involved in such projects (e.g., mapping invasive plant species in an urban forest; Hawthorne et al. 2015) gain experience in teamwork and policy processes in a real-world context, addressing conservation issues that are important to the local community.

Recommendation 6

Implementing any of the previous recommendations will require a commitment of time and energy on the part of CPB faculty. Likewise, the mentoring necessary to prepare students for careers (Campbell et al. 2005) in CPB will require effort beyond that normally involved in advising graduate students within a single discipline. Thus, our final recommendation is that faculty incentive systems will need to be modified to promote cooperation among departments and to reward individual faculty for real-world problem solving at local, national, and international levels. Such cultural changes cannot be accomplished by the small number of CPB faculty alone; faculty in other disciplines who are engaged in applied community work will need to be enlisted as allies in influencing institutional culture.

The feasibility of implementing this strategy will vary among institutions. University structure can either hinder or facilitate the work we have described. A precedent for more applied community or regional engagement exists in the mission of some land-grant institutions. For instance, Florida Gulf Coast University, with a mission that "promotes and practices environmental sustainability, …

nurtures community partnerships, values public service, [and] encourages civic responsibility,"[11] is actively involved in environmental management and restoration in southwest Florida. Conservation paleobiology plays an essential role in the university's involvement in these activities, with faculty working across disciplines and with government agencies (e.g., Army Corps of Engineers, South Florida Water Management District, USGS, US Fish & Wildlife, Everglades National Park) and NGOs, including the Conservancy of Southwest Florida, the Audubon Society, Caloosahatchee River Watch, and the Southwest Florida Watershed Council (Boyer et al. 2017).

Some institutions also already have cross-disciplinary institutes or centers that could be the locus of CPB activities. The existence of such units within a university means that systems are in place to facilitate communication and cooperation among disciplines. For instance, the Atkinson Center for a Sustainable Future[12] is a cross-disciplinary center at Cornell University (of which Dietl is a faculty fellow) focusing on economic development, energy, and the environment; it involves faculty from the natural sciences, social sciences, agriculture, medicine, engineering, arts, and humanities. The Center supports students and postdoctoral scholars through grants and fellowships (including some current CPB projects) and provides internships at environmental organizations as well as engagement with policy makers and other external stakeholders. For instance, the NatureNet Science Fellows Program[13] is offered in collaboration with the Nature Conservancy, and the Atkinson Postdoctoral Fellows in Sustainability[14] must have a co-advisor external to the university (e.g., Environmental Defense Fund). The Center's activities thus align well with recommendations made by the Conservation Paleobiology Workshop (2012) concerning partnerships with stakeholders, postdoctoral opportunities, and integration with the social sciences. When such cross-disciplinary units already exist within a university, involvement by faculty in cross-disciplinary work is more likely to be rewarded by the institution.

Collaborative research tends to be more costly in terms of time, effort, and infrastructure (National Research Council 2005). The availability of funding thus has been a major factor in developing cross-disciplinary conservation programs; the Pew Charitable Trust's initiative on "Integrated Approaches to Training in Conservation and Sustainable Development" provided a significant impetus to such programs (Jacobson 1990). Integrative Graduate Education and Research Traineeship (IGERT) grants from the National Science Foundation (now the NSF Research Traineeship program[15]) have supported successful interdisciplinary training programs (National Research Council 2005; Morse et al. 2007; Moslemi et al. 2009); the Tropical Conservation and Development program at University

[11] http://www.fgcu.edu/info/mission.asp.

[12] https://www.acsf.cornell.edu/index.php.

[13] http://www.nature.org/science-in-action/naturenet-science-fellowship.xml.

[14] http://www.atkinson.cornell.edu/grants/postdoc/.

[15] https://www.nsf.gov/funding/pgm_summ.jsp?pims_id=505015.

of Florida has been supported by grants from the MacArthur Foundation, Ford Foundation, and state of Florida, among others (Kainer et al. 2006). Likewise, service learning at IUPUI is supported by both corporate and civic sponsors (Tedesco and Salazar 2006). Research-funding agencies are also making cross-disciplinary work more attractive financially (National Research Council 2005). For instance, the "Dynamics of Coupled Natural and Human Systems" program[16] at the NSF is cross-directorate (Biological Sciences; Geosciences; and Social, Behavioral, and Economic Sciences), supporting research on the interactions of human and natural systems by teams with expertise in the natural, social and behavioral sciences. In our experience, university administrators tend to be more enthusiastic about initiatives that can attract external funding. Thus, continuing to work with agencies such as the NSF (Conservation Paleobiology Workshop 2012) to enhance funding opportunities for CPB work should remain a priority.

4 Okay, But . . .

We recognize that the approach we have recommended is not without challenges. Our recommendations that CPB training be cross disciplinary, with an emphasis on real-world experience, raise a variety of issues.

Cross-disciplinary programs suffer from institutional constraints (Jacobson and Robinson 1990). Tenure and promotion systems, as well as vehicles for program evaluation, occur within a departmental context. Rewards and incentives for faculty to improve performance, to the extent that they exist, are distributed primarily at the departmental level, and reward systems tend to value products of individual scholarship more than team efforts (National Research Council 2005). Cross-disciplinary research, and faculty participation in cross-disciplinary programs, can be difficult to evaluate (Klein 2008); disciplines vary in their paradigms, terminology, research methods, philosophical approaches (Jacobson and Robinson 1990; Morse et al. 2007), and opportunities for publication and funding. Most faculty are uncomfortable venturing beyond a narrow range of research topics (whether it be taphonomy or brachiopod systematics or drilling predation). We prefer to stay within our comfort zone and avoid taking risks, especially when the rewards are uncertain. Soulé and Press (1998) also noted that applied research, especially when used for advocacy, is often seen as subjective or biased (see also Flessa 2017; Dietl and Flessa 2018). Such work may suffer when subject to peer review for publication, funding, tenure, and promotion. For instance, candidates for tenure and promotion may find that their institution classifies such applied research as service or instruction (if students are involved) rather than as a research contribution.

Departments are often placed in adversarial roles with one another, as they compete for limited resources (Jacobson and Robinson 1990). With state support

[16]http://www.nsf.gov/funding/pgm_summ.jsp?pims_id=13681.

for higher education declining, at public universities even modest departmental contributions to cross-disciplinary initiatives (e.g., to bring in a speaker relevant to multiple departments) require a sacrifice of resources. Even cooperation by offering cross-departmental co-taught courses can be problematic in terms of sharing costs, faculty salaries, and allocation of credit for the students enrolled, despite general acknowledgment that students may benefit from observing how faculty with different expertise work together (e.g., Kuban 2016). Joint appointments of faculty in more than one department can help alleviate such problems but may be difficult to negotiate. Thus, departments may have few incentives to cooperate with one another (Kelley, personal experience as department chair). Programs that depend on the good will of participating departments for resources and faculty may suffer (thus Soulé and Press 1998 argued for autonomy of environmental studies programs).

In addition, allocation of resources is often based on generation of student credit hours by a department. Courses that do not attract a minimum number of students are threatened with cancellation. As a result, department chairs may discourage faculty from advising students to take courses outside their home department. Even in the absence of these institutional barriers to cross-disciplinary education, attitudinal barriers may inhibit implementation of our recommendations. "Hard" sciences may be seen as superior in some way to "softer" disciplines (social sciences, humanities), which may discourage natural science faculty from advising students to take courses representing the "human dimension" of conservation.

In addition, philosophical disagreements occur regarding the appropriate balance between depth and breadth of training—a common concern in conservation biology (Jacobson 1990; Jacobson and Robinson 1990; Kainer et al. 2006; Morse et al. 2007; Moslemi et al. 2009). Soulé and Press (1998) argued that environmental studies programs are plagued by a problem of "hyper-diverse shallow curricula" (p. 397), resulting in "multidisciplinary illiteracy" (p. 402) and a lack of quality and rigor. Thus, a key question for CPB is whether breadth of training can be increased without sacrificing depth of training in the core discipline of paleontology.

Undoubtedly, tradeoffs will be involved in developing programs in CPB (see Muir and Schwartz 2009). Conservation paleobiologists need to be trained as paleobiologists; the core science, including training in how to do paleobiological research, must remain. In addition, extending time to degree is undesirable (Kainer et al. 2006; Blickley et al. 2012). Time to degree is one of the metrics used to evaluate program productivity (Kelley, personal experience as department chair); programs that take longer to complete are at a disadvantage in terms of recruitment and cost (to students and to the institution).

Muir and Schwartz (2009) recognized that training cannot be added to conservation biology requirements without cutting other aspects of training and concluded that not all skills need to be taught to all students. We may need to rethink current training programs for CPB and not automatically funnel our students into PhD programs focused only on research skills. The need for conservation paleobiologists with PhDs will remain, in order to train students, but we should also be preparing students to enter the workforce at the MS level. Indeed, MS programs may be more amenable to implementing such aspects as internships. For example, at UNCW the MS Geoscience degree[17] has both a thesis and a non-thesis option. Students in the non-thesis option have the opportunity to do a final project, an internship, or both, and the non-thesis option requires 33 credit hours as compared to 30 credit hours in the thesis option. Similarly, at Georgia State University the MS in Geosciences[18] non-thesis option usually incorporates a project, often with a partner organization, and many are related to sustainability or conservation. The greater flexibility of the non-thesis option could be an excellent venue for taking a broader range of courses (our Recommendations 1 and 2) and providing real-world experience (Recommendation 3). Such students may not become experts in taphonomy, brachiopod systematics, or even drilling predation, but they would be better prepared for conservation careers. Practice-oriented degree programs (e.g., the Master of Professional Studies at Cornell,[19] which requires a problem-solving project) serve a similar function.

5 In the Meantime . . .

Programmatic changes (Recommendations 1, 2, and 3) and modifications to institutional attitudes and cultures (Recommendation 6) will require time to implement. Although informal program elements described in Recommendations 4 and 5 can be introduced more quickly, is there more that students currently in the CPB pipeline can do to develop the skills needed for a conservation career?

Current students who aspire to a career in conservation paleobiology would do well to follow the recommendations of Blickley et al. (2012). Their analysis of advertisements for conservation science positions and interviews of conservation professionals revealed that skills desired for new hires vary from governmental to nonprofit to the private job sector (see also Lucas et al. 2017). Government and nonprofit sectors showed preference for interpersonal, networking, oral communication, program leadership, and project-management skills; nonprofits also valued fundraising skills. Technical, disciplinary, written communication, and field skills were priorities for the private sector. Because student time is limited for development of skills beyond those needed for academic success in the discipline, a student who is able to identify a particular career goal can target acquisition of the relevant skills. CPB students should assess strategically whether their program of study can provide these skills. If the answer is "no," students should take charge of their future by making their own opportunities (Perez 2005; Blickley et al. 2012): mentoring undergraduates in research, planning and participating in K-12 outreach activities, seeking an internship or volunteering with an organization in the desired job sector, asking a practitioner from an outside organization to be a thesis/dissertation

[17]http://uncw.edu//msgeoscience/index.html.

[18]http://geosciences.gsu.edu/grad-programs/m-s-degree-in-geosciences/.

[19]https://cals.cornell.edu/mps.

committee member, meeting with alumni, planning departmental events, organizing a seminar or asking a faculty member to offer a course or an independent study on professional skills (e.g., Kelley co-developed a "Professional Skills Practicum" course at the University of North Dakota; the Network of Conservation Educators and Practitioners at the American Museum of Natural History[20] offers open-access teaching modules that include communication skills). The possibilities are limited only by student energy, time, and creativity; advisors can suggest opportunities and help students assess feasibility of ideas and achieve a balance between completing program requirements and acquiring non-academic skills.

6 A Bright Future

We are convinced that, if we are clever, conservation paleobiologists can find ways to collaborate with existing cross-disciplinary programs in the conservation sciences. Working within the framework of existing cross-disciplinary units holds particular promise. Admittedly this approach is not without obstacles, including convincing other conservation workers of the applicability of the fossil record to biodiversity conservation (Durham and Dietl 2015; Flessa 2017; Smith et al. 2018). Regardless of issues in overcoming institutional hurdles to formalizing CPB programs, implementing any of the informal approaches proposed in Recommendations 4 and 5 will better prepare tomorrow's conservation paleobiologists to function effectively in the conservation world.

We stand at the brink of exciting changes and progress in CPB education. If we continue along our present trajectory, CPB will be increasingly ineffective in its primary mission. However, if we help our students blaze a new path, the payoff is clear: we will do a better job of training our CPB students, which will produce a more effective workforce doing conservation on the ground.

Acknowledgments We thank Carrie Tyler and Chris Schneider for the opportunity to contribute to this volume. Karl Flessa and Frank Forcino provided helpful reviews of the manuscript.

References

Blickley JL, Deiner K, Garbach K et al (2012) Graduate student's guide to necessary skills for nonacademic conservation careers. Conserv Biol 27:24–34

Boyer AG, Brenner M, Burney D et al (2017) Conservation paleobiology roundtable: from promise to application. In: Dietl GP, Flessa KW (eds) Conservation paleobiology: science and practice. University of Chicago Press, Chicago, pp 291–302

[20] http://www.amnh.org/our-research/center-for-biodiversity-conservation/capacity-development/network-of-conservation-educators-and-practitioners-ncep/.

Brewer C (2001) Cultivating conservation literacy: "trickle-down" education is not enough. Conserv Biol 15:1203–1205
Brewer C (2002) Outreach and partnership programs for conservation education where endangered species conservation and research occur. Conserv Biol 16:4–6
Campbell SP, Fuller AK, Patrick DAG (2005) Looking beyond research in doctoral education. Front Ecol Environ 3:153–160
Ciannelli L, Hunsicker M, Beaudreau A et al (2014) Transdisciplinary graduate education in marine resource science and management. ICES J Mar Sci 71(5):1047–1051. https://doi.org/10.1093/icesjms/fsu067
Clark TW (2001) Developing policy-oriented curricula for conservation biology: professional and leadership education in the public interest. Conserv Biol 15:31–39
Conservation Paleobiology Workshop (2012) Conservation paleobiology: opportunities for the earth sciences. Report to the Division of Earth Sciences, National Science Foundation. Paleontological Research Institution, Ithaca
Dietl GP (2016) Brave new world of conservation paleobiology. Front Ecol Evol 4:21
Dietl GP, Flessa KW (2018) Should conservation paleobiologists save the world on their own time? In: Tyler CL, Schneider CL (eds) Marine conservation paleobiology. Springer, Cham, pp 11–22
Dietl GP, Kidwell SM, Burney DA et al (2015) Conservation paleobiology: leveraging knowledge of the past to inform conservation and restoration. Annu Rev Earth Planet Sci 43:79–103
Durham SR, Dietl GP (2015) Perspectives on geohistorical data among oyster restoration professionals in the United States. J Shellfish Res 34:227–239
Flessa KW (2017) Putting the dead to work: translational paleoecology. In: Dietl GP, Flessa KW (eds) Conservation paleobiology: science and practice. University of Chicago Press, Chicago, pp 283–289
Gruenewald DA (2003) Foundations of place: a multidisciplinary framework for place-conscious education. Am Educ Res J 40:619–654
Hansen SS, Fortner S (2016) Geoscience service-learning literature themes. Document prepared for the National Academy of Sciences (NAS) Service-Learning in Undergraduate Geosciences Workshop, 20–21 April 2016. http://sites.nationalacademies.org/cs/groups/dbassesite/documents/webpage/dbasse_171831.pdf
Hawthorne TL, Elmore V, Strong A et al (2015) Mapping non-native invasive species and accessibility in an urban forest: a case study of participatory mapping and citizen science in Atlanta, Georgia. Appl Geogr 56:187–198
Jacobson SK (1990) Graduate education in conservation biology. Conserv Biol 4:431–440
Jacobson SK, McDuff MD (1998) Training idiot savants: the lack of human dimension in conservation biology. Conserv Biol 12:263–267
Jacobson SK, Robinson JG (1990) Training the new conservationist: cross-disciplinary education in the 1990s. Environ Conserv 17:319–327
Jarvis RM, Borrelle SB, Breen BB et al (2015) Conservation, mismatch and the research-implementation gap. Pac Conserv Biol 21:105–107
Kainer KA, Schmink M, Covert H et al (2006) A graduate education framework for tropical conservation and development. Conserv Biol 20:3–13
Kelley PH, Dietl GP (2012) The "Research Experiences for Undergraduates in Biodiversity Conservation" program: training the first generation of conservation paleobiologists. Geol Soc Am Abstr Programs 44:76
Klein JT (2008) Evaluation of interdisciplinary and transdisciplinary research: a literature review. Am J Prev Med 35:S116–S123
Knight AT, Cowling RM, Rouget M et al (2008) Knowing but not doing: selecting priority conservation areas and the research–implementation gap. Conserv Biol 22:610–617
Kroll AJ (2007) Integrating professional skills in wildlife student education. J Wildl Manag 71:226–230
Kuban A (2016) Transdisciplinary collaboration: an introduction. In The Trenches 6:1–2
Kurilla B (2016) The evolution of the 21st-century scientist. Am Sci Macroscope. August 12

Lucas J, Gora E, Alonso A (2017) A view of the global conservation job market and how to succeed in it. Conserv Biol 31:1223–1231
Meffe GK (1998) Softening the boundaries. Conserv Biol 12:259–260
Morse W, Nielsen-Pincus M, Force J et al (2007) Bridges and barriers to developing and conducting interdisciplinary graduate-student team research. Ecol Soc 12:1–14
Moslemi JM, Capps KA, Johnson MS et al (2009) Training tomorrow's problem solvers: an integrative approach to graduate education. Bioscience 59:514–521
Muir MJ, Schwartz MW (2009) Academic research training for a nonacademic workplace: a case study of graduate student alumni who work in conservation. Conserv Biol 23:1357–1368
National Research Council (2005) The geological record of ecological dynamics: understanding the biotic effects of future environmental change. National Academy Press, Washington DC
Noss R (1999) Is there a special conservation biology? Ecography 22:113–122
Ort MH, Anderson DE, Ostergren DM (2006) Integrating policy and land management issues into a natural sciences education: teaching environmental sciences on the lower San Juan River, Utah. J Geosci Educ 54:116–122
Perez HE (2005) What students can do to improve graduate education in conservation biology. Conserv Biol 19:2033–2035
Rosenfield P (1992) The potential of transdisciplinary research for sustaining and extending linkages between the health and social sciences. Soc Sci Med 35:1343–1357
Semken S, Butler Freeman C (2008) Sense of place in the practice and assessment of place-based science teaching. Sci Educ 92:1042–1057
Semken S, Butler Freeman C, Watts NB et al (2009) Factors that influence sense of place as a learning outcome and assessment measure of place-based geoscience teaching. Electron J Sci Educ 13:1–25
Shriberg M, Harris K (2012) Building sustainability change management and leadership skills in students: lessons learned from "Sustainability and the Campus" at the University of Michigan. J Environ Stud Sci 2:154–164
Smith JA, Durham SR, Dietl GP (2018) Conceptions of long-term data among marine conservation biologists and what conservation paleobiologists need to know. In: Tyler CL, Schneider CL (eds) Marine conservation paleobiology. Springer, Cham, pp 23–54
Sobel D (2004) Place-based education. Connecting classrooms and communities. The Orion Society, Great Barrington
Soulé ME, Press D (1998) What is environmental studies? Bioscience 48:397–405
Tedesco LP, Salazar KA (2006) Using environmental service-learning in an urban environment to address water quality issues. J Geosci Educ 54:123
White R, Fleischner TL, Trombulak SC (2000) The status of undergraduate education in conservation biology. Education Committee of the Society for Conservation Biology Report

A Conceptual Map of Conservation Paleobiology: Visualizing a Discipline

Carrie L. Tyler

Abstract Disciplinary boundaries frame the basic questions and central issues of research, providing the context for the evolution of prevailing theories or paradigm shifts. This chapter aims to outline the development and scope of conservation paleobiology using bibliometrics. Publication records relating to research on conservation paleobiology were downloaded from Web of Science to generate two datasets, one aimed at producing a more conservative representation of conservation paleobiology, and the other more expansive. Bibliographic maps were created to provide insight into the development and structure of the discipline for both characterizations of conservation paleobiology research (conservative versus expansive). Bibliographic maps indicated that individual researchers working on conservation paleobiology specialize in several fields. Regardless of how conservation paleobiology is defined, research involving both paleontology and conservation appears to be highly multidisciplinary, including at least three main research domains broadly categorized as: (1) environmental history and conservation archeobiology, (2) genetics and evolutionary biology, and (3) ecology. Furthermore, paleontological publications did not form a distinct cluster, but rather were integrated within conservation science. This supports the proposition that, in practice, conservation paleobiology is a field of study within conservation science, and not a sub-discipline within paleontology. Analyses also revealed emerging research fronts in several topics and confirmed the need for long-term data that pre-dates human activities.

Keywords Science mapping · Conservation science · Paleoecology · Bibliometrics · Discipline structure

---

C. L. Tyler (✉)
Department of Geology and Environmental Earth Science, Miami University, Oxford, OH, USA
e-mail: tylercl@miamiOH.edu

C. L. Tyler, C. L. Schneider (eds.), *Marine Conservation Paleobiology*,
Topics in Geobiology 47, https://doi.org/10.1007/978-3-319-73795-9_10

1 Determining the Current State and Structure of Conservation Paleobiology

Conservation paleobiology is a young and emerging discipline, rapidly developing in response to the urgent need for long-term data in conservation science, in the face of climate change and biodiversity loss (Smith et al. 2018). However, conservation paleobiology has yet to rally together to formalize goals, or to establish whether the goals of conservation paleobiology are consistent with, or independent of those of conservation science (expanding our understanding of the role of biodiversity in ecosystem health and functioning, and the conservation of biodiversity). Meine et al. (2006) asked *"Is [the field of conservation biology] performing its core function—providing reliable and useful scientific information on biological diversity and its conservation—in the most effective manner possible? Is that information making a difference on the ground?"*. Given that the Earth system is facing unprecedented environmental pressures and reduction in biodiversity, it is worth considering how conservation paleobiology can similarly aim to make the most difference. The creation and organization of a common conceptual framework is vital to identify central issues within the field, particularly in young disciplines (Torraco 2005; van der Have and Rubalcaba 2016), as disciplinary boundaries frame the basic questions and central issues of research, providing the context for the evolution of prevailing theories, or paradigm shifts. Given the complexity and size of the scientific literature, personal knowledge and experience are no longer sufficient for decision-making, or understanding and identifying such research trends.

Scientific contributions form a dynamic and self-organizing system of knowledge, and research subfields share a common knowledge base in the form of article references. A research field can, therefore, be defined as a network of publications covering a set of research questions and methodologies, referring to overlapping literature (Van Den Besselaar and Heimeriks 2006). Journals with the same research fields are expected to have aggregated citation patterns, and inter-journal citations can be used to map journals with similar citing patterns (Van Den Besselaar and Heimeriks 2006). These patterns can be analyzed to produce a visualization of the bibliometric landscape of the field. Through network modeling and visualization, the intellectual landscape can be explored to reveal current discipline structure and dynamics, by identifying prominent scholars, seminal articles, collaborators, and prevalent research areas. In other words, depicting the current set of questions that the scientific community is addressing. Emergent research trends and frontiers can be identified, or pre-eminent individuals and publications in these cutting-edge research areas. Bibliometrics can be used to assess such shifts in research focus using citation data (Small 1977), and networks of citations can be traced to explore the history and evolution of article chains on particular topics. Bibliometrics is the quantitative analysis of publications and their citations and employed globally to quantitatively evaluate research performance by scientists, university administrators, government labs, funding agencies (including the National Science Foundation), and policymakers. These analyses can be advantageous, gathering objective information necessary for decision-making, and providing a comprehensive perspective on research activity, as citation data provide an objective and quantitative indicator for evaluating research performance. Weighted measures, such as papers per researcher or citations per paper, remove characteristics such as the place of production, or reputation, that can color human perceptions of quality. For example, when thinking of "the best," it is not hard to think automatically of the biggest producers such as individuals, labs, and universities. But these locations may not be the source of the most impactful work. Bibliometric analyses can therefore be used to identify top performers, balancing human perceptions of reputation. Here bibliometrics was employed to map conservation paleobiology, in an effort to characterize the structure and dynamics of the discipline.

2 Mapping a Discipline

Creating a science map is the first step towards exploring and understanding scientific frontiers in conservation paleobiology. Here bibliometric networks were constructed to create graphic representations of conservation paleobiology to: (1) conceptualize intellectual structure and dynamics and (2) visualize scientific advancement and emerging research topics.

Bibliometric data were downloaded from Web of Science (WoS) which contains over 90 million records covering a wide range of natural science disciplines as far back as the 1900s and includes thorough coverage of records within the fields of conservation and paleontology. Despite known inaccuracies with assigning articles to WoS categories, the overall network structures observed when using WoS are robust to changes in classifications, degree of aggregation using journals rather than subject categories, and over time periods studied herein (Rafols and Leydesdorff 2009). WoS also allows expansion of searches beyond the Web of Science™ Core Collection's nine indexes, to simultaneously search 15 different content sets, by employing the "All Databases" search option. Although the latter does not yield data for all record fields, excluding several types of bibliometric analyses. WoS was searched by "Topic," which searches the title, abstract, author keywords, and Keywords Plus® fields within a record. This ensured that the searched phrases are likely to be a significant component of the publication and not merely briefly mentioned in the body of the text. To conservatively capture records fulfilling the narrowest definition of conservation paleobiology, the words "conservation paleobiology" were searched in "All Databases." However, the search term "paleobiology" may not accurately capture the full breadth of the discipline, given that conservation paleobiology includes studies using many types of paleontological data, such as paleogeography or paleoecology. Therefore, to more broadly visualize the conservation paleobiology literature as a whole, a second search was conducted using the WoS "Core Collection" and the terms "conservation paleo*." The addition of a wildcard following "paleo" expanded the topic search to include all records using variations with the prefix "paleo." This increased inclusiveness, capturing any studies combining research on paleontology and conservation, but not specifically defined using the term "paleobiology." Use of the "Core Collection" searched records in the nine most relevant databases and expanded available record fields, facilitating additional bibliometric analyses.

VOSViewer was used to construct and visualize bibliometric networks (www.vosviewer.com; Van Eck and Waltman 2010, 2014), which uses an algorithm analogous to multidimensional scaling to position data in a multidimensional vector space and minimize stress (Van Eck et al. 2010; Leydesdorff and Rafols 2012). Data downloaded from citation databases such as WoS can be input directly into VOSViewer to construct relationships between citations, keywords, documents, and authors in the form of a science map (Boyack and Klavans 2010; Small et al. 2014). Two types of mapping approaches were applied here: visualizations based on bibliographic data (co-authorship and co-citation) and text data (co-occurrence). Highly connected portions within networks, or modules, were grouped by VOSViewer. Modularity is assigned based on the degree to which nodes in the network can be divided into groups. Nodes within the same group are connected more densely than nodes between different groups. Modularity can thus be used to identify the integration and strength of relationships between groups of authors, documents, topics, and fields of research.

Here bibliometric networks were constructed using a variety of approaches, relying on different units of analyses necessary for the examination of both discipline structure and dynamics and emerging research areas. What is the current state of the field of marine conservation paleoecology, and what have been our major contributions to date? As the application of a combination of bibliometric analyses can provide a deeper understanding of discipline structure and dynamics (Yan and Ding 2012; Chang et al. 2015), conservation paleobiology was explored using three types of bibliometric networks: co-authorship, text co-occurrence, and author and document co-citation.

Bibliographic Co-Authorship Visualizations

To map discipline structure and dynamics, it is important to first examine the individual participants and researchers that form the discipline. Bibliographic co-authorship analyses are widely used to investigate the structure of research fields (White and Griffith 1981; Ramos-Rodríguez and Ruíz-Navarro 2004; Eom 2008; Raasch et al. 2013), as significant collaborations are expected to result in co-authored publications. Co-authorship is thus viewed as evidence of scientific collaboration and interactions among scientists and research teams (Melin and Persson 1996). Although not all collaborations result in co-authored publications, the number of collaborations that do not produce co-authored publications is typically considered negligible (Melin and Persson 1998).

Documents that have more than one author are considered co-authored, and links between co-authors form networks. The strength of the relationship is based on the number of co-authored documents. In all co-authorship analyses, full counting was employed, meaning that each occurrence has equal weight. For example, if a publication had three co-authors, it was assigned to each author with a weight of one. Authors with fewer than five documents were excluded. Temporal overlays were then applied to identify shifts in productivity over time and to determine which authors and research groups are currently contributing most to the field.

Text Co-Occurrence Visualizations

Co-occurrence analysis is a well-established method in bibliometrics (Callon et al. 1983), employing the co-occurrence of keywords and mapping proximity between keywords in scientific documents (Boyack and Klavans 2010). The more frequently words co-occur, the stronger the relationships between them, as they reside in similar research sub-fields. Emerging topics can therefore be identified by examining the frequency of the use of specific terms and the body of literature that scientists are *actively* citing (de Solla Price 1965). By detecting and visualizing research fronts, we can ask how conservation paleobiology started, recognize critical paths in its evolution, and identify state-of-the-art research. Temporal overlays were applied to maps produced by text co-occurrence analyses, to identify fast-growing, emerging areas of research.

VOSViewer extracted terms from titles and abstracts, terms are defined as a sequence of nouns and adjectives ending with a noun. Binary counting was employed, counting the presence or absence of a word and not the number of times a word occurs in a document. A relevance score was calculated for each word, and using these scores, the 60% most relevant terms were selected for mapping. VOSViewer used the extracted terms to make a map, where distances between two terms indicates number of co-occurrences of the terms. The smaller the distance between two terms, the larger the number of co-occurrences of the terms.

Bibliographic Co-Citation Visualizations

Citation analysis employs one of the most crucial indicators of scholarship: citations (Small 1977). Author co-citation analysis records indirect interactions between scientists through the scientific literature, by identifying and counting the number of times author pairs are cited among a set of articles (White and Griffith 1981; McCain 1990; White and McCain 1998). In other words, co-citation occurs when two authors are cited by a third author, regardless of which articles are being cited (Small 1973; White and Griffith 1981; White and McCain 1998). When more authors cite a pair of authors, higher co-citation strength results (Small 1973), i.e., the more cited authors two documents have in common, the greater the co-citation strength. This is often used as an indication of similarity in subject specialties between authors (White and McCain 1998). Co-citation measures the association between concepts represented by highly cited papers (Marshakova 1973; Small 1973) and can be used to identify areas of common interest. Altogether, these authors constitute the "invisible college," bringing to light documents in the research network that refer to each other without being linked by recognizable organizational ties (de Solla Price 1965; Crane 1972; Lievrouw 1989; Gmür 2003). Authors of highly cited papers in co-citation analysis are thought to constitute the leading scientists in a discipline (White and McCain 1998). Author co-citation analysis can therefore also be used to identify highly cited, and presumably influential researchers, and their specialties.

Journal co-citation was also employed to identify the proportions of documents appearing in journals assigned to multiple subject categories and the prevalence of boundary crossing co-citations. Large numbers of publications in interdisciplinary journals would indicate that authors consider their research relevant for audiences in multiple disciplines.

As only the WoS Core Collections data contain full records, necessary for co-citation analyses, all co-citation analyses were performed using the more conservative "Core Collection" dataset (see below). Co-citation analyses were applied using authors, and sources (journals) as the units of analyses. The full-counting method was applied assigning each occurrence equal weight. Disciplinary affiliations of authors were assigned based on authors' institutional web pages to view the structure and membership of the science.

Bibliographic Coupling Visualizations

Bibliographic coupling can be used to represent a disciplines knowledge base (Garfield et al. 2003). When two documents cite the same reference, they are bibliographically coupled, with higher values indicating a strong relationship between the two documents (Kessler 1963). Bibliometric coupling measures similarity between papers and the number of references two papers have common (Kessler 1963). To determine which journals currently publish high impact and/or cutting-edge conservation paleobiology research, journal names in the references were used as the unit of analysis, and a temporal overlay was applied. Similar to co-citation analysis, full records were required, restricting analysis to the more conservative "Core Collection" dataset (see below). The full-counting method was applied assigning each occurrence equal weight.

3 Bibliometric Networks

WoS searches resulted in two datasets consisting of bibliographic records (Table 1): The topic search for the phrase "conservation paleo*" in the WoS "Core Collection" (CP-CC) yielded 833 records, and "conservation paleobiology" in "All Databases" (CPb-AD) yielded 3418 records. Although number of records varies dramatically between these datasets, they record similar publication trends, and the number of records produced per year in both increases substantially from 1990 to 2000 to present (Fig. 1).

Table 1 Bibliographic datasets

WoS database	Topic search terms	Records
Core collection	Conservation paleo*	833
All databases	Conservation paleobiology	3418

Number of records downloaded from WoS for each database and search term. Topic searches included the title, abstract, author keywords, and Keywords Plus® fields within a record. Records downloaded from WoS on October 21st, 2016

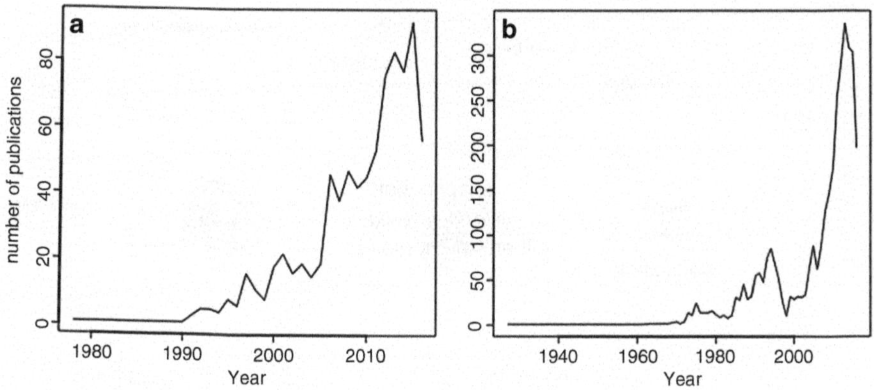

Fig. 1 Publication trends through time. Number of publications per year from first usage to 2016. Panels plotting data generated using two search terms and either WoS "Core Collection" or "All Databases". (**a**) Number of records per year resulting from search terms "conservation paleo*" in WoS "Core Collection"; (**b**) Number of records per year resulting from search terms "conservation paleobiology," in WoS "All Databases"

Bibliographic Co-Authorship Networks

Bibliographic co-authorship networks were generated using both datasets, however, as the CP-CC dataset did not yield a sufficiently populated map for meaningful interpretation (only 14 authors met the minimum requirement of five documents[1], creating 14 isolated nodes), only the CPb-AD network is presented here.

The CPb-AD dataset contained 10,318 authors, of which 69 met the minimum threshold of five documents. Twenty-three clusters of authors were identified, of which ten reveal strong relationships between authors, representing working groups, frequent collaborations between authors, or shared lines of inquiry (Fig. 2). The most influential groups of authors were identified using number of citations as an estimate of research impact. Four groups including authors with average citation rates above 40 are apparent in Fig. 2, while the remainder of authors has an average

[1] Authors ordered by number of citations, from highest to lowest: S. T. Jackson, D. R. Foster, K. J. Willis, C. Saiz-Jimenez, J. M. Gonzalez, M. C. Portillo, J. C. Svenning, V. Rull, J. Salse, T. Vegas-Vilarrubia, G. P. Dietl, E. Montoya, J. M. Pandolfi, and J. L. McGuire.

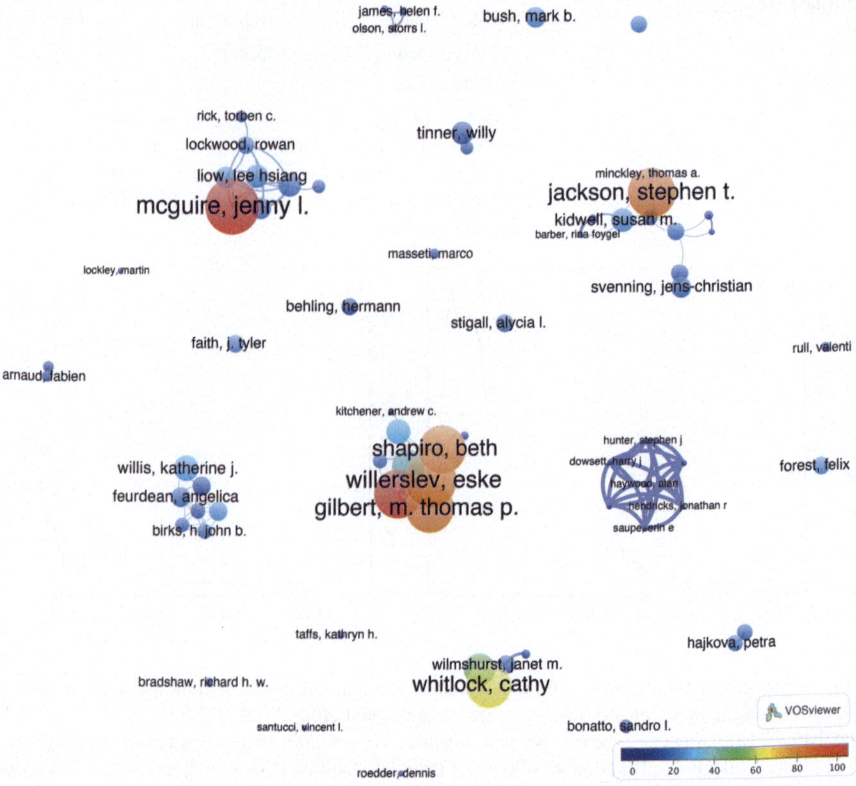

Fig. 2 Citation rates of research groups and frequent collaborators for CPb-AD. Visualization of CPb-AD bibliographic co-authorship analysis. Clusters generated by strong relationships between authors, indicating research groups and frequent collaborations. Nodes were weighted by number of citations, with larger nodes and warmer colors signifying higher numbers of citations. Colors correspond to the scale on the bottom right, numbers on scale are average number of citations

citation rate of 20. The largest group of related authors consisted of a cluster of 11 authors, the next largest of nine authors, and clusters three and four both consisting of eight authors (Fig. 3a–d). Clusters of authors are highly interdisciplinary (Table 2). When nodes were weighted by documents, and viewed with a temporal overlay, individuals within groups that made their most significant contributions in the previous decade could be distinguished from those actively publishing (Fig. 4). All clusters containing more than one author include at least one individual with on average more recent publications (after 2013), and three of the four largest research groups contained members with average publication dates prior to 2010 (Figs. 3 and 4).

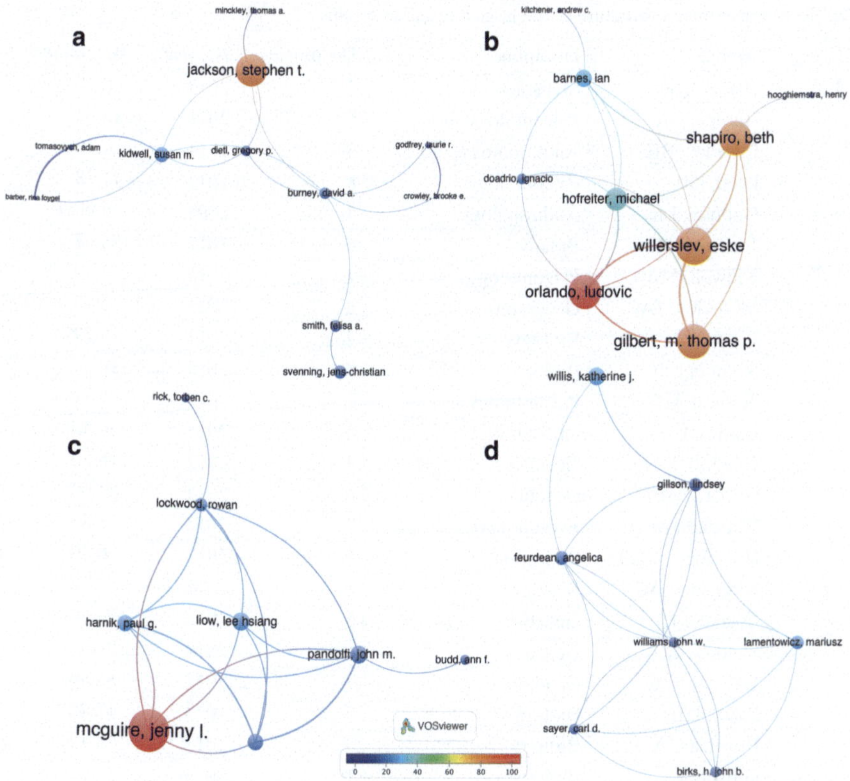

Fig. 3 Largest research groups and most frequently cited contributors for CPb-AD. The four largest research groups identified in the CPb-AD bibliographic co-authorship analysis (from Fig. 2). (**a**) The largest research group consisting of 11 collaborating authors, (**b**) second largest cluster consisting of 9 authors, (**c, d**) tied for third position, both consisting of eight authors. Nodes were weighted by number of citations, with larger nodes and warmer colors signifying higher numbers of citations. Colors correspond to the scale on the bottom right, numbers on scale are average number of citations. When compared with Fig. 4, it is apparent that the second largest cluster (**b**) was the only cluster lacking an author whose average contributions predate 2010

Text Co-Occurrence Networks

VOSViewer extracted 69,407 terms from the CPb-AD dataset, of which 1523 met the minimum threshold of 10 occurrences. Relevance scores were then calculated for the 914 terms making up the 60% most relevant terms. The resulting terms list was carefully inspected and terms not related to research topics, such as geographic locations or species names were removed (the following are examples of removed terms: "usage," "English," "book," "page," "chapter," "bibliography," "illustration," "Balkan peninsula," "Americas," "southern Brazil"). The text co-occurrence analysis identified three modules (Fig. 5). All three modules included references to conservation and anthropogenic impacts in various forms, however, modules

Table 2 Disciplinary structure of the largest research groups

Group	Author	Discipline	Documents	Av. year	Av. citations
1	Barber, RF	Statistics	5	2015	0.20
	Burney, DA	Paleontology	6	2011	18.33
	Crowley, BE	Anthropology	5	2014	4.40
	Dietl, GP	Paleontology	8	2013	11.38
	Godfrey, LR	Anthropology	5	2008	8.00
	Jackson, ST	Botany	6	2012	84.67
	Kidwell, SM	Paleontology	7	2013	23.29
	Minckley, TA	Geography	6	2013	5.17
	Smith, FA	Biology	7	2014	15.29
	Svenning, JC	Biology	8	2015	16.50
	Tomasovych, A	Paleontology	7	2014	5.71
2	Barnes, I	Paleontology	6	2013	29.50
	Doadrio, I	Genetics	5	2012	16.80
	Gilbert, MPT	Genetics	6	2013	84.00
	Hofreiter, M	Evolutionary biology	7	2012	38.86
	Hooghiemstra, H	Palynology	5	2012	6.80
	Kitchener, AC	Biology	5	2011	2.60
	Orlando, L	Genetics	5	2012	99.20
	Shapiro, B	Genetics	6	2014	81.67
	Willerslev, E	Genetics	7	2012	82.86
3	Birks, HJB	Biology	5	2013	17.40
	Feurdean, A	Biology	7	2013	18.14
	Gillson, L	Biology	10	2013	11.80
	Lamentowicz, M	Paleontology	5	2011	24.00
	Sayer, CD	Geography	5	2013	16.20
	Williams, JW	Geography	6	2012	14.17
	Willis, KJ	Ecology	7	2009	26.43
4	Budd, AF	Paleontology	5	2005	13.80
	Harnik, PG	Paleontology	6	2012	24.33
	Liow, LH	Ecology and evolution	7	2013	24.00
	Lockwood, R	Paleontology	5	2014	21.60
	Mcguire, JL	Biology	6	2014	103.17
	Pandolfi, JM	Historical ecology	9	2012	18.00
	Rick, TC	Archaeology	6	2013	9.50
	Simpson, C	Paleobiology	8	2013	17.88

Authors in the four largest research groups (Figs. 2 and 3) and their disciplines. Authors classified into generalized disciplines based on research interests. Documents—number of publications, average publication year, and average citation—average scores of the documents by author. Each working group contains collaborators from a minimum of four disciplines, indicating widespread interdisciplinary exchange and integration

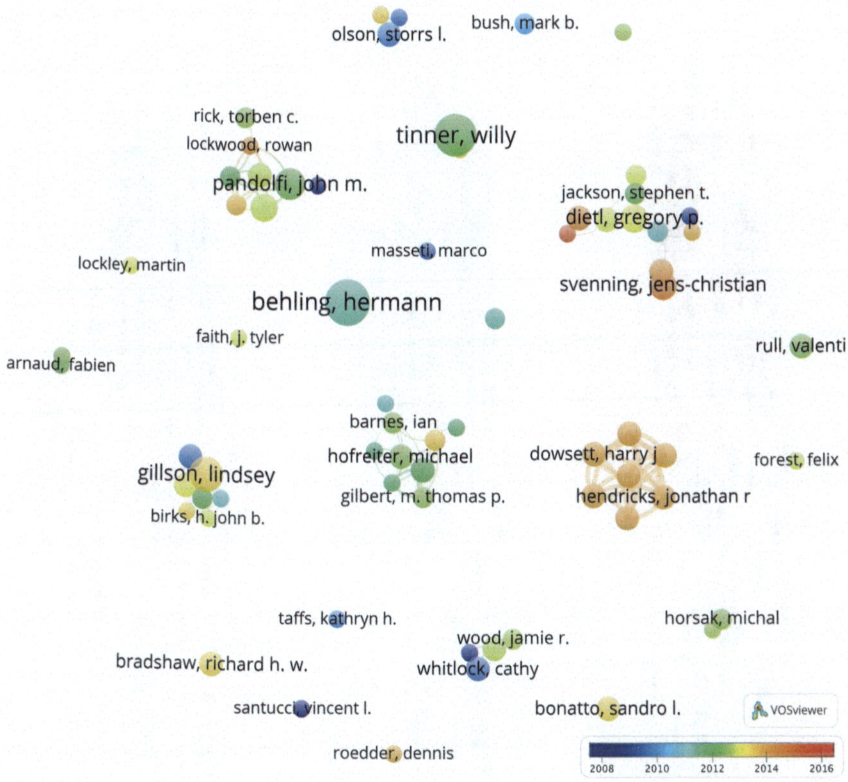

Fig. 4 Publication rates of research groups and frequent collaborators for CPb-AD. Visualization of CPb-AD bibliographic co-authorship analysis as in Fig. 2. Nodes were weighted by number of documents, with larger nodes signifying higher numbers of documents. Warmer colors indicate more recent publications, and colors correspond to the scale on the bottom right where numbers on the scale are average publication year

appeared to define boundaries grouping relatively more cohesive sub-disciplines with greater intellectual overlap (Fig. 5a). Module one consisted of 134 items and included terms describing environmental history and conservation archeobiology such as "anthropogenic activity," "anthropogenic disturbance," "anthropogenic impact," "baseline," "death assemblage," "drought," "disturbance," "European settlement," "fire," "little ice age," "macrofossil," "microfossil," "palaeoecology," "palynology," "pollen," and "sea level rise." Module two consisted of 118 phrases and was dominated by references to genetics and evolutionary biology such as "allele," "base pair," "biodiversity hotspot," "conservation concern," "dispersal," "divergence," "DNA," "endemic," 12 phrases beginning with "gene" (e.g., gene flow, genetic drift), "invasive species," "molecular clock," "niche," "population," and "refuge," with no direct references to paleontology. Module three consisted of 57 terms describing ecology and paleontology including "biodiversity

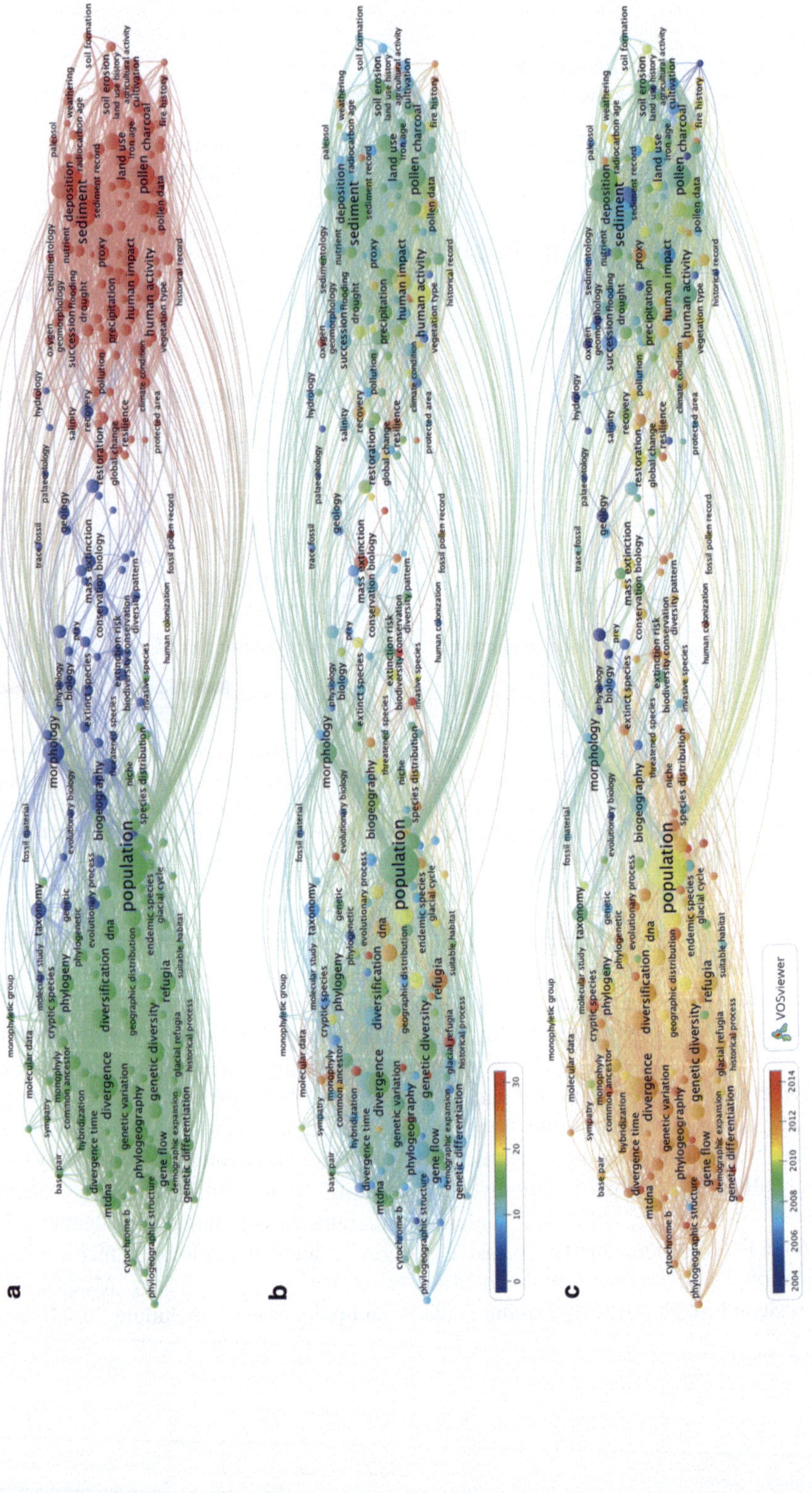
a
b
c
0
10
20
30
2004
2006
2008
2010
2012
2014
VOSviewer
population
morphology
biogeography
taxonomy
phylogeny
diversification
divergence
phylogeography
genetic diversity
gene flow
refugia
dna
mtdna
sediment
deposition
land use
pollen
charcoal
human impact
human activity
proxy
precipitation
restoration
mass extinction
conservation biology
soil erosion

Fig. 5 Most commonly occurring terms in the CPb-AD dataset. Size of nodes indicates occurrence frequency, with words associated with larger nodes occurring more frequently. (**a**) Colors differentiate the three modules identified by VOSViewer, based on strength of inter-group relationships. Modules could be classified into three field of research: environmental history and conservation archeobiology (red, module one), genetics and evolutionary biology (green, module two), and ecology and paleontology (blue, module three). Paleontological terms occurred in both the historical ecology module (red) and ecology and paleontology (blue), but not in genetics and evolutionary biology. (**b**) Citation frequency overlay, warmer colors represent more highly cited topics. Scale on bottom left represents average number of citations. Commonly occurring terms (larger nodes and corresponding text) did not have the highest average citation frequencies, and common terms plotted with cooler colors. (**c**) Temporal overlay with warmer colors denoting more recent publications, colors correspond to the scale on the bottom left indicating average publication year

loss," "biotic interaction," "conservation biology," "conservation paleobiology," "extinction risk," "geographic range," "morphology," "paleontology," "paleobiology," "paleoecology," "paleoenvironment", "predation," "recovery," "reintroduction," "taphonomy," and "trace fossil."

Colors signifying average citation frequency scores for the documents in which a term occurs were overlaid on the text co-occurrence map to identify major areas of research (Fig. 5b). The average publication year of documents in which common terms occur are prior to, or during, 2012 for the ten most common terms in each module (Table 3). Similarly, average publication years of which documents in which

Table 3 Ten most frequently occurring terms by module within the CPb-AD dataset

Module	Terms	Occurrences	Av. year	Av. citations
1	Sediment	304	2008	14.39
	Erosion	206	2009	11.53
	Pollen	180	2008	13.99
	Soil	146	2005	13.64
	Human activity	143	2008	18.41
	Fire	132	2010	17.11
	Deposition	113	2008	10.38
	Human impact	107	2010	21.24
	Land use	98	2007	14.32
	Charcoal	87	2010	12.46
2	Population	614	2010	13.97
	Divergence	173	2012	15.29
	Genetic diversity	155	2012	15.54
	Diversification	141	2011	16.16
	Refugia	131	2011	25.25
	Phylogeography	116	2012	15.48
	Evolutionary history	110	2012	10.99
	Dispersal	108	2010	18.21
	Genetic structure	106	2012	11.08
	Gene flow	103	2011	14.50
3	Morphology	142	2008	13.68
	Biogeography	98	2008	16.49
	Trait	71	2011	18.39
	Taxonomy	67	2009	7.72
	Mass extinction	58	2010	28.43
	Recovery	51	2010	22.41
	Conservation biology	48	2009	11.54
	Paleoecology	47	2008	16.51
	Biology	45	2006	10.56
	Paleontology	41	2001	11.20

The ten most common terms in each module and the frequency with which they occur (Fig. 5). Average publication year and average citation are average scores of the documents in which the term occurs. Module numbers correspond with the colors in Fig. 5a (see figure caption)

a term occurs were overlaid on the text co-occurrence map to identify emerging research fronts (Fig. 5c). Terms within genetics and evolutionary biology were overwhelmingly more recent, likely reflecting technological advances in genetics. Although terms with more recent average publication scores did not have high occurrence frequencies (Table 4). The origin of conservation paleobiology in 2006 and its subsequent rise is apparent in the temporal overlay, as the average publication year for terms such as "paleoenvironment" (2005), "palaeontology" (2006), "taphonomy" (2007), "microfossil" (2007), "paleoecology" (2008), "paleoecological data" (2008), "paleobiology" (2009), "paleoecological record" (2009), "death

Table 4 Ten most recently occurring terms by module within the CPb-AD dataset

Module	Terms	Av. year	Occurrences	Av. citation
1	Anthropocene	2015	17	18.24
	Ecosystem service	2013	19	13.42
	Charcoal record	2012	17	12.94
	Palaeoecological data	2012	17	20.77
	Grain size	2012	18	8.72
	Anthropogenic impact	2012	25	9.84
	Baseline	2012	39	16.31
	Proxy	2012	71	11.37
	Palaeoecological record	2012	15	10.47
	Anthropogenic disturbance	2012	24	5.75
2	Haplotype network	2014	11	2.18
	High genetic diversity	2014	11	7.36
	Ecological niche model	2014	16	6.69
	Species distribution model	2014	18	9.11
	Lineage diversification	2014	12	6.67
	Nuclear gene	2014	19	7.63
	Cryptic diversity	2013	15	6.47
	CP-DNA	2013	11	12.00
	Distribution range	2013	22	9.09
	Cryptic species	2013	45	7.42
3	Biodiversity loss	2013	19	26.42
	Evolutionary process	2013	28	19.96
	Extinction risk	2012	38	13.47
	Geographical range	2012	22	11.55
	Conservation biologist	2012	10	9.70
	Body size	2012	29	14.55
	Molecular study	2012	16	11.00
	Reintroduction	2012	13	9.46
	Biotic response	2012	13	12.77
	Biodiversity conservation	2012	20	33.30

The ten most recent terms in each module based on average publication year of the documents in which the term occurs, and the frequency with which they occur (Fig. 5). Occurrences are listed for each term and average citation scores (average scores of the documents in which the term occurs). Module numbers correspond with the colors in Fig. 5a (see figure caption)

assemblage" (2011), "fossil evidence" (2011), and "conservation paleobiology" (2011) appear over time.

VOSViewer extracted 26,856 terms from the CP-CC dataset, of which 442 met the minimum threshold of ten occurrences. Relevance scores were then calculated for the 265 terms making up the 60% most relevant terms. The resulting terms list was carefully inspected, and terms not related to research topics, such as geographic locations or species names were removed. The text co-occurrence analysis identified three modules (Fig. 6). Similar to the larger CPb-AD dataset, references to conservation and anthropogenic impacts occur in all modules; however, modules are grouped by sub-disciplines with greater intellectual overlap (Fig. 6a). Module one consisted of 42 items and included terms describing paleontology, conservation archeobiology, and environmental history such as "archaeology," "condition," "core," "deposit," "fossil," "human impact," "paleontology," "sediment," and "soil." Module two consisted of 41 phrases and was dominated by references to genetics and evolutionary biology such as "biogeography," "biology," "dispersal," "divergence," "DNA," "gene," "gene flow," "genetic diversity," "genetic variation," "haplotype," "mtDNA," and "phylogeography." Module three consisted of 22 terms describing climate science and ecology including "biotic response," "climate change," "climatic condition," "conservation status," "extinction risk," "expansion," "fragmentation," "geographic range," "global warming," "last glacial maximum," "migration," "species distribution," and "temperature."

A color overlay was applied to visualize average citation frequency scores to identify major areas of research (Fig. 6b). The average publication year of documents in which common terms occur was prior to, or during, 2013 for the ten most common terms in each module (Table 5). The average publication year overlay identified emerging research fronts (Fig. 6c). Terms within a combination of genetics and ecology were overwhelmingly more recent, many of which pertained to ecological response to climate change. Terms with more recent average publication scores, again, did not have high occurrence frequencies (Table 6). The origin of conservation paleobiology in 2006 and its subsequent rise were apparent in the temporal overlay, and terms such as "paleobiology" (average publication year 2011), "fossil" (2010), "paleontology" (2009), "paleoecology" (2009), and "paleoecological data" (2006) appeared over time.

Fig. 6 Most commonly occurring terms in the CP-CC dataset. Size of nodes indicates occurrence frequency, with words associated with larger nodes occurring more frequently. (**a**) Colors differentiate the three modules identified by VOSViewer, based on strength of inter-group relationships. Modules could be classified into three field of research: paleontology, conservation archeobiology and environmental history (red, module one), genetics and evolutionary biology (green, module two), and climate science and ecology (blue, module three). Paleontological terms occur in the conservation biology module. (**b**) Citation frequency overlay, warmer colors represent more highly cited topics. Scale on bottom left represents average number of citations. Commonly occurring terms (larger nodes and corresponding text) did not have the highest average citation frequencies, and common terms plotted with cooler colors. (**c**) Temporal overlay with warmer colors denoting more recent publications, colors correspond to the scale on the bottom left indicating average publication year

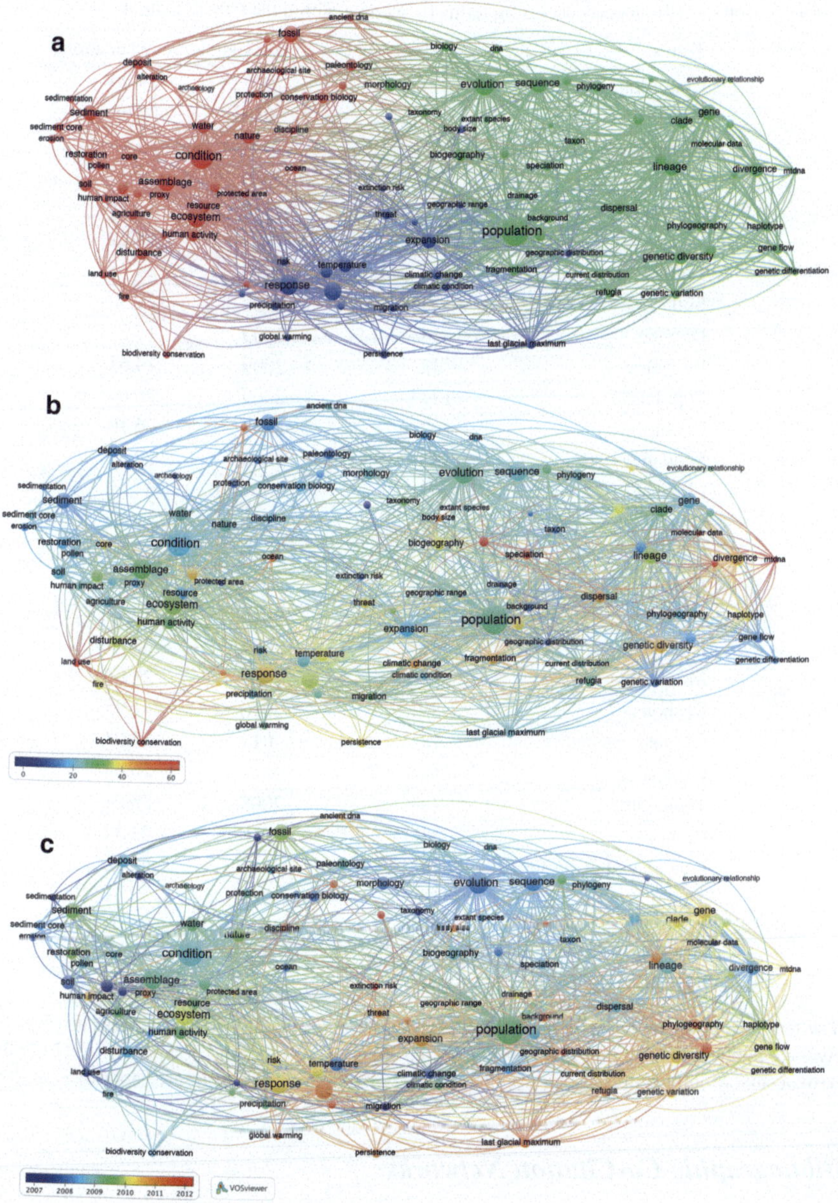
a
b
c
ancient dna
fossil
biology
dna
deposit
alteration
archaeological site
paleontology
archaeology
protection
conservation biology
morphology
evolution
sequence
phylogeny
evolutionary relationship
sedimentation
sediment
sediment core
erosion
water
nature
discipline
taxonomy
extant species
body size
taxon
clade
gene
molecular data
restoration
core
pollen
condition
biogeography
speciation
lineage
divergence
mtdna
soil
assemblage
human impact
proxy
ocean
protected area
resource
agriculture
ecosystem
human activity
extinction risk
threat
geographic range
drainage
background
dispersal
phylogeography
haplotype
expansion
population
geographic distribution
genetic diversity
gene flow
disturbance
risk
temperature
fragmentation
current distribution
genetic differentiation
land use
response
climatic change
climatic condition
refugia
genetic variation
fire
precipitation
migration
global warming
last glacial maximum
biodiversity conservation
persistence
0
20
40
60
2007
2008
2009
2010
2011
2012
VOSviewer

Table 5 Ten most frequently occurring terms by module within the CP-CC dataset

Module	Terms	Occurrences	Av. year	Av. citations
1	Condition	145	2009	19.70
	Ecosystem	94	2010	32.98
	Assemblage	85	2009	29.61
	Fossil	61	2010	14.90
	Sediment	57	2010	13.11
	Water	50	2009	25.24
	Nature	48	2010	24.96
	Soil	44	2008	21.30
	Deposit	43	2009	15.21
	Resource	42	2009	14.67
2	Population	173	2009	26.60
	Evolution	102	2008	26.16
	Sequence	83	2008	22.37
	Lineage	67	2009	34.37
	Gene	58	2010	19.83
	Genetic diversity	50	2010	43.42
	Divergence	48	2008	44.42
	Clade	47	2010	29.17
	Biogeography	41	2009	38.44
	Morphology	40	2008	23.68
3	Response	86	2010	41.95
	Climate change	78	2011	35.04
	Expansion	58	2011	16.88
	Temperature	49	2008	38.04
	Richness	41	2010	22.34
	Last glacial maximum	29	2011	22.66
	Threat	28	2011	39.61
	Precipitation	26	2009	22.08
	Climatic change	24	2008	46.79
	Risk	23	2010	23.57

The ten most common terms in each module and the frequency with which they occur (Fig. 6). Average publication year and average citation are average scores of the documents in which the term occurs. Module numbers correspond with the colors in Fig. 6a (see figure caption)

Bibliographic Co-Citation Networks

Bibliographic co-citation networks were created using the WoS CP-CC data for cited authors, and cited sources. In the author co-citation network, out of 28,839 cited first authors, 167 met the minimum threshold of 25 documents per author. Isolated nodes which did not share any citations in common with other papers in the dataset were excluded in the visualization for enhanced comprehensibility, which shows relationships between 153 of the authors. Four modules were present (Fig. 7),

Table 6 Ten most recently occurring terms by module within the CP-CC dataset

Module	Terms	Av. year	Occurrences	Av. citation
1	Paleobiology	2011	17	15.90
	Discipline	2011	19	19.64
	Proxy	2011	17	15.72
	Ancient DNA	2011	17	23.36
	Human impact	2011	18	24.70
	Protected area	2011	25	6.23
	Ecosystem	2010	39	32.98
	Fossil	2010	71	14.90
	Restoration	2010	15	18.41
	Nature	2010	24	24.96
2	Genetic structure	2012	11	15.81
	Historical biogeography	2011	11	12.90
	Background	2011	16	9.81
	Phylogeography	2011	18	20.29
	Evolutionary history	2011	12	14.17
	Refugia	2011	19	26.48
	Genetic variation	2011	15	12.77
	Genetic diversity	2010	11	43.42
	mtDNA	2010	22	59.42
	Gene	2010	45	19.83
3	Extinction risk	2013	19	8.00
	Conservation status	2012	28	3.50
	Conservation planning	2012	38	25.56
	Geographic distribution	2011	22	8.35
	Last glacial maximum	2011	10	22.66
	Global warming	2011	29	26.47
	Species distribution	2011	16	33.07
	Climate change	2011	13	35.04
	Body size	2011	13	49.44
	Persistence	2011	20	39.76

The ten most recent terms in each module based on average publication year of the documents in which the term occurs, and the frequency with which they occur (Fig. 6). Occurrences are listed for each term, and average citation scores (average scores of the documents in which the term occurs). Module numbers correspond with the colors in Fig. 6a (see figure caption)

broadly representing the following research domains based on the disciplinary affiliations of authors: genetics and evolutionary biology (red), ecology (green), paleontology (blue), and conservation ecology and paleobiology (yellow). Multiple highly cited and influential authors were present in all four specialties (Fig. 7): (1) genetics and evolutionary biology—J. C. Avise (70 citations), C. Moritz (67 citations), A. J. Drummond (63 cites); (2) ecology—K. J. Willis (155 citations), V. Rull (134 citations), S. T. Jackson (96 cites); (3) paleontology—S. M. Kidwell (104 citations), D. Jablonski (85 citations), J. B. C. Jackson (71 cites); and (4)

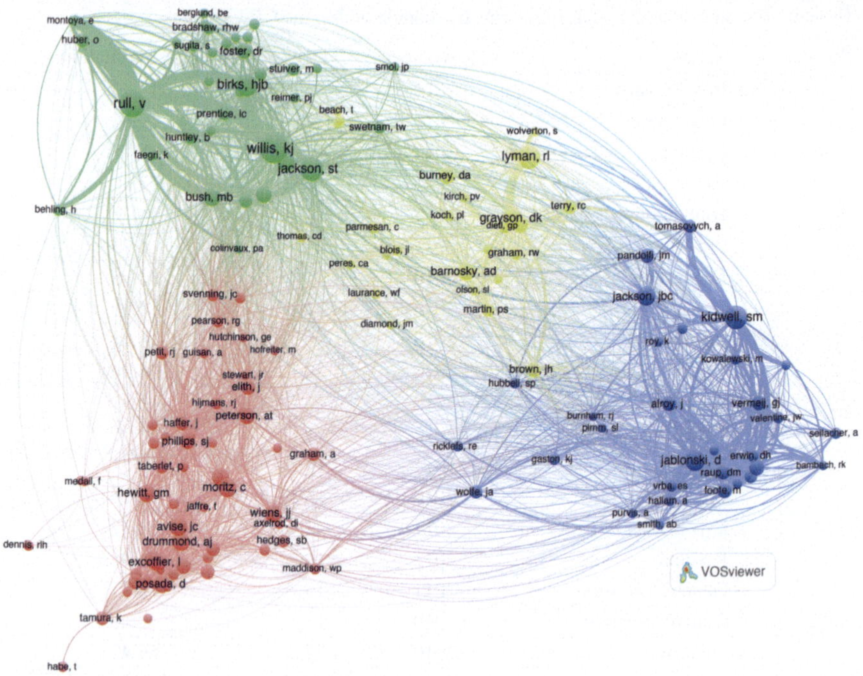

Fig. 7 The invisible college of conservation paleobiology crosses research domains. Visualization of bibliographic co-citation author analysis of the CP-CC dataset. Links indicate relationships between authors (nodes) that have published a minimum of 25 documents cited in the CP-CC dataset. Greater line width corresponds to stronger relationships. Invisible colleges are based on documents in a research network that refer to each other without being linked by recognizable organizational ties. Authors' last names and first initials are shown over nodes. Size of nodes represents relative number of citations, with larger nodes identifying greater numbers of citations. Colors denote four distinct modules representing disciplinary affiliations of authors with the fields of genetics and evolutionary biology (red), ecology (green), paleontology (blue), and conservation ecology and conservation paleobiology (yellow)

conservation ecology and paleobiology—D. K. Grayson (83 cites), A. D. Barnosky (64 cites), G. P. Dietl (58 cites).

Bibliometric Coupling Networks

In the network visualization of bibliographic coupling using sources (Fig. 9a), modules did not clearly define journal disciplines, mainly due to the influence of high impact interdisciplinary journals such as *PNAS* and *Science*. The five journals with the largest numbers of citations included *Science* with 1762 citations, *PNAS* with 1142, *Quaternary Science Review* with 547, *Biodiversity and Conservation* with 461, and *Molecular Phylogenetics and Evolution* with 453 (Fig. 9a).

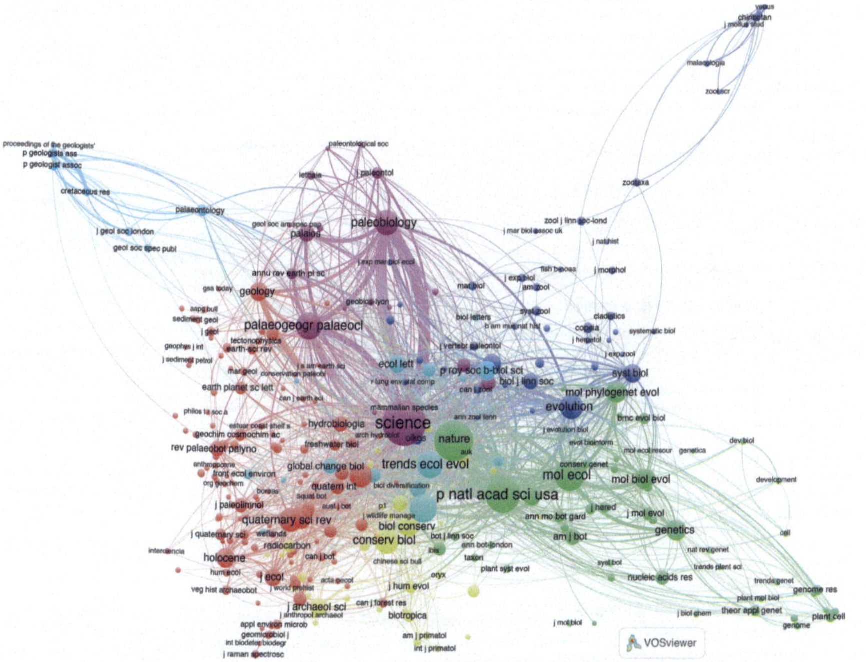

Fig. 8 The journal landscape of conservation paleobiology. Visualization of the bibliographic co-citation journal sources analysis using the CP-CC dataset. Links indicate relationships between journal sources (nodes) that have a minimum of 20 citations of the sources. Greater line width corresponds to stronger relationships. Size of nodes represents relative number of citations, with larger nodes identifying greater numbers of citations. Colors denote seven modules, loosely approximating journal disciplines such as earth sciences (red), paleontology (pink), and evolutionary biology (green). The highest impact interdisciplinary journals plotted in the center of the network and had a greater number of links with multiple modules

Journals more recently publishing articles on conservation paleobiology (Fig. 9b) included *Ecology and Evolution* (average publication year 2015), *PLOS One* (2013), *Ecography* (2013), *Global Change Biology* (2012), and *Quaternary International* (2012). More highly cited journals had older average years of publication, and the average publication years for the top five most highly cited journals ranged from 2008 to 2010 (e.g., *Science* average is 2009). The largest numbers of articles (Fig. 9c) were published in *PNAS* and *PLOS One* (26 documents), followed by *Molecular Phylogenetics and Evolution* and *Quaternary International* (17), *Biological Conservation* (15), and *Quaternary Science Reviews* (13). Of these, *PLOS One* had the most recent average publication year.

Out of 15,946 journal sources in co-citation network, 312 journals met the minimum requirement of 20 citations of a source. Seven modules were present (Fig. 8), which loosely approximated journal disciplines such as earth sciences (red), paleontology (pink), and evolutionary biology (green). Although dominated by geological journals, the largest module (in red), for example, also included

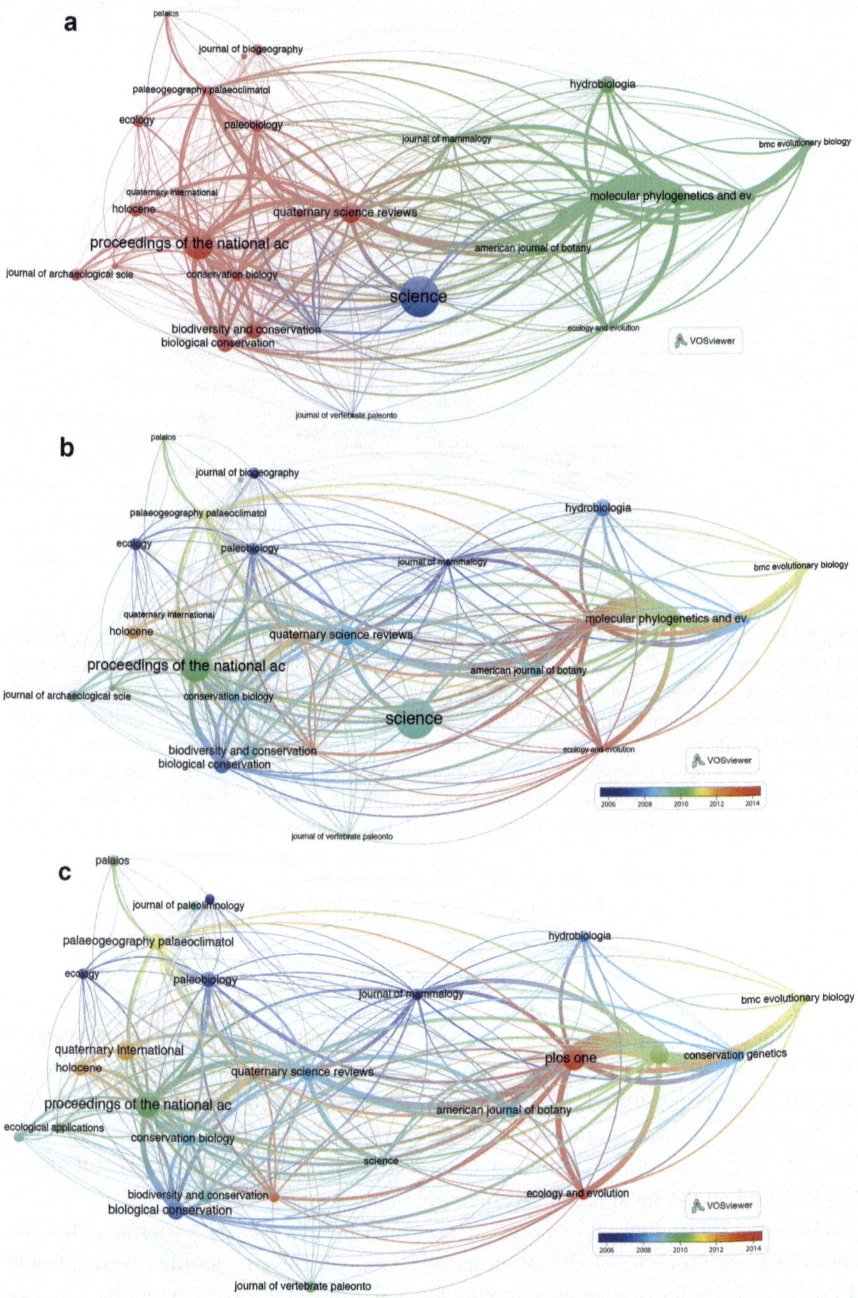
a
palaios
journal of biogeography
palaeogeography palaeoclimatol
ecology
paleobiology
journal of mammalogy
hydrobiologia
bmc evolutionary biology
quaternary international
holocene
molecular phylogenetics and ev
quaternary science reviews
proceedings of the national ac
american journal of botany
journal of archaeological scie
conservation biology
science
biodiversity and conservation
biological conservation
ecology and evolution
VOSviewer
journal of vertebrate paleonto
b
palaios
journal of biogeography
palaeogeography palaeoclimatol
hydrobiologia
ecology
paleobiology
journal of mammalogy
bmc evolutionary biology
quaternary international
holocene
molecular phylogenetics and ev
quaternary science reviews
proceedings of the national ac
american journal of botany
journal of archaeological scie
conservation biology
science
biodiversity and conservation
biological conservation
ecology and evolution
VOSviewer
2006
2008
2010
2012
2014
journal of vertebrate paleonto
c
palaios
journal of paleolimnology
palaeogeography palaeoclimatol
hydrobiologia
ecology
paleobiology
journal of mammalogy
bmc evolutionary biology
quaternary international
holocene
plos one
conservation genetics
quaternary science reviews
proceedings of the national ac
american journal of botany
ecological applications
conservation biology
science
biodiversity and conservation
biological conservation
ecology and evolution
VOSviewer
2006
2008
2010
2012
2014
journal of vertebrate paleonto

journals from ecology, paleontology, and conservation such as *Ecology, Human Ecology,* and *Climatic Change*. The top ten journals with the highest citations in conservation paleobiology were: *Science* (1602 citations), *Nature* (1289), *PNAS* (1047), *Molecular Ecology* (744), *Journal of Biogeography* (643), *TREE* (582), *Palaeogeography Palaeoclimatology Palaeoecology* (577), *Conservation Biology* (524), and *Paleobiology* (508). These high impact interdisciplinary journals, plotted towards the center of the network, and had a greater number of links to multiple modules.

4 The Intellectual Landscape

The approximate origin of the concept of conservation paleobiology appears to have been around 2006, and its subsequent rise was visible in the rapid increase in references including the phrase (Fig. 1), and in the temporal overlays (Figs. 5 and 6). After this phase of rapid ramping-up, publications on conservation paleobiology increased exponentially peaking at 92 in 2015 in the CP-CC dataset, and 336 in 2010 the CP-AD dataset (Fig. 1). These findings are consistent with other assessments of the growth and development of conservation paleobiology (Tyler and Schneider 2018).

Twenty-three interconnected groupings including researchers pursuing similar lines of inquiry, intellectual collaborations, or research "teams" are evident in the co-authorship network, many of which are interdisciplinary (Fig. 2). Interdisciplinary co-authorship indicates a dependence on methods and knowledge from other scientific domains (Moody 2004), which is to be expected in a rapidly growing new discipline, where ideas, publications, and core-literature have not yet reached a critical mass. Disciplines bring researchers with similar interests together, and

---

Fig. 9 Trending journals for cutting-edge research in conservation paleobiology. Bibliographic coupling visualization of sources. Link width indicates strength of relationships. (**a**) Colors differentiate the three modules identified by VOSViewer, based on strength of inter-group relationships. Modules do not clearly define journal disciplines, mainly due to the influence of high impact interdisciplinary journals such as *PNAS* and *Science*. Node sizes indicate numbers of citations, the largest of which is *Science* with 1762 citations, *PNAS* with 1142, *Quaternary Science Review* with 547, *Biodiversity and Conservation* with 461, and *Molecular Phylogenetics and Evolution* with 453. (**b**) Temporal overlay, warmer colors represent more recent average publication years (scale on bottom left). Size of nodes indicates number of citations. Journals more recently publishing articles on conservation paleobiology plotted with warmer colors include *Ecology and Evolution* (average publication year 2015), *PlosOne* (2013), *Ecography* (2013), *Global Change Biology* (2012), and *Quaternary International* (2012). (**c**) Node sizes indicate number of documents published with temporal overlay (colors correspond to the scale on the bottom left) indicating average publication year. The largest numbers of articles were published in *PNAS* and *PlosOne* (26 documents), followed by *Molecular Phylogenetics and Evolution* and *Quaternary International* (17), *Biological Conservation* (15), and *Quaternary Science Reviews* (13)

citations may be motivated by a combination of intellectual and interpersonal ties between authors (White et al. 2004). Groups are therefore likely the result of a combination of factors in addition to direct collaboration, including intellectual ties with former advisees, and academic social networks. Three of the four largest schools of thought contained at least one highly influential researcher, and one more "senior" researcher with contributions averaging in 2010.

Researchers working on the topic of conservation paleobiology were clearly not constrained to paleontology, and disciplinary affiliations of authors included genetics and evolutionary biology, ecology, paleontology, and conservation ecology and paleobiology (Fig. 7). Furthermore, authors working predominantly on conservation included both paleontologists and conservation scientists, while authors with more traditional paleontological contributions formed a distinct intellectual cluster (Fig. 7), likely predominantly contributing foundational science (e.g., advances in taphonomy and the fidelity of the fossil record) facilitating the expansion of research in conservation paleobiology.

Applications of conservation paleobiology have been broad (Figs. 5, 6 and 7), and research on the topics of conservation and paleontology has been conducted by scholars from a wide range of disciplines, with significant intellectual exchange between sub-disciplines. Conservation research conducted by paleontologists did not form a distinct sub-discipline (i.e., conservation research within paleontology), but was substantively coherent and interrelated with environmental history and conservation archeobiology (Figs. 5 and 6), within conservation science. These findings indicate a strong unity and inter-relatedness of content despite intellectual sub-communities, suggesting that conservation science is one research field with at least three major research themes: (1) environmental history and conservation archeobiology, (2) genetics and evolutionary biology, and (3) ecology. Regardless of the dataset used, conservation paleobiology did not form a distinct module in either dataset, and content is integrated into environmental history and conservation archaeobiology. Increased interdisciplinary exchange and integrating problem solving by scholars from different disciplines could have many potential benefits, as poorly understood problems benefit from the greater complexity afforded by interdisciplinary approaches, often yielding superior results (Birnbaum 1981). Although not currently a sub-discipline, it is possible that over time as literature accrues, paleontological research on conservation paleobiology may become an independent self-sustaining structure.

Although many perceive conservation paleobiology to fall under the *sensu stricto* domain of paleontology, given that conservation paleobiology frequently draws from expertise and methods in conservation sciences, these results suggest that when multiple types of paleontological data are considered, conservation paleobiology may in practice be a sub-discipline within conservation science. These results also indicate that conservation paleobiology is not a closed circuit, i.e., we are not only talking to ourselves. If conservation paleobiology identifies with the values and ethical norms steering conservation science (Soule 1985; Noss 1999; Meine et al. 2006), and we accept conservation paleobiology as a field of research within conservation science (Dietl 2016), perhaps our goals should similarly be to forge relationships with economists, educators, environmental policy makers, and conservation scientists (Savarese 2018; Dietl and Flessa 2018). Recognizing conservation paleobiology as a sub-discipline of conservation science would allow us to hook into pre-existing infrastructure and professional societies. For example, attending professional conferences outside of paleontology, such as the "Society for Conservation Biology" meeting, as suggested by Savarese (2018). Intra-disciplinary research is typically favored by individual experience and expertise, communication skills, career incentives and rewards, public outlets, and institutional organizational structures (Klein and Porter 1990). Attempts to integrate with conservation science may therefore be less than straightforward for researchers already trained and networked within paleontology, and typically housed in geosciences departments. However, simply recognizing the intellectual landscape of the field could be beneficial, and appreciating the breadth of the conservation community could prompt researchers to actively seek out exposure to topics and tools across disciplines.

While authors publishing cutting-edge research are likely to target well-established high impact interdisciplinary journals such as *Science,* or *PNAS*, most articles were published in journals within the authors' specialty. Given the interdisciplinary nature of the science, authors should carefully consider the target audience, a large part of which lies outside of their specific discipline. Authors may wish to submit their research to interdisciplinary or conservation science journals to reach the broader community working on conservation paleobiology, or journals publishing an increasing number of conservation paleobiology papers.

5 Emerging Frontiers

Thorough reviews of potential research areas to which conservation paleobiology has contributed are provided elsewhere (Willis et al. 2010; Dietl et al. 2011, 2015; Dietl and Flessa 2011; Fordham et al. 2016). Here instead, I encourage researchers to consider potential paleontological contributions within the discipline framework identified by bibliometrics. Paleontological and historical terms appeared within various sub-disciplines (Tables 3, 4, 5 and 6). On average, terms within genetics and evolutionary biology were overwhelmingly more recent, likely reflecting technological advances and breakthroughs (Figs. 5 and 6). Although terms with more recent average publication scores did not have high occurrence frequencies (Tables 3 and 5), this is likely a product of the novelty of the topic. The newest research themes and topics have not been present long enough to accumulate high occurrences, nor are they likely to be widely recognized by the scientific community and extensively published upon. Some of these emerging topics can be expected to dissipate, but significant research areas should persist and grow. The terms "paleobiology," "fossil," "Anthropocene," "palaeoecological record," and "last glacial maximum" are included among the most recent topics, consistent with the rise of this emerging sub-discipline, and the increasing awareness in conservation science of the need for long-term data that pre-dates human activities

(Smith et al. 2018). However, an increased awareness and expressed need for long-term or paleontological data does not necessarily indicate conservation paleobiology in practice. Furthermore, terms relating to explicit policy-oriented research were less common (e.g., "conservation status," "conservation planning," "protected area"). Several trending topics are apparent (Tables 4 and 6), and while many related to broad problems that need to be addressed (e.g., "climate change"), several topics likely to represented emerging frontiers to which paleobiology has much to contribute, including extinction risk, refugia (e.g., see Schneider 2018), identifying evolutionary relationships between populations, species and genetic diversity, ecosystem services, establishing baselines (e.g., Kusnerik et al. 2018), and biotic responses to perturbation (e.g., Roopnarine and Dineen 2018).

6 Conclusions

We have lacked a common organizing framework to identify central questions in the field, which is particularly important for the emergence of new fields of research (Torraco 2005). Here bibliometrics was used to promote a clearer understanding of the intellectual structure of the field and its current development, to remove some of the ambiguity surrounding the scope and meaning of conservation paleobiology. The conservation paleobiology community is arguably at a critical juncture, either moving forward predominantly within paleontology and geology, or actively identifying with conservation science (Dietl 2016). Results showed that conservation paleobiology is a relatively young emerging field of research that is highly interdisciplinary. Paleontological research did not form a cohesive sub-discipline, but was assimilated within other historical approaches to conservation. Although research communities appear clustered into several sub-disciplines, research is highly integrated across communities. Interdisciplinary collaborations and approaches will therefore likely play an increasingly important role in the future, as innovate and impactful research is likely to be interdisciplinary, and many of the challenges we face today require a broader approach.

Acknowledgments Thank you to Chris Schneider and Lindsey Leighton for reviewing this manuscript, and to Anthony Giuffre for thoughtful discussions, which helped shape this contribution.

References

Birnbaum PH (1981) Contingencies for interdisciplinary research: matching research questions with research organizations. Manage Sci 27:1279–1293

Boyack KW, Klavans R (2010) Co-citation analysis, bibliographic coupling, and direct citation: which citation approach represents the research front most accurately? J Am Soc Inf Sci Technol 61:2389–2404

Callon M, Courtial J-P, Turner WA, Bauin S (1983) From translations to problematic networks: an introduction to co-word analysis. Soc Sci Inf 22:191–235
Chang Y-W, Huang M-H, Lin C-W (2015) Evolution of research subjects in library and information science based on keyword, bibliographical coupling, and co-citation analyses. Scientometrics 105:2071–2087
Crane D (1972) Invisible colleges: diffusion of knowledge in scientific communication. University of Chicago Press, Chicago
Dietl G (2016) Brave new world of conservation paleobiology. Front Ecol Evol 4:21
Dietl GP, Flessa KW (2011) Conservation paleobiology: putting the dead to work. Trends Ecol Evol 26:30–37
Dietl GP, Flessa KW (2018) Should conservation paleobiologists save the world on their own time? In: Tyler CL, Schneider CL (eds) Marine conservation paleobiology. Springer, Cham, pp 11–22
Dietl GP, Kidwell SM, Brenner M et al (2011) Conservation paleobiology: opportunities for earth science. Ithaca, New York
Dietl GP, Kidwell SM, Brenner M et al (2015) Conservation paleobiology: leveraging knowledge of the past to inform conservation and restoration. Annu Rev Earth Planet Sci 43:79–103
Eom S (2008) All author cocitation analysis and first author cocitation analysis: a comparative empirical investigation. J Informetr 2:53–64
Fordham DA, Akçakaya HR, Alroy J et al (2016) Predicting and mitigating future biodiversity loss using long-term ecological proxies. Nat Clim Chang 6:909–916
Garfield E, Pudovkin AI, Istomin VS (2003) Why do we need algorithmic historiography? J Am Soc Inf Sci Technol 54:400–412
Gmür M (2003) Co-citation analysis and the search for invisible colleges: a methodological evaluation. Scientometrics 57:27–57
van der Have RP, Rubalcaba L (2016) Social innovation research: an emerging area of innovation studies? Res Policy 45:1923–1935
Kessler MM (1963) Bibliographic coupling between scientific papers. Am Doc 14:10–25
Klein JT, Porter AL (1990) Preconditions for interdisciplinary research. In: Birnbaum-More PH, Rossini FA, Baldwin DR (eds) International research of management: studies in interdisciplinary methods from business, government, and academia. Oxford University Press, New York, pp 11–19
Kusnerik KM, Lockwood R, Grant AN (2018) Using the fossil record to establish a baseline and recommendations for oyster mitigation in the mid-Atlantic U.S. In: Tyler CL, Schneider CL (eds) Marine conservation paleobiology. Springer, Cham, pp 75–103
Leydesdorff L, Rafols I (2012) Interactive overlays: a new method for generating global journal maps from web-of-science data. J Inf Secur 6:318–332
Lievrouw LA (1989) The invisible college reconsidered. Commun Res 16:615–628
Marshakova IV (1973) System of document connections based on references Sci Technical Inf Ser. VINITI 3–8
McCain KW (1990) Mapping authors in intellectual space: a technical overview. J Am Soc Inf Sci 41:433–443
Meine C, Soule M, Noss RF (2006) "A mission-driven discipline": the growth of conservation biology. Conserv Biol 20:631–651
Melin G, Persson O (1996) Studying research collaboration using co-authorships. Scientometrics 36:363–377
Melin G, Persson O (1998) Hotel cosmopolitan: a bibliometric study of collaboration at some European universities. J Am Soc Inf Sci 49:43–48
Moody J (2004) The structure of a social science collaboration network: disciplinary cohesion from 1963 to 1999. Am Sociol Rev 69:213–238
Noss R (1999) Is there a special conservation biology? Ecography (Cop) 22:113–122
Raasch C, Lee V, Spaeth S, Herstatt C (2013) The rise and fall of interdisciplinary research: the case of open source innovation. Res Policy 42:1138–1151

Rafols I, Leydesdorff L (2009) Content-based and algorithmic classifications of journals: perspectives on the dynamics of scientific communication and indexer effects. J Am Soc Inf Sci Technol 60:1823–1835

Ramos-Rodríguez A-R, Ruíz-Navarro J (2004) Changes in the intellectual structure of strategic management research: a bibliometric study of the strategic management journal, 1980–2000. Strateg Manag J 25:981–1004

Roopnarine PD, Dineen AA (2018) Coral reefs in crisis: the reliability of deep-time food web reconstructions as analogs for the present. In: Tyler CL, Schneider CL (eds) Marine conservation paleobiology. Springer, Cham, pp 105–139

Savarese M (2018) Effectively connecting conservation paleobiological research to environmental management: examples from Greater Everglades' restoration of southwest Florida. In: Tyler CL, Schneider CL (eds) Marine conservation paleobiology. Springer, Cham, pp 55–73

Schneider CL (2018) Marine refugia, past, present, and future: lessons from ancient geologic crises for modern marine ecosystem conservation. In: Tyler CL, Schneider CL (eds) Marine conservation paleobiology. Springer, Cham, pp 161–206

Small H (1973) Co-citation in the scientific literature: a new measure of the relationship between two documents. J Am Soc Inf Sci 24:265–269

Small HG (1977) A co-citation model of a scientific specialty: a longitudinal study of collagen research. Soc Stud Sci 7:139–166

Small H, Boyack KW, Klavans R (2014) Identifying emerging topics in science and technology. Res Policy 43:1450–1467

Smith JA, Durham SR, Dietl GP (2018) Conceptions of long-term data among marine conservation biologists and what conservation paleobiologists need to know. In: Tyler CL, Schneider CL (eds) Marine conservation paleobiology. Springer, Cham, pp 23–54

de Solla Price DJ (1965) Networks of scientific papers. Science 149(3683):510–515

Soule ME (1985) What is conservation biology? Bioscience 35:727–734

Torraco RJ (2005) Writing integrative literature reviews: guidelines and examples. Hum Resour Dev Rev 4:356–367

Tyler CL, Schneider CL (2018) An overview of conservation paleobiology. In: Tyler CL, Schneider CL (eds) Marine conservation paleobiology. Springer, Cham, pp 1–10

Van Den Besselaar P, Heimeriks G (2006) Mapping research topics using word-reference co-occurrences: a method and an exploratory case study. Budapest Sci 68:377–393

Van Eck NJ, Waltman L (2010) Software survey: VOSviewer, a computer program for bibliometric mapping. Scientometrics 84:523–538

Van Eck NJ, Waltman L (2014) Visualizing bibliometric networks. In: Ding Y, Rousseau R, Wolfram D (eds) Measuring scholarly impact: methods and practice. Springer, New York, pp 285–320

Van Eck NJ, Waltman L, Dekker R, van den Berg J (2010) A comparison of two techniques for bibliometric mapping: multidimensional scaling and VOS. J Assoc Inf Sci Technol 61:2405–2416

White HD, Griffith BC (1981) Author cocitation: a literature measure of intellectual structure. J Am Soc Inf Sci 32:163–171

White HD, McCain KW (1998) Visualizing a discipline: an author co-citation analysis of information science. J Am Soc Inf Sci 49:327–355

White HD, Wellman B, Nazer N (2004) Does citation reflect social structure? Longitudinal evidence from the "Globenet" interdisciplinary research group. J Am Soc Inf Sci Technol 55:111–126

Willis KJ, Bailey RM, Bhagwat SA, Birks HJB (2010) Biodiversity baselines, thresholds and resilience: testing predictions and assumptions using palaeoecological data. Trends Ecol Evol 25:583–591

Yan E, Ding Y (2012) Scholarly network similarities: how bibliographic coupling networks, citation networks, cocitation networks, topical networks, coauthorship networks, and coword networks relate to each other. J Am Soc Inf Sci Technol 63:1313–1326

Index

C. L. Tyler, C. L. Schneider (eds.), *Marine Conservation Paleobiology*, Topics in Geobiology 47, https://doi.org/10.1007/978-3-319-73795-9

R

S

The manufacturer's authorised representative in the EU is Springer Nature Customer Service Centre GmbH, Europaplatz 3, 69115 Heidelberg, Germany. If you have any concerns regarding our products, please contact ProductSafety@springernature.com

Printed and bound by CPI Group (UK) Ltd, Croydon, CR0 4YY

05/10/2026

02228139-0006